Communications in Computer and Information Science 1976

Rationale

The CCIS series is devoted to the publication of proceedings of computer science conferences. Its aim is to efficiently disseminate original research results in informatics in printed and electronic form. While the focus is on publication of peer-reviewed full papers presenting mature work, inclusion of reviewed short papers reporting on work in progress is welcome, too. Besides globally relevant meetings with internationally representative program committees guaranteeing a strict peer-reviewing and paper selection process, conferences run by societies or of high regional or national relevance are also considered for publication.

Topics

The topical scope of CCIS spans the entire spectrum of informatics ranging from foundational topics in the theory of computing to information and communications science and technology and a broad variety of interdisciplinary application fields.

Information for Volume Editors and Authors

Publication in CCIS is free of charge. No royalties are paid, however, we offer registered conference participants temporary free access to the online version of the conference proceedings on SpringerLink (http://link.springer.com) by means of an http referrer from the conference website and/or a number of complimentary printed copies, as specified in the official acceptance email of the event.

CCIS proceedings can be published in time for distribution at conferences or as post-proceedings, and delivered in the form of printed books and/or electronically as USBs and/or e-content licenses for accessing proceedings at SpringerLink. Furthermore, CCIS proceedings are included in the CCIS electronic book series hosted in the SpringerLink digital library at http://link.springer.com/bookseries/7899. Conferences publishing in CCIS are allowed to use Online Conference Service (OCS) for managing the whole proceedings lifecycle (from submission and reviewing to preparing for publication) free of charge.

Publication process

The language of publication is exclusively English. Authors publishing in CCIS have to sign the Springer CCIS copyright transfer form, however, they are free to use their material published in CCIS for substantially changed, more elaborate subsequent publications elsewhere. For the preparation of the camera-ready papers/files, authors have to strictly adhere to the Springer CCIS Authors' Instructions and are strongly encouraged to use the CCIS LaTeX style files or templates.

Abstracting/Indexing

CCIS is abstracted/indexed in DBLP, Google Scholar, EI-Compendex, Mathematical Reviews, SCImago, Scopus. CCIS volumes are also submitted for the inclusion in ISI Proceedings.

How to start

To start the evaluation of your proposal for inclusion in the CCIS series, please send an e-mail to ccis@springer.com.

Anban Pillay · Edgar Jembere · Aurona J. Gerber
Editors

Artificial Intelligence Research

4th Southern African Conference, SACAIR 2023
Muldersdrift, South Africa, December 4–8, 2023
Proceedings

Springer

Editors
Anban Pillay 🆔
University of KwaZulu-Natal
Durban, South Africa

Edgar Jembere 🆔
University of KwaZulu-Natal
Durban, South Africa

Aurona J. Gerber 🆔
University of the Western Cape
Cape Town, South Africa

ISSN 1865-0929   ISSN 1865-0937 (electronic)
Communications in Computer and Information Science
ISBN 978-3-031-49001-9   ISBN 978-3-031-49002-6 (eBook)
https://doi.org/10.1007/978-3-031-49002-6

This Springer imprint is published by the registered company Springer Nature Switzerland AG
The registered company address is: Gewerbestrasse 11, 6330 Cham, Switzerland

Paper in this product is recyclable.

Preface

Message from the General Chairs

Dear authors and readers,

It is with great pleasure that we write this foreword to the proceedings of the Fourth Southern African Conference for Artificial Intelligence Research (SACAIR 2023), held as a hybrid online and in-person event from the 4th to 8th December 2023. The program included an unconference for students on the 4th December 2023 (a student-driven event allowing students to interact with each other and with sponsors and potential employers), a day of tutorials on the 5th December, and the main conference from 6–8 December 2023.

SACAIR 2023 was the fourth international conference focused on Artificial Intelligence hosted by the SACAIR Steering Committee[1], an affiliate of the Centre for AI Research (CAIR), South Africa. The Centre for AI Research (CAIR)[2] is a South African distributed research network established in 2011 that aims to build world-class Artificial Intelligence research capacity in South Africa. CAIR conducts foundational, directed, and applied research into various aspects of AI through its nine research groups based at six South African universities (the University of Pretoria, the University of KwaZulu-Natal, the University of Cape Town, Stellenbosch University, University of the Western Cape, and North-West University).

Although still a young conference, SACAIR is quickly establishing itself as a premier artificial conference in the Southern African region. The fourth conference builds on the success of previous conferences. The inaugural CAIR conference, the Forum for AI Research (FAIR 2019), was held in Cape Town, South Africa, in December 2019, SACAIR 2020 was held in February 2021 after being postponed due to the Covid pandemic and SACAIR 2021 was an online event hosted by the University of KwaZulu-Natal, in Durban, in December 2021. The 2022 edition of the conference was held in Stellenbosch, Western Cape in December.

We are pleased that this, our fourth annual Southern African Conference for Artificial Intelligence Research (SACAIR), continued to enjoy the confidence of the South African artificial intelligence research community. The conference attracted support from both authors, who submitted high-quality research papers, as well as researchers who supported the conference by serving on the international program committee. The conference, held under the theme of Human-Centered AI, brought together a diverse group of researchers and practitioners in the fields of Computer Science, Information Systems, Knowledge Representation and Reasoning, Law, and Philosophy of AI.

As AI continues to affect all aspects of the human experience, an inter-disciplinary conference of this nature is essential as it serves as a platform for experts from various fields to exchange knowledge, ideas, and insights. This conference enables a holistic

---

[1] https://sacair.org.za.

[2] https://www.cair.org.za/.

approach to AI, integrating diverse perspectives and disciplines. Such an exchange is vital to address the complex challenges posed by a rapidly evolving field, with massive disruptive potential, like AI. An inter-disciplinary conference plays a significant role in advancing knowledge, driving innovation, and shaping the future of this transformative technology. The papers presented in this proceedings volume represent a rich tapestry of research and ideas, showcasing cutting-edge advancements in various aspects of AI and its impact on society.

The conference was organized as a multi-track conference that would cover broad areas of Artificial Intelligence namely:

- Algorithmic, Data Driven and Symbolic AI (Computer Science & Engineering)
- Socio-technical and human-centered AI (Information Systems)
- Responsible and Ethical AI (Philosophy and Law)

The accepted papers show a healthy balance between contributions from logic-based AI and those from data-driven AI, as the focus on knowledge representation and reasoning remains an important ingredient of studying and extending human intelligence. In addition, important contributions from the fields of socio-technical and human-centred AI and responsible and ethical AI are reported in this volume.

A conference of this nature is not possible without the hard work and contributions from many stakeholders. We extend our sincere gratitude to our sponsors: the Artificial Intelligence Journal (AIJ), the National Institute of Computational Sciences (NiTHeCS), the Centre for Artificial Intelligence Research (CAIR), Entelelect, and the National Research Foundation. These sponsors have made it possible to offer generous scholarships to students and emerging academics to participate in the conference. We sincerely thank the technical chairs for their work in overseeing technical aspects of the conference and the publication of the two volumes of the proceedings, the international panel of reviewers, our keynotes, authors, and participants for their contributions. Finally, we extend our gratitude to the track chairs, the local organizing committee, student organizers, and the conference organizer for their substantive contributions to the success of SACAIR 2023.

We encourage all the readers of this proceedings volume to embrace the diversity of ideas presented in these papers.

October 2023

Terence van Zyl
Dustin van der Haar

Message from the Program Chairs

This, volume 1976 of the Springer Communications in Computer and Information Science (CCIS) series, contains the revised accepted papers of the South African Conference for Artificial Intelligence, 2023 (SACAIR 2023). The second SACAIR 2023 proceedings volume is published as an online proceedings that will be available on the SACAIR 2023 website.

The inter- and trans-disciplinary nature of the SACAIR series of conferences in Artificial Intelligence is unique in providing a venue for researchers from a diverse set of disciplines that include Computer Science, Engineering, Information Systems, Law, Philosophy, and the Humanities. The organization of such a conference has to carefully consider the differing research methods, interests, publication standards, and cultures of these disciplines. The conference was thus organized around the following three tracks: Algorithmic, Symbolic and Data-Driven AI (Computer Science and Engineering - CSE), Socio-technical and human-centered AI (Information Systems - IS) and Responsible and Ethical AI (Philosophy and Law - PHIL).

1. **Algorithmic, Symbolic and Data-Driven AI** (*Computer Science & Engineering*) track. Topics in this track included:

   - Deep Learning and Machine Learning
   - Computer Vision and Image Processing
   - Natural Language Processing
   - Pattern Recognition
   - Evolutionary Computing
   - Biometrics and Cybersecurity in AI
   - Explainable AI
   - Knowledge Representation and Reasoning
   - Ontologies
   - Computational Logic
   - Multi-agent Systems
   - Agent-Based Modelling
   - Quantum Artificial Intelligence

2. **Socio-technical and Human-Centred AI** (*Information Systems*) track. Topics in this track include:

   - AI Information Systems
   - AI and Socio-Technical Systems
   - Human-Centred Artificial Intelligence
   - AI for and in Business, including AI Adoption
   - AI Supporting Sustainable Development and the Circular Economy
   - AI for Development and Social Good
   - Society 5.0

3. **Responsible and Ethical AI** (*Philosophy and Law*) track. Topics in this track included:

- Data Ethics
- Machine Ethics
- Ethics of Socio-robotics
- Neuro Ethics
- AI and the Law
- Responsible AI Governance

The program committee comprised 126 members (representing some 53 research institutions), 27 of whom were from outside Southern Africa. Each paper was reviewed by at least two members of the program committee in a rigorous, double-blind process. Great care was taken to ensure the integrity of the conference including careful attention to avoid conflicts of interest. The following criteria were used to rate submissions and to guide decisions: relevance to SACAIR, significance, technical quality, scholarship, and presentation, which included quality and clarity of writing.

We received 90 abstracts, and after submission and a first round of evaluation, 66 submissions were sent to our SACAIR program committee for review. The papers consisted of 47 in the CSE track, 7 in the IS track and 12 in the PHIL track. Twenty-two full research papers were selected for publication in this Springer CCIS volume (which translates to an acceptance rate of 33.3%), whilst a further 14 papers were accepted for inclusion in the online volume (21.2%). The total acceptance rate for publication in the two SACAIR 2023 proceedings volumes was 54.5% for reviewed submissions. In total, eight papers from the Responsible and Ethical AI track, three papers from the Socio-technical and Human-Centered AI track, and 25 papers from the Algorithmic, Symbolic and Data-Driven AI track were accepted for publication in the two volumes.

Thank you to all the authors who submitted work of an exceptional standard to the conference and congratulations to the authors whose work was accepted for publication. We place on record our gratitude to the Program Committee members, whose thoughtful and constructive comments were well received by the authors.

October 2023

Anban Pillay
Edgar Jembere
Aurona J. Gerber

Organization

General Chairs

Terence van Zyl   University of Johannesburg, South Africa
Dustin van der Haar   University of Johannesburg, South Africa

Technical Committee Chairs

Aurona Gerber   University of the Western Cape, South Africa
Edgar Jembere   University of KwaZulu-Natal, South Africa

Proceedings Chair

Anban Pillay   University of KwaZulu-Natal, South Africa

Publicity Chair

Margaux Bowditch   University of Johannesburg, South Africa

Unconference Chairs

Tevin Moodley   University of Johannesburg, South Africa
Sibonelo Dlamini   University of KwaZulu-Natal, South Africa
Asad Jeewa   University of KwaZulu-Natal, South Africa
Yuvika Singh   University of KwaZulu-Natal, South Africa

Tutorial Chairs

Senekane Makhamisa   University of Johannesburg, South Africa
Hima Vadapalli   University of Johannesburg, South Africa

Program Committee

Algorithmic, Data-Driven and Symbolic AI Track

Track Chairs

Alta de Waal	BMW IT Hub South Africa, South Africa & University of Pretoria, South Africa
Edgar Jembere	University of KwaZulu-Natal, South Africa
Clement Nyirenda	University of the Western Cape, South Africa
Anban Pillay	University of KwaZulu-Natal, South Africa
Giovanni Casini	Istituto di Scienza e Tecnologie dell'Informazione, Italy
Jesse Heyninck	Open Universiteit, The Netherlands

Program Committee

Richard Booth	Cardiff University, UK
Colin Chibaya	Sol Plaatje University, South Africa
Laura Giordano	Universitá del Piemonte Orientale, Italy
Ricardo Gonçalves	Universidade NOVA de Lisboa, Portugal
Louise Leenen	University of the Western Cape, South Africa
Thomas Meyer	University of Cape Town, South Africa
Pere Pardo	University of Luxembourg, Luxembourg
Iliana M. Petrova	National Institute for Research in Digital Science and Technology (Inria), France
Ramon Pino Perez	Université d'Artois, France
Umberto Straccia	Istituto di Scienza e Tecnologie dell'Informazione (CNR-ISTI), Italy
Bubacarr Bah	African Institute for Mathematical Sciences, South Africa
Shakuntala Baichoo	University of Mauritius, Mauritius
Mncedisi Bembe	University of Mpumalanga, South Africa
Mutanga Bethel	Mangosuthu University of Technology, South Africa
Sangeeta Biswas	University of Rajshahi, India
Anna Sergeevna Bosman	University of Pretoria, South Africa
Willie Brink	Stellenbosch University, South Africa
Dane Brown	Rhodes University, South Africa
Jan Buys	University of Cape Town, South Africa

Richard Chilipa   Malawi University of Science and Technology,
                                   Malawi
Olawande Daramola   Cape Peninsula University of Technology,
                                   South Africa
Marelie Davel   North-West University, South Africa
Allan De Freitas   University of Pretoria, South Africa
Pieter De Villiers   University of Pretoria, South Africa
Febe de Wet   Stellenbosch University, South Africa
Sibonelo Dlamini   University of KwaZulu-Natal, South Africa
Roald Eiselen   North-West University, South Africa
Inger Fabris-Rotelli   University of Pretoria, South Africa
Mandlenkosi Gwetu   University of KwaZulu-Natal, South Africa
Charis Harley   University of Johannesburg, South Africa
Albert Helberg   North-West University, South Africa
Frederic Isingizwe   University of the Western Cape, South Africa
Steven James   University of the Witwatersrand, South Africa
Asad Jeewa   University of KwaZulu-Natal, South Africa
Herman Kamper   Stellenbosch University, South Africa
Bennet Kankuzi   North-West University, South Africa
Z. Karim   University of the Witwatersrand, South Africa
Gift Khangamwa   University of the Witwatersrand, South Africa
Richard Klein   University of the Witwatersrand, South Africa
Paul Okuthe Kogeda   University of KwaZulu-Natal, South Africa
Eduan Kotzë   University of the Free State, South Africa
Hope Mauwa   University of Mpumalanga, South Africa
Thipe Modipa   University of Limpopo, South Africa
Abiodun Modupe   University of Pretoria, South Africa
Deshendran Moodley   University of Cape Town, South Africa
Tevin Moodley   University of Johannesburg, South Africa
Rudzani Mulaudzi   University of the Witwatersrand, South Africa
Kondwani Munthali   University of Malawi, Malawi
Fred Nicolls   University of Cape Town, South Africa
Oluwakorede Oluyide   University of the Western Cape, South Africa
Thembinkosi Semwayo   University of the Witwatersrand, South Africa
Jules-Raymond Tapamo   Univesity of KwaZulu-Natal, South Africa
Hima Vadapalli   University of Johannesburg, South Africa
Mohamed Variawa   University of Johannesburg, South Africa
Hairong Wang   University of the Witwatersrand, South Africa
Michael Zimba   Malawi University of Science and Technology,
                                   Malawi

Socio-Technical and Human-Centered AI Track

Track Chairs

Knut Hinkelmann   FHNW University of Applied Sciences and Arts
                                   Northwestern Switzerland, Switzerland
Sunet Eybers   University of Pretoria, South Africa

Program Committee

Johan Breytenbach   University of the Western Cape, South Africa
Massimo Callisto De Donato   University of Camerino, Italy
Fabrizio Fornari   University of Camerino, Italy
Pret Gerber   University of Pretoria, South Africa
Ridewaan Hanslo   University of Pretoria, South Africa
Felix Härer   Digitalization and Information Systems Group,
                                   University of Fribourg, Germany
Bernhard Humm   Hochschule Darmstadt - University of Applied
                                   Sciences, Germany
Emanuele Laurenzi   FHNW University of Applied Sciences and Arts
                                   Northwestern Switzerland, Switzerland
Zola Mahlaza   University of Cape Town, South Africa
Andreas Martin   FHNW University of Applied Sciences
                                   Northwestern Switzerland, Switzerland
Machdel Matthee   University of Pretoria, South Africa
Douglas Parry   Stellenbosch University, South Africa
Marco Piangerelli   University of Camerino, Italy
Henk Pretorius   University of Pretoria, South Africa
Catherine S. Price   University of KwaZulu-Natal, South Africa
Danie Smit   University of Pretoria, South Africa
Riana Steyn   University of Pretoria, South Africa
Corné Van Staden   University of South Africa, South Africa
Bruce Watson   Stellenbosch University, South Africa
Hans Friedrich Witschel   Fachhochschule Nordwestschweiz, Switzerland

Responsible and Ethical AI Track

Track Chairs

Emma Ruttkamp-Bloem	University of Pretoria, South Africa
Fabio Tollon	Stellenbosch University, South Africa

Program Committee

Paige Benton	University of Pretoria, South Africa
Jan Bergen	University of Twente, The Netherlands
Ivan Bock	Stellenbosch University, South Africa
Mary Carman	University of the Witwatersrand, South Africa
Kristy Claassen	University of Twente, The Netherlands
Ashley Coates	University of the Witwatersrand, South Africa
Vasundhra Dahiya	Indian Institute of Technology Jodhpur, India
Tanya de Villiers-Botha	Stellenbosch University, South Africa
Arzu Formánek	University of Vienna, The Netherlands
Cindy Friedman	Utrecht University, The Netherlands
James Garforth	University of Edinburgh, UK
Joris Graff	Utrecht University, The Netherlands
Zach Gudmunsen	University of Leeds, UK
Johan May	Stellenbosch University, South Africa
Isaac Olouch	University of Twente, The Netherlands
Andrea Palk	Stellenbosch University, South Africa
Helen Sarah Robertson	University of the Witwatersrand, South Africa
Achim Rosemann	De Montfort University, UK
Ethan Vorster	University of Pretoria, South Africa
Shenao Yan	University of Connecticut, USA

Our Sponsors

Contents

Responsible and Ethical AI Track

Emerging AI Discourses and Policies in the EU: Implications for Evolving AI Governance

Ana Paula Gonzalez Torres(✉) , Kaisla Kajava , and Nitin Sawhney

Department of Computer Science, Aalto University, Espoo, Finland
ana.gonzaleztorres@aalto.fi

Abstract. With the emergence of powerful generative AI technologies and the increasing prevalence of high-risk AI systems in society, the conversation around the regulation of AI has gained critical global traction. Meanwhile, policymakers and regulators are struggling to stay on top of technological advances in AI. Our work uses an interdisciplinary approach combining computational linguistics, law, and sociology to examine the developments in policy discourses around the AI Act (AIA) in the European Union (EU) and their implications for global AI policy. We base our analysis on findings from an ongoing study of multi-stakeholder feedback to the AI Act, leveraging Natural Language Inference (NLI) to examine the language used by diverse stakeholders to justify viewpoints for and against the AIA. Based on the outcomes of that analysis and engaging an AI policy perspective, we examine how those initial discourses are reflected in the amendments to the AI Act during the EU co-legislative process. The study is anchored on identified trends of contentious points in the regulation, such as the definition of AI, general principles, prohibited practices, a tiered approach to foundation models, general-purpose and generative AI, high-risk categorization as well as measures supporting innovation, such as AI regulatory sandboxes. We reflect on the implications of emerging discourses, regulatory policies, and experimental frameworks for global AI governance. Our take is that experimental regulation, such as regulatory sandboxes, can offer a window of opportunity to incorporate a global perspective and produce better-informed regulation and governance of AI.

Keywords: AI Act · AI Discourses · Natural Language Inference · AI Policy · Justifications Analysis (JA) · Regulatory Sandboxes

1 Introduction

With the emergence of powerful generative AI technologies and the increasing prevalence of high-risk AI systems in society, the conversation around the regulation of AI has gained critical global traction. The discourse around AI continues to be vibrant, given the technology's proliferation across societal sectors. AI is a unique example of a technological phenomenon that is rapidly starting to encompass virtually all corners of society, affecting citizens across demographic groups at a global scale. The magnitude of the changes that AI development and applications have already caused and are expected to cause in the future have prompted a wide range of stakeholders to contest over the regulation of AI.

A. Pillay et al. (Eds.): SACAIR 2023, CCIS 1976, pp. 3–17, 2023.
https://doi.org/10.1007/978-3-031-49002-6_1

In April 2021, the European Commission (EC) proposed the first regulation on artificial intelligence (AI) systems, the AI Act (AIA) [1]. It marked a regulatory turn in the AI narrative space, shifting the discursive attention of a previously largely ungoverned technological sector. Furthermore, it added a legal dimension to the AI discourse between stakeholders with diverse domains of knowledge, emphasizing the multidisciplinarity of AI as a technology in today's world [2]. Thus, ever since, policymakers, regulators, and stakeholders have been struggling to stay on top of regulatory discourses as well as technological advances in AI. It signals that the language used around and about AI technologies both reflects and influences the current state of AI, its regulation and development.

In this paper, we aim to understand how the current regulatory discourses and policy choices pursued within the European Union (EU) legislative process have been shaped by initial reactions to the AIA as portrayed in the feedback submitted by multiple stakeholders. The feedback period opened after the publication of the EC proposal, from 26 April 2021 to 6 August 2021, during which stakeholders could submit their views and critiques through an online portal[1]. Our study aims to draw qualitative correlations between discursive phenomena and the progress of changes to the AIA proposal from the multistakeholder feedback period to subsequent amendments to the legal text. We acknowledge that the legislative process is complex and multi-faceted, driven both by customs and procedural rules as well as deliberations between diverse parties. Thus, we consider the relationship between multi-stakeholder feedback to the proposed regulation and the modifications applied to the legal text one of correlation rather than causation. Therefore, many changes are likely to have been directly influenced by stakeholder input through diverse fora, and most of the influential stakeholders, such as technology companies, have shared their official views on the AIA with the EC. Many conversations between regulators and stakeholders occur behind the scenes, but that discourse is out of reach to us other than by mediation of media platforms. Therefore, we consider the feedback submitted through the EC online portal as a discursive indicator of the climate surrounding regulation, allowing researchers to study perceptions of AI governance.

Our work uses an interdisciplinary approach combining computational linguistics, law, and sociology to examine the developments in policy discourses around the regulation of AI in the European Union and their implications globally. We base our analysis on findings from an ongoing study of multi-stakeholder feedback to the AI Act leveraging the pragmatic sociological theory of justification [3] and Natural Language Inference (NLI) to examine the language and discourse of justifications by diverse stakeholder groups [4]. From the outcomes of that study, we take an AI policy perspective to examine the legal implications of these regulatory discourses to the European Union legislative process through central points of contestation emerging as trends in the development of the regulation: (a) the definition of AI; (b) general principles; (c) prohibited practices; (d) a tiered approach to foundation models, general-purpose AI and generative AI; (e) high-risk categorization; and (f) measures supporting innovation, such as AI regulatory sandboxes. The policy analysis aims to understand how diverse stakeholders' feedback

---

[1] All stakeholder feedback can be found and downloaded at: https://ec.europa.eu/info/law/bet ter-regulation/have-your-say/initiatives/12527-Artificial-intelligence-ethical-and-legal-requir ements/feedback_en?p_id=24212003.

has impacted current trends in legislative amendments to the original text of the AIA. Our stance is that the study on discourses and justifications can offer grounds for legislative changes and reflection on the implications of the EU legislative debate for future AI policy. In our understanding, these approaches could aid in analysing discourses around AI governance at a global scale, provide an analytical framework for contentious points in AI policy, and shed light on stakeholders' justifications supporting policy choices. Our goal is to reflect on the impact that policies emerging in the EU, such as regulatory sandboxes, have for the global context in incentivizing international collaboration and potential transnational approaches to responsible AI.

2 AI Act and Policy Discourses

In the European policy context, AI has been presented as a transformational force, either with redeeming of "salvific" qualities [5, 6] drawing from techno-solutionist discourse [7], or through mystified lens [5, 8, 9] with allusion to dystopian narratives [10]. These discourses often occur in conjunction with each other, serving either to promote or criticize AI. Meanwhile, the discussion regarding regulation and AI technologies occurs between narratives from technical experts working in various research and development subfields of AI, as well as legislators and policy experts. The latter are seen as lacking technical expertise in AI, while the technical experts have been found to view wider democratic involvement in AI development as a complicating factor, especially in the case of AGI (Artificial General Intelligence) [11]. However, the deliberations converge on the fundamentals of what AI *is* and what it can *do* or be *used for*. This complicated conjunction can be seen in the AIA [1], which has resulted in a complex piece of legislation due both to its complex subject matter and its relevance to multiple stakeholders as well as society at large. Nonetheless, the AIA is a landmark proposal as it recognizes the potential of AI technologies as enablers of innovation and their potential risks of causing harm.

At the time of writing, the AI Act is at the last stage of the 'co-decision' procedure of the European Union legislative process, the so-called "trilogue" between the European Commission (EC), Council of the European Union (CE), and European Parliament (EP) [12]. This means that members of the European Parliament (MEPs) engage in interinstitutional negotiations with the EU Council of Ministers, representing European governments and the European Commission in trilogues that will define the final compromised version of the AIA, which then can be voted into law. In the context of our study, the CE and EP proposed amendments to the EC AIA allow us to trace the proposed changes to the initial proposal for AI [1] in comparison to the justifications for or against the AIA used by diverse stakeholders during the feedback period. Thus, AIA and policy discourses go hand in hand as we examine how initial reactions to the AIA have been reflected in the trends of amendments to the EU regulation.

3 Grounding the Analysis

This paper is based on our ongoing study [4] examining how different stakeholder groups justified their views on the first draft of the AIA proposal. That study aims to pinpoint discursive patterns in diverse stakeholder groups' opinions on AI and the AIA, explicitly focusing on acts of justification for or against the proposal. The researchers leveraged the theory of justification by Boltanski and Thévenot [3] coupled with its empirical operationalization called Justifications Analysis (JA) [13]. Boltanski and Thévenot defined a set of six *worlds of justification*, also known as *orders of worth*, to describe the main ways stakeholders justify their arguments in public debates. Boltanski and Thévenot's theory places most arguments under the *civic, domestic, fame, industrial, inspired*, and *market* worlds of justification, and several additions have been made to the theory thereafter [14–16]. These worlds of justification are used by stakeholders in argumentation for or against changes to the social order in a process of legitimization. That is to say, the theory posits that claims made by stakeholders regarding a particular issue in a public debate draw from one or several of these orders of worth.

Based on this analytical framework, Kajava *et al.* [4] studied a dataset of EC proposal feedback consisting of 128 English-language documents from stakeholders in academic and research institutions, the technology industry, non-profit and non-governmental organizations, and the public sector. From this set, 20 documents were selected to be manually coded based on the JA framework. The researchers then leveraged a zero-shot RoBERTa-based [17] Natural Language Inference (NLI) model for assisted reading to extend their analysis to the remaining 108 feedback documents. The model, fine-tuned on the Multi-Genre Natural Language Inference Corpus (MNLI) [18], can be used for entailment tasks, where the user provides a set of classes, a hypothesis, and an instance of text to be classified as inputs, and receives a probability score for logical entailment (e.g., whether the hypothesis entails, contradicts, or has a neutral logical relationship to each sentence to be classified).

As the classification of justifications is an interpretative and extensive task to be solved as a machine learning problem, Kajava *et al.* [4] created a proxy task based on observed features common to a justification in the dataset. Justifications were found to typically allude either to a *challenge* or to a *benefit*, corresponding to Boltanski and Thévenot's [3] idea that criticism and justification tend to go hand in hand. The task was thus formulated as a sentence-level classification setting to detect the entailment of either a *challenge* or a *benefit*. Sentences scoring over a set probability threshold were selected as working hypotheses and manually analyzed for justifications. The set of 20 documents, coded entirely by hand, was used as a testing set for the model's performance. The model usefully captured 88.5% of the sentences containing a justification. The method was thus used for assisted reading, considerably speeding up manual coding of justifications by allowing researchers to focus on parts of the text that likely had higher relevance to the analytical task at hand. The results indicated a stakeholder-wide trend of grounding justifications both for and against articles of the AIA (and attitudes towards AI in a general sense) on values of a) human rights, democracy, and fairness, b) efficiency, best practices, and expertise, c) innovation and technological progress, d) investment, competitiveness, and cost, and e) agility, cooperation, and participation. However, the perspectives varied somewhat between stakeholder groups.

In this paper, while we draw from the results of an analysis of justifications based on the worlds of Boltanski & Thévenot [3], for the purposes of the policy analysis, we focus on the content of the justifications present in regulatory amendments rather than discussing the specific *worlds* of justification. Leveraging the assistive NLI approach provides a reading aid, allowing us to cover more documents. JA enables us to examine how the trends in policy discourses have been justified by stakeholders, and policy analysis reveals how those initial discursive developments have played into the ongoing state of legislation. In the legislative process, each of the three co-legislating institutions are negotiating to keep their position on the AIA afloat in the final version of the AIA. While the quantitative analysis of discourse using an NLI model provided distributions of stakeholder justifications [4] following the framework of Boltanski & Thévenot [3] and JA [13], we draw qualitative insights based on the results of the NLI model to correlate stakeholder justifications with legislative points of contestation.

Finally, we reflect on how the discourses in multi-stakeholder feedback indicate a need for a global approach to the regulation of AI. Reflecting on the Brussels effect of the AIA, the last section of the paper examines the current initiatives and trends that the EU AI law has sparked in the global context of AI governance. We examine which stakeholder justifications were and are still prevalent in the EU legislative process. In the policy analysis, we examine the needs and values that emerge in the different discourses and policies around global AI governance for responsible AI while envisioning a way forward by means of AI regulatory sandboxes.

4 Policy Analysis

From the policy perspective, we aim to compare the original EC AIA proposal and its related feedback to the current trends in the legislative process, meaning the amendments that the Council of the European Union (CE) and the European Parliament (EP) seek to consolidate, through trilogue negotiations, in the final compromised legislative text of the AIA. Based on an analysis of the legislative developments as portrayed in two AI policy journalism media outlets[2,3], we determined the following trends as main points of contestation in the regulatory negotiations: the definition of AI, general principles, prohibited practices, a tiered approach to foundation models, general-purpose AI, and generative AI, high-risk categorization, and measures supporting innovation, such as AI regulatory sandboxes. These trends were corroborated by examining the parts of the AIA that have been subjected to the most legislative debate and amendments in the legislative negotiation [1, 19, 20]. In the following, we will discuss, from a policy perspective, the relation between the results of the ongoing study on the feedback to the EC AIA proposal [4] and the current trends in the changes, as positioned by the EU co-legislators.

As a preliminary consideration, we reiterate that our study does not claim a consequential link between feedback and legislative amendments. We acknowledge that the complex political nature of the legislative process is subject to power struggles. For instance, "[m]ajor decisions take place in smaller groups (such as the trilogues) that allow for the achievement of consensus with other institutional actors such as the Council" and that "[t]he conclusion of complex (package) deals obscures, however, who has won or lost on particular issues" [12]. Nonetheless, our study shows that certain perspectives revealed in the analysis of discourse and justifications in the feedback documents [4] are reflected in subsequent amendments to the AIA proposal.

---

[2] https://www.euractiv.com/sections/artificial-intelligence/.

[3] https://futureoflife.org/newsletters/.

4.1 The Definition of AI

In the multi-stakeholder feedback, we see recurring calls for a definition of AI that is more closely aligned with "global definitions" as the EC proposal is considered "misaligned" with other used AI definitions (Facebook feedback). The feedback reflects a desire for closer alignment with the Organization for Economic Cooperation and Development (OECD) definition of an "AI system" as "a machine-based system that can, for a given set of human-defined objectives, make predictions, recommendations, or decisions influencing real or virtual environments. AI systems are designed to operate with varying levels of autonomy" [21]. For instance, as proposed by the Information Technology Industry Council in their feedback, "while continuously building on the OECD's proposed definition of AI, this regulation should have a more targeted scope by clarifying this definition and excluding traditional software and control systems".

This trend in the feedback documents can be seen in the CE and EP positions. From an analysis of different legislative drafts by the co-legislators, we gather that the EP and CE have adopted some elements of the OECD definition of an "artificial intelligence system" [21] while maintaining elements of the EC proposal. The CE's amendment combines the OECD's and EC proposal's definition to limit it to "machine learning and/or logic - and knowledge-based approaches" [19], whilst the Parliament opted for maintaining the OECD "machine-based system" definition but opted to include the issue of "explicit or implicit objective" [20]. The calls for a clearer definition of AI can be seen as closely connected to an ideal of efficiency, which was found to recur in stakeholders' justifications. As seen in the feedback, there is a notion that a good definition of AI will entail better-aligned legislation, which in turn will boost efficient practices by reducing legal uncertainty and allowing preferred best practices in technology development.

4.2 General Principles

The EP position introduces a newly added article 4a, "General principles applicable to all AI systems," which is established to be applicable in the "development and use" of "AI systems or foundation models" [20]. It is meant to promote a "human-centric European approach" and to encompass a) "human agency and oversight," b) "technical robustness and safety," c) "privacy and data governance," d) "transparency," e) "diversity, non-discrimination and fairness," f) "social and environmental well-being". We note that the principles resemble requirements established in the original AIA, such as articles 10, 13, 14, 15 and mentioned in article 69's codes of conduct.

We can see this reflected in the feedback, for example, by VDMA, "the requirements should better be formulated as essential principles, leaving detailed prescriptions to standards or guidelines". Similar feedback has echoed from industry stakeholders such as ORANGE "[a] right balance between the need to ensure that AI applications implemented and used in the EU follow clear ethical principles, including strong governance and transparency rules, and the necessity to ensure that EU players thrive and innovate in the global race for AI leadership without undue burdens". The request for general principles from industry as a measure for certainty is better exemplified by the feedback of Norwegian Open AI Lab when they mention the "missed opportunity with respect to defining a set of common regulatory principles across applications of technology for autonomous systems" or the call to "align AI usage standards with EU principles" based on Glovo feedback and in line with other stakeholders such as Standard Chartered and Twilio. Whilst the technology industry furthers "general principles" because of a need for certainty, there is simultaneously a wide understanding that such principles should follow fundamental rights, such as in the feedback from Digitalcourage e.V., AI Austria, or Artificial Intelligence Association of Lithuania (AIAL). Moreover, there are instances calling for specific principles that can be reflected in the EP position draft. For example, the European Consumer Organization states, "[a]ll AI (and not only high-risk) are [to be] subject to a minimum set of rules (starting with basic principles of transparency, fairness, accountability, non-discrimination, security, etc.)".

4.3 Prohibited Practices

The amendments make it clear that the CE takes a chirurgical approach, refining the original EC proposal to remove specifications related to enforcement authorities and public authorities [19], which now would apply to both the private and public sectors. Meanwhile, the EP has taken major revisions to include in prohibited practices "purposefully manipulative or deceptive techniques", "biometric categorization systems", "systems for risk assessment", "untargeted scraping of facial images", "systems to infer emotions" in law enforcement, border management, workplace, and education. Also, modifications include 'no real' remote biometric identification and remove exceptions but allow it for ex-post investigations on serious crimes [20].

The stakeholder feedback shows similar trends, for instance, the recognition that "emotion recognition systems and biometric categorization systems are not included in the list of prohibited AI practices under title II but are instead – partially – included in the category of 'high-risk' AI systems under Article 6(2). Stakeholders stated "disappoint[ment] given the unacceptably high risks such systems pose to human rights in many contexts" (Amnesty International feedback) and that "the proposed Regulation also should include, but not be limited to prohibitions on: non-consensual facial recognition; the storage of facial images; and the retention of other personal data incidentally captured by an AI system but not critical to its safe and intended operation" (ACM). Consistency remains from different stakeholders. For instance, in the Microsoft feedback, there is a call to further limit the types of crimes for which the system can be used, requiring each individual use of the technology to be subject to prior judicial authorization and creating additional fundamental rights protections, especially in issues of law enforcement use of real-time remote biometric identification systems in publicly accessible spaces. The changes of the co-legislator institutions are consistent with justifications found in the feedback, relating to values such as fairness, equality, end-user and citizen agency, and human rights, and focusing on collective good and well-being.

4.4 A Tiered Approach to Foundation Models, General-Purpose AI, and Generative AI

The CE has introduced in its position draft the definition of a "general purpose AI system" as an "AI system that - irrespective of how it is placed on the market or put into service, including as open source software - is intended by the provider to perform generally applicable functions such as image and speech recognition, audio and video generation, pattern detection, question answering, translation and others; a general purpose AI system may be used in a plurality of contexts and be integrated in a plurality of other AI systems" [19]. Meanwhile the EP position draft introduced, in the newly drafted recital 60e, the interplay between foundation models as "algorithms designed to optimize for generality and versatility of output" while "general purpose AI systems can be an implementation of a foundation model" [20]. The EP defines a "foundation model" as "an AI system model that is trained on broad data at scale, is designed for generality of output, and can be adapted to a wide range of distinctive tasks" [20]. The EP position also introduced a different definition of "general purpose AI system" as "an AI system that can be used in and adapted to a wide range of applications for which it was not intentionally and specifically designed" [20].

The original EC proposal does not explicitly mention foundation models, general-purpose AI, or generative AI. Nonetheless, the feedback shows that several stakeholders raised the issue of general-purpose models, as exemplified by Microsoft: "AI models are often made available as general-purpose AI services that are put to use in customer-defined applications", IBM: "the definition of AI systems should be made more explicit regarding the distinction between general-purpose AI tools and AI systems", or OpenAI: "encourage the Commission to examine how general-purpose AI systems will be effectively evaluated by the proposed AIA". Furthermore, the Leverhulme Centre for the Future of Intelligence (University of Cambridge) recommended to "consider how to identify and regulate high-risk uses of general-purpose systems, now before these systems become more dominant in the market".

Thus, although the discourse might suggest that the recent advances in the field of AI (especially generative) have been disruptive to legislators, it seems that part of the phenomena and issues were already widely known at the time of the first AIA proposal and during the subsequent feedback period. It is seen in the tiered approach developed in the amendments of the EU co-legislators. For foundation models[4], the EP subjects them to a respect for the introduced "general principles", revised article 28 on the "responsibilities along the AI values chain of providers, distributors, importers, deployers or other third parties", and specific newly drafted article 28b "obligations of the provider of a foundation model", as such answering to some of the calls seen in the feedback. For general-purpose AI systems, the CE introduced a newly drafted "Title Ia" to establish compliance requirements and obligations. For generative AI, the CE included them in their definition of AI. At the same time, the EP subjected them to further transparency obligations and implemented adequate safeguards against generating content in breach of EU law, such as copyright law. This implementation seems to align with observed justifications, as legal certainty for compliance leads to support for good processes. This step in the regulation can be seen in the Center for Human-Compatible Artificial Intelligence (CHAI, University of California, Berkeley) call to update the EC proposed AI Act to "address increasingly generalized AI systems that have multiple purposes, such as OpenAI's Generative Pre-Trained Transformer 3 and DeepMind's AlphaFold system".

---

[4] The term "foundation model" originated in the academic paper by Bommasani, R, et al.: On the Opportunities and Risks of Foundation Models. arXiv:2108.07258 (2021).

4.5 High-Risk Categorization

Initially, the draft law automatically classified high-risk AI applications that fell into a list of use cases in Annex III. Both co-legislators removed this automatism and introduced an "extra layer". For the Council, this layer concerns the significance of the output of the AI system in the decision-making process, with purely accessory outputs kept out of the scope. The EP position draft established the introduction of "significant risk," meaning "a risk that is significant as a result of the combination of its severity, intensity, probability of occurrence, duration of its effects, and its ability to affect an individual, a plurality of persons or to affect a particular group of persons" [20]. Thus, an extra layer was added for AI applications to fall in the high-risk category, whilst the list of high-risk areas and use cases was made more precise and extended in the areas of law enforcement and migration control. Finally, the EP modified Annex III regarding biometrics, critical infrastructure, education and vocational training, private and public services, and benefits whilst introducing considerations for recommender systems of the largest social media systems that might influence electoral outcomes, AI used in dispute resolutions, and cross-border management. The CE maintained chirurgical interventions in further detailing the provisions in Annex III and introduced considerations for AI systems for risk assessment in case of life decisions and health insurance.

We can trace the issue of significant risk to the early feedback. For instance, as raised by Avaaz Foundation, "[w]e are concerned about the limit the AI Act places on the categories of industry that are deemed as high risk" as "[t]his would allow services that are not in Annex III, but which do pose significant risks of harm to the health and safety or a risk of adverse impact on fundamental rights, to slip out of any regulatory oversight". In general, Annex III has been seen as containing "potential loopholes in the list provided in Annex III" (Centre for Commercial Law, University of Aberdeen) whilst others state it as "overly broad and would encompass AI applications that are not intended to be covered by the Regulation" (techUK). It could explain the trend on expanding Annex III and adding further details. We see hints of a need for an extra layer: "strongly advise not to consider AI in critical infrastructure automatically as High-Risk but only when health, safety & security are at risk" (Siemens AG). Stakeholders have been concerned about the overregulation of AI systems that do not pose a legitimate risk, which would lead to a slowdown in the process of AI development. Meanwhile, the expansion in Annex III can be traced to multiple stakeholders. Concerns for the impact on elections were raised by Amnesty International and for recommender systems by KU Leuven. While stakeholders maintain that biometric categorization is an overly broad area and imposes unnecessary regulation on AI (Getty Images UK Ltd). As seen in these feedback examples, this trend in modifications can be said to further the line of justifications advancing legal certainty (and, by proxy, the value of efficiency) and promoting increased citizen protection by regulating more cases as high risk.

4.6 Measures Supporting Innovation: AI Regulatory Sandboxes

In recognition of the complex regulatory framework posed by the provisions of the AI Act, the EC proposal includes "AI regulatory sandboxes" as a "measure in support of innovation", which aims "to reduce the regulatory burden" [1]. Likewise, the explanatory memorandum of the AIA states that the measure derives from a willingness to "create a legal framework that is innovation-friendly, future-proof and resilient to disruption". Thus, article 53 on "AI regulatory sandboxes" establishes them as "a controlled environment that fosters innovation and facilitates the development, testing and validation of innovative AI systems". It can be seen as a measure addressing legal uncertainties deriving from the new regulatory framework as AI systems within this regulatory environment are "under the direct supervision and guidance by the competent authorities with a view to ensuring compliance with the requirements of this Regulation and, where relevant, other Union and Member States legislation supervised within the sandbox" [1].

In the initial EC proposal, it was voluntary for the member states' competent authorities to implement AI regulatory sandboxes. Nonetheless, the EP amendments foresee this provision as mandatory to "at least one AI regulatory sandbox at national level", with the option for joint implementation with other member states [20]. The Parliament's position also entails providing AI developers that exit a sandbox the presumption of conformity for their systems. However, the CE position established AI regulatory sandboxes and real-world testing as voluntary measures to be decided by countries individually, while the exit report "could be taken into account during conformity assessment procedures or market surveillance checks" [19].

An analysis of the feedback shows the popularity of sandboxes among different stakeholders (stated, among many others, by Future of Life Institute, The Federation of Finnish Enterprises, City of Stockholm, Region Västra Götaland, Ministry of Local Government and Modernisation, Norway). Stakeholders supporting the use of regulatory sandboxes justified them as enablers of innovation but expressed concerns about discrimination in access between small/large companies (Centre for Information Policy Leadership (CIPL), German Insurance Association). In the case of real-world testing, it has been suggested that "[u]nless there are real humans inside the sandbox, it is hard to see how this could reveal (say) potential psychological harms. We suggest that Article 13 recognizes the limits of sandboxes for unknown risks and takes a prudent approach in dealing with psychological harms" (CHAI, University of California, Berkeley). However, other justifications supporting the measure are cautious due to concerns about market fragmentation. "For instance, we believe the deployment and outcomes of the regulatory sandboxes should also be aligned and leveraged at the EU level through the appropriate mechanisms [..] to avoid fragmentation across countries" (Johnson & Johnson). Thus, stakeholders suggested the "creation of a European-level, coordinated approach, following consultation with all stakeholders" (Nokia) or even "international cooperation between different sandboxes" (DeepMind). There are also warnings against the view of this measure as *panacea*. For instance, the differences in national legal systems "while Norway already has introduced regulatory sandboxes in their legal code, Sweden prohibits them" (CLAIRE). Thus, there are calls for precaution against the use of AI regulatory sandboxes "for dealing with problematic AI systems which push the boundaries of the AIA. Instead, they should be used as a precautionary step for the regulators to keep up with new regulatory challenges that future AI systems could present" (Amnesty International). The different perspectives regarding this popular but contested measure are also portrayed in the prudent approach by the CE and more support for the measure as a driver of innovation by EP.

5 Global AI Policy Implications

In terms of global AI policy, in the feedback, we can see early traces of stakeholders raising that AI "opportunities and benefits will be not limited to one country or government" (DeepMind) and the need for a "global dialogue about AI governance" (Facebook), especially in the form of international standards to avoid market fragmentation (Fujitsu, Splunk, IDEMIA, ITI - Information Technology Industry Council, DeepMind, Norwegian Open AI Lab). While it is recognized that the EU has the opportunity to "shape the global debate on AI" (RELX, Civil Liberties Union for Europe), it is sustained (Splunk) that the EU and US should maintain a dialogue to align AI regulation if not a global regulatory alignment (European Federation of Pharmaceutical Industries and Associations). In terms of solutions, DeepMind favored a global approach using platforms such as the "Global Partnership on AI" or fora for international cooperation such as "EU-US Trade and Technology Council".

In this light, at the time of writing, we see three different forms that AI policy has taken to address the need for a global approach. Firstly, in line with the need for a global perspective, in 2020, the United Nations (UN) raised the need for a multi-stakeholder advisory body on AI. As established in the UN Secretary-General's Roadmap for Digital Cooperation, currently in 2023, it is being formed to provide global steerage on AI[5]. The initiative aims to enhance multi-stakeholder efforts on global AI cooperation as a diverse forum to share and promote best practices, standardization, and compliance efforts [22]. Secondly, following the willingness of the EU to shape the global debate, we see the AI Pact. It is an initiative by EU officials to engage with industry actors and align with G7 goals. Under the justification of legal certainty, it is meant to be a voluntary measure that anticipates the AIA by means of raising "awareness to the principles and democratic process underlying the EU AI law"[6]. Thirdly, as seen in the feedback, following a 2023 EU-US Trade and Technology Council meeting, there is an agreement for transatlantic cooperation for responsible AI[7]. It is presented as a joint EU-US initiative to produce a set of voluntary commitments for industry and to eventually involve G7 countries under the disguise of the G7 Hiroshima AI Process[8].

The importance of the AIA relates to the "Brussels effect" [23] in setting an initial agenda, influencing, and sparking international appetite for global AI regulation. From the feedback, we recognized approaches that have been formalized as current trends in global policy scenarios as soft law attempts at establishing steering committees and voluntary measures. Whilst the trends signal a move by the EU to set the attempts at internationally coordinated policy moves, there is a lack of cohesion in what degree of global AI policy is necessary. Attempts from the US-EU, G7, or UN levels signal different degrees of influence that countries could have in shaping the regulation of a technology that affects every corner of society [24].

---

[5] https://www.un.org/techenvoy/content/artificial-intelligence.

[6] https://ec.europa.eu/commission/presscorner/detail/en/statement_23_3344.

[7] https://ec.europa.eu/commission/presscorner/detail/en/statement_23_2992.

[8] https://www.whitehouse.gov/briefing-room/statements-releases/2023/05/20/g7-hiroshima-leaders-communique/.

It is worth noticing that while there are concurrent interests in setting the regulatory narrative, it remains unclear how to translate responsible AI into practical measures. Thus, there is a need for regulation that is based on empirical experimentation with regulatory measures touching upon the practical nature of technical compliance. A move forward in the way towards global AI policy could entail the use of experimental regulation by means of regulatory sandboxes. As seen in our policy analysis, this measure is meant to support innovation while alleviating regulatory burdens and providing space for experimentation under regulatory guidance. While it is important to acknowledge the risk of discriminatory market advantages and unforeseeable liability (Facebook feedback), regulatory sandboxes could provide a way for improved regulation, leaning on an international effort to support experimental frameworks based on different approaches to regulation [25, 26]. For instance, technologies that have global implications because of their global coverage (e.g., ChatGPT) could be the subject of study in an international AI regulatory sandbox. Following the line of the UN efforts, this experimental framework could be overseen by an international steering committee supervising its impact, with the participation of diverse national regulatory bodies overseeing AI-based systems across borders. This configuration could address the national impacts of AI systems while enhancing governmental abilities to mitigate international harms. On the one hand, it could be a tool for reducing uncertainty deriving from translating regulatory provisions to practical technical compliance. On the other hand, it could provide co-learning opportunities from experimentation in a controlled environment, especially if under global supervision, potentially leading to the co-creation of cohesive international standards and a global AI policy direction.

6 Conclusions

While the AI Act is still under negotiation and thus bound to changes beyond our current analysis, we believe that studying the initial discourses around AI legislation based on different stakeholders and their justifications provides a window of opportunity to understand the implications for the direction of global AI policy. We focused on analyzing stakeholders' proposed solutions or suggested changes to the EC proposed AI Act raised in the feedback documents. The analysis of multi-stakeholder feedback covered a significant amount of documents from several sectors. It was targeted to analyze discourses belonging to a specific point in time and to compare them to the ensuing legislative changes. The stakeholder justifications for and against the AIA were compared to legislative trends, as portrayed by AI policy journalism media outlets, in the amendments to the EC proposed AIA by the EP and CE in the trilogue negotiations.

The comparison allowed us to produce a backward policy analysis, which suggests that a significant part of the changes advocated by stakeholders in the feedback are reflected in the trends of amendments to the EC AIA proposal as well as new considerations based on the moving state of the world. The first conclusion is that several contested amendments, which have been introduced in the AIA proposal, are present in the stakeholder responses. These include, among others, issues related to the OECD definition of AI, general-purpose AI, AI regulatory sandboxes, and prohibiting practices related to elections. Secondly, some of the newly introduced regulatory amendments related to the regulation of foundation models, general-purpose AI, and generative AI are reflected in the feedback. The discourses concurrent to the introduction of the AIA preceded the release of large generative models, such as ChatGPT, and the rise in public attention around generative AI triggered a discursive pivot point in time. This determined the need to introduce entirely novel regulatory provisions. Issues which were to become especially relevant for the regulation of these technologies were already raised by diverse stakeholders in their feedback to the EC.

Recognizing the emerging need for ad hoc regulatory provisions that tackle emerging issues in AI governance, we focus on implications for evolving AI governance. The feedback documents show a call for a global approach to the regulation of AI, and current efforts lean towards aligning responsible AI policies and regulations internationally, which may vary in the degree of influence that countries aim to have in the AI regulatory discourse. These encompass broader aims from the UN and other impulses by the EU trying to draw resources to build on these initial efforts. In this instance, there are crucial challenges of engaging wider global concerns while leveraging relevant expertise in setting the agenda for harmonized global AI policies and regulations. Our analysis of stakeholder feedback reveals the need for a coherent international approach, while the analysis of amendments to the AIA proposal highlights the emerging interest in measures that support innovation and ease the burden of regulatory compliance.

We believe that some of these concerns and challenges could be addressed by establishing AI regulatory sandboxes [26]. This sandbox approach could be envisioned as a way forward in the global regulation of AI by implementing responsible oversight. It could provide access to regulators (or outcomes from sandbox experiments) within or across national jurisdictions, and an opportunity to co-learn from experimenting with different compliance measures based on the impacts of an AI-based system. Thus, future AI policy and regulation could be better informed by experimentation as it could allow regulators in different countries to take distinct approaches to regulation and compliance in their own national contexts while facilitating mutual learning and oversight. For instance, AI regulatory sandboxes could support directed regulation (as proposed by the EU) or voluntary codes of conduct (e.g., a laissez-faire approach such as in the US), whereby more or less stringent compliance measures are only subjected to certain AI systems and in specific circumstances of use. This approach has the potential to be used as a mechanism for piloting and validating novel AI systems in conjunction with citizens, stakeholders in public and private sectors, policymakers, and international regulators throughout the AI lifecycle. Our position is that there is a need for a future-proof measure for regulation that does not ultimately rely on multi-stakeholder feedback as a measure to fill in regulators' blind spots. This framework could set the agenda for global AI policy based on experimentation and co-creation of knowledge, potentially informing emerging standards or best practices. Sandboxes could function as an assistive approach in democratizing how AI policies and regulations are devised both within countries and in transnational contexts.

References

1. European Commission (EC).: Proposal for a Regulation of the European Parliament and of the Council Laying Down Harmonised Rules on Artificial Intelligence (artificial intelligence act) and amending certain union legislative acts, COM/2021/206 final', Brussels (2021)
2. Dignum, V.: AI is multidisciplinary. AI Matters 5(4), 18–21 (2020)
3. Boltanski L., Thévenot L.: On justification: Economies of worth. Princeton University Press, Princeton (2006)
4. Kajava, K., Gonzalez Torres, A. P., Rannisto, A., Sakai, S.: Justifying AI & Its Regulation: Examining Multi-Stakeholder Perspectives on the AI Act (2023). (Manuscript in preparation)
5. Ossewaarde, M., Gulenc, E.: National varieties of artificial intelligence discourses: myth, utopianism, and solutionism in west European policy expectations. Computer 53(11), 53–61 (2020)
6. Kajava, K., Sawhney, N.: Language of algorithms: agency, metaphors, and deliberations in AI discourses. In: Lindgren, S. (ed.) Handbook of Critical Studies in Artificial Intelligence. Edward Elgar Publishing (2023). (In press)
7. Katzenbach, C.: "AI will fix this" – the technical, discursive, and political turn to AI in governing communication. Big Data Soc. 8(2), 20539517211046182 (2021)
8. Musa Giuliano, R.: Echoes of myth and magic in the language of artificial intelligence. AI & Soc. 35(4), 1009–1024 (2020). https://doi.org/10.1007/s00146-020-00966-4
9. Elish, M.C., Boyd, D.: Situating methods in the magic of big data and AI. Commun. Monogr.. Monogr. 85(1), 57–80 (2018)
10. Köstler, L., Ossewaarde, R.: The making of AI society: AI futures frames in German political and media discourses. AI & Soc. 37, 249–263 (2022)
11. Graham, R.: Discourse analysis of academic debate of ethics for AGI. AI Soc. 37, 1519–1532 (2022)https://doi.org/10.1007/s00146-021-01228-7
12. Neuhold, C.: The "Legislative Backbone" keeping the Institution upright? The Role of European Parliament Committees in the EU Policy-Making Process. European Integration Online Papers (2001)
13. Ylä-Anttila, T., Luhtakallio, E.: Justifications analysis: understanding moral evaluations in public debates. Sociol. Res. Online 4(21), 1–15 (2016)
14. Boltanski, L., Chiapello, E.: The new spirit of capitalism. Int. J. Politics, Culture, Soc. 18, 161–188 (2006)
15. Lafaye, C., Thévenot, L.: An ecological justification? Conflicts in the development of nature in justification, evaluation and critique in the study of organizations. In: Cloutier, C., Gond, J.-P. (eds.) Research in the Sociology of Organizations 52, pp. 273–300. Emerald Publishing Limited, Bingley (2017)
16. Sharon, T.: When digital health meets digital capitalism, how many common goods are at stake? Big Data Soc. 5(2), 2053951718819032 (2018)
17. Liu, Y. et al.: RoBERTa: A Robustly Optimized BERT Pretraining Approach. arXiv preprint arXiv:1907.11692 (2019)
18. Williams, A., Nangia, N., Bowman, S.: A Broad-Coverage Challenge Corpus for Sentence Understanding through Inference. In: Proceedings of the 2018 Conference of the North American Chapter of the Association for Computational Linguistics: Human Language Technologies 1 (Long Papers), pp. 1112–1122. Association for Computational Linguistics. New Orleans, Louisiana (2018)
19. Council of the European Union (CE).: Proposal for a Regulation of the European Parliament and of the Council laying down harmonised rules on artificial intelligence (Artificial Intelligence Act) and amending certain Union legislative acts—General approach. Interinstitutional File: 2022/0085(COD) 14128/22 (2022)

20. European Parliament (EP).: Draft compromise amendments on the draft report. Proposal for a regulation of the European Parliament and of the Council on harmonised rules on Artifcial Intelligence (Artificial Intelligence Act) and amending certain Union Legislative Acts (COM(2021)0206—C9 0146/2021—2021/0106(COD)) (2023)
21. OECD.: Recommendation of the Council on Artificial Intelligence, OECD/LEGAL/0449 (2019)
22. United Nations General Assembly.: Road map for digital cooperation: implementation of the recommendations of the High-level Panel on Digital Cooperation. Report of the Secretary-General (2020)
23. Bradford, A.: The Brussels Effect: How the European Union Rules the World. Oxford University Press, New York (2020)
24. D'Ignazio, C.: Data Feminism. MIT Press, Cambridge (2022)
25. Ranchordas, S.: Experimental regulations for AI: sandboxes for morals and mores (SSRN Scholarly Paper No. 3839744). SSRN Electronic Journal (2021). https://doi.org/10.2139/ssrn.3839744
26. Gonzalez Torres, A.P., Sawhney, N.: Role of regulatory sandboxes and MLOps for AI-enabled public sector services. Rev. Socionetwork Strat. **17**, 297–318 (2023). https://doi.org/10.1007/s12626-023-00146-y (2023)

Intergenerational Justice as Driver for Responsible AI

Emma Ruttkamp-Bloem[1,2]([✉]) [iD]

[1] Department of Philosophy, University of Pretoria, Pretoria, South Africa
`emma.ruttkamp-bloem@up.ac.za`
[2] Centre for Artificial Intelligence (CAIR), Pretoria, South Africa

Abstract. I show that considering intergenerational justice as a core AI ethics value is an effective way to ensure responsible AI debates are hooked to the real world. I argue that an integral part of addressing concerns about the actionability of AI ethics regulation should be concern for the effectiveness of its defense of social rights on the one hand, and its engaging with core groupings to affect such a defense on the other. Current responsible AI debates lack sufficient consideration of the rights of future generations. This seems odd as AI technology is changing the world in which future generations will live, and therefore responsible AI practices should surely demonstrate interaction with issues of intergenerational justice. The paper highlights the manner in which AI ethics concerns relate in general to future social stability through three lenses - environmental justice (harm to the environment and ecosystems), structural justice (harm from identity prejudice and structural bias), and moral justice (harm from value change). The argument is not that AI technology holds only bad news for future generations, but rather that framing responses to threats to social justice and human flourishing potentially posed by AI technology in intergenerational terms can contribute significantly to establishing concrete measures to underpin AI ethics and governance discourses.

Keywords: Intergenerational Justice · AI Ethics · Data Activism · Responsible AI · Social Justice

1 Introduction

Over the past decade concern about potential threats to human rights and human autonomy posed by AI technology, validated and fueled by the actual realization of some of these threats [2] has led to a proliferation of regulations. Over 200 regulations saw the light, coming from different sources. Intergovernmental, national, regional, and private and public sector entities all in various ways reacted to the challenge of realizing responsible governance of AI technology [3].

The regulation set to have the most far-reaching impact of any current legislation, apart from the UNESCO global Recommendation on the ethics of AI [54] is

A. Pillay et al. (Eds.): SACAIR 2023, CCIS 1976, pp. 18–30, 2023.
https://doi.org/10.1007/978-3-031-49002-6_2

the European Union's AI Act. This legislation takes a risk-based approach to AI, dividing AI technology up into different categories based on their risk to society and placing varying legal obligations on the use of this technology depending on which risk category is applicable [16]. Recently various writers have pointed out the ineffectiveness of most current regulation for reasons that range from being too abstract and over-focused on the 'what' of regulation, rather than the 'how', not speaking to the tech community's expertise or training, being top-down and too rigid, lacking sufficient normative power, etc [23,28,38,40]. In this paper I argue that an integral part of addressing concerns about the actionability of AI ethics regulation should be concern for the effectiveness of its defense of social rights on the one hand, and its engaging with core groupings in order to affect such a defense on the other. In this regard, I focus here on the lack of sufficient engagement with the rights of future generations and of placing intergenerational concerns at the center when values such as inclusivity and non-discrimination are considered.

In the domain of intergenerational ethics, the debate is about whether current generations should be concerned with the wellbeing of future generations in the sense of acknowledging a moral obligation to take the needs of both near- and far future generations seriously in political, economic, and social decision-making. While this debate originated in climate change settings [41], I argue here that this debate should be engaged with in AI ethics contexts as well, and that the moral obligation exists in this context, as the issue of intergenerational justice is just as core to the sustainability of responsible AI and related technologies such as the Internet of Things as it is to the nature of climate engineering decisions in climate change debates.

I start by briefly describing the domain of intergenerational ethics and how it might link to AI ethics. I unpack core AI ethics concerns relating in general to future social stability through three lenses - environmental justice (harm to the environment and ecosystems), 'structural justice' (harm from identity prejudice and structural bias), and 'moral justice' (harm from value change). I then consider current and future social justice concerns in AI ethics in terms of data activist debates as the focus of these debates on future benefit to communities suggests a concrete way to achieve engagement with intergenerational concerns. I move on to show how intergenerational justice could be a core driver for concrete and actionable responsible AI. I conclude with a brief plea for linking intergenerational issues more strongly to social justice issues in AI ethics debates such that a more central role is ascribed to social justice issues in these debates and concrete measures can be put in place to underpin AI ethics and governance discourses.

2   AI Ethics as Intergenerational Ethics

While AI technology has the potential to bring untold benefit to humanity - from democratizing healthcare and education, to informing measures addressing climate change and sustainability concerns - it also holds the potential to do harm.

Concerns range from environmental damage and widening the digital divide, to the fact that AI technology challenges humans' unique experiences in various ways, and include concerns about limiting human agency, and threats to the right to mental integrity (to not be manipulated). These issues raise additional concerns about human self-understanding; social, cultural, and environmental interaction; autonomy; agency; worth; and dignity [54].

The aim of this paper is to highlight the manner in which these concerns relate to social stability in general through three lenses - environmental justice (harm to the environment and ecosystems), 'structural justice' (harm from identity prejudice and structural bias), and 'moral justice' (harm from value change). In considering how these concerns might impact future societies, an argument is made for including intergenerational justice as a value in AI ethics regulation. Intergenerational justice as an AI ethics value, might then be concretized by principles such as the right to privacy and data protection, transparency and explainability, sustainability, and fairness and inclusivity in novel ways which would contribute significantly to hooking AI ethics discourse to the real world.

Given that the driver of current AI technologies is human data, and that decisions generated by these technologies impact society in concrete manners from decisions about who gets bail, who is stopped and frisked by police, who gets into which educational institution and has access to which level of healthcare; when engaging in AI ethics debates, in one way or another, one comes up against the challenge of openness to the call for a "futurist mindset" [50]. Steele [50] writes that due to the social and economic changes humanity will have to face for various reasons in the near future, there is a clear need "to compose a new civilizational story line that will guide the evolution of our species". I suggest that the narrative needed to guide our species into the future in the context of AI technologies is a social justice narrative. The power of AI technologies to disrupt social stability is clear from many concrete examples including the Cambridge Analytical scandal and potential threats around technology such as AI systems that create deep fakes and automated hacking tools, as well as from the obstacles it holds for finding environmental equilibrium. I pick up this point in Sect. 3.

Ulrick Beck [6] writes that the 'risk society' in which we live is based on the fact that the risks we face are global and 'democratized' in the sense that they affect all of humanity. And very often, these risks that include climate change, geopolitical tensions, and inequality [59], are brought about by human economic-technological actions. The kinds of threats that AI technologies hold for humanity are global risks as they affect all of humanity in some way, and they also have the potential to create 'wicked problems' [45][1], among many other reasons, because of their tendency to latch onto diverse social and geo-political patterns. Given the global and 'wicked' nature of AI ethics concerns, and the

---

[1] Design theorists Rittel and Weber [45] coined the notion of wicked problems that are problems that are difficult to solve and typically cannot be solved in only one 'correct' way, but only in better or worse ways given the inherent contradictions in the factors causing the problems.

likelihood that a hope of finding 'final' solutions seems misguided given the continuous exponential progress of AI technologies and the complexity of the factors leading to these concerns, intergenerational thinking should play a core role in AI ethics debates.

We will now first consider the nature of intergenerational ethics and justice to understand these claims better. Solum [49] writes that intergenerational ethics relates to at least environmental policy, health policy, intellectual property law, international development policy, social security policy, and telecommunications policy. More concretely, in the intergenerational ethics context the main question related to justice for future generations is to consider what such generations are owed by us [22]. When considering a response to this question, one first has to determine whether talk of an ethical obligation to future generations makes sense at all in the first place. Given that the sustainability of AI technologies depends on the technologies being fully adopted and that adoption in its turn is based on the fact that the technology is trusted, it is clear that the trust of future generations is also at issue both for future societal success of these technologies in terms of real benefit (as opposed to the economic profit of Big Tech) and the sustainability of AI technologies. Thus, in the AI ethics context, there certainly is some moral obligation to engage with the rights of future generations.

Having established the moral obligation for engaging with intergenerational concerns in AI ethics debates, let us now briefly consider the current status of discussion of intergenerational ethics in AI ethics discourses. There is a well-known perception that the tech world belongs to the young, and recent statistics confirm that the median age of a software developer in the US is 29 [5]. Apart from the fact that this is not the whole story, as for instance, the median age of the top 0.1% of tech start-ups in terms of growth is 45 [59], the young age of tech teams on the whole seems, however, not to be leading to any serious wholescale intergenerational analyses of the impact of AI technology on human society in the tech world. That said, UNICEF recently developed a suite of documents focused on AI and Children [55], and similarly, there is a recent European Commission's JTC publication on AI and the Rights of the Child [17]. These documents focus on near-future generations and are helpful in terms of thinking about mitigating harm to these groups but perhaps less helpful in terms of empowering near-future generations such that they can flourish in tandem with AI technologies. In its turn, the WEF recently developed a report on 'AI Regulation through an Intergenerational Lens' [58], which, while offering valuable insights as to the diversity of generations impacted by AI technologies, focuses more on the important issue of navigating between collectivist and individualist ethics in striving to find global solutions to AI ethics concerns than specifically on intergenerational justice as an AI ethics value.

There has been some engagement in the literature with intergenerational ethics, but on the whole very little, if one considers the major impact AI technologies will have on the nature and structure of future societies. One instance of such engagement is Klockmann et al. [30] who undertook an online study in which they tested whether individuals who know that their actions induce externalities on the future decision-making of AI systems and who know that such externalities will affect future participants, would adjust their behavior to ensure minimal harm to future participants. They found that only when these individuals themselves stood to be harmed by future algorithmic decisions, did they adjust their actions. This study is important as it highlights and illustrates the need for further engagement with how seriously tech teams as well as policy makers take intergenerational concerns, especially given that they might not be looking too far into the future when considering harm, going by this study.

While acknowledging that the above is a far cry from a thorough literature review, the samples mentioned combined with the fact that intergenerational justice features in no significant AI regulations as a core value, at the very least make it interesting to raise the question whether intergenerational justice is taken seriously enough in AI ethics communities. Reflecting then on the moral obligation AI ethicists have to engage with intergenerational justice relates to the so-called 'non-identity problem' which is the problem of determining on the one hand whether and in which ways the character of future generations depends on decisions and actions of present generations, and on the other to what extent present generations are in fact able to take interests of future generations into account in their thinking and actions (see, e.g., [26,35,36]). For the latter, see Sects. 3 and 4, but now let us first consider through three lenses - which I term environmental justice, structural justice, and moral justice - the ways in which the well-being of future generations is impacted on by AI technologies[2].

In the first instance, in terms of environmental justice, the environmental threats that efficient AI technological processes hold concretely add to existing climate change concerns, which obviously impacts future generations. The environmental cost of AI technologies in terms of energy cost is well-known [34], for instance one roundtrip flight by way of New York and San Francisco leaves a carbon footprint of 1 984 lbs of CO_2 equivalent, while a transformer (213M parameters) with neural architecture search leaves a footprint of 626 155 lbs of CO_2 equivalent [51]. In addition, there are serious concerns around water use for cooling in big data centers, for instance, there are estimates that the evaporation tower-cooled freshwater cost of ChatGPT is around 10 ml per prompt [53]. And so, the environmental dark side of AI technology is real and might exacerbate existing climate concerns for future societies.

The fact that many private sector companies do all they can to minimize these harms [47], is encouraging, and, at the same time proof that there is acknowledgement that future harm is relevant when engaging with AI technologies [27]. Of course, I also recognize the strong promise for positive use of AI technology in the energy (e.g., [12]) and sustainability domains (e.g., [19]). However, the environmental impact of AI systems and potential contribution to harm to

---

[2] Note that while the focus here is on negative impact, the intent is not at all to imply that there is no positive impact in terms of the benefits that AI technologies offer humanity. Far from it. The intent is simply to ensure that AI technologies are governed responsibly such that their potential benefit to current and future generations can be realized.

future societies in this regard should not be underestimated. In addition, from a slightly more negative angle, in the environmental sustainability context, Halsband [24] found, for instance that uses of AI technologies in decision support in climate change contexts included problematic assumptions "about social discounting and future persons' preferences", which seems to point to further work needed to cement intergenerational justice concerns in this context in sensible ways.

Turning now to the second lens through which to consider how AI ethics concerns relate to future social stability, in terms of what I term structural justice, I consider the issue of fairness[3]. Fairness usually refers broadly to accessible (findable) and reusable (interoperable) data - the so-called FAIR principles refer to findable, accessible, interoperable, and reusable data (see, e.g., [44]). But there is more. Data is sensitive because of concerns about the right to privacy, and also, because it may contain various kinds of bias based on ideological, historical, racial, epistemic, or other kinds of stereotyping linked to social identities. In the context of this paper the concern is that if humans' inherent prejudices - that already ensure social instability for present generations - are amplified or at least solidified by data-driven processes leading to predictions and decisions, future generations will be impacted too. This impact might also be more severe in future contexts than in current ones given additional factors such as the creation of echo chambers by online platforms and accompanying conformation bias[4], both of which amplify and perpetuate social divides and might contribute to future social and political instability.

Concerns about AI technologies' impact on fairness is perhaps best described in terms of what Crawford has identified as representation and allocation harm from data-driven AI [11]. Representation harm relates to identity prejudice and structural bias in society and the threat it brings is that existing social and political - and epistemic - stereotyping is amplified by machine learning algorithms' inherent tendency to latch onto perceived patters in its decision-making processes. Allocation harm is more concrete and immediate as here latching onto patterns plays out in terms of access to resources in domains such as banking, education, and healthcare. These kinds of harm are determining the outlines of future societies in a concrete manner, in the sense of establishing the boundaries between the haves and the have nots, and of individual autonomy to address such categorization.

This brings us to the third lens through which to consider including intergenerational justice in AI ethics and governance debates, which I have termed

---

[3] In the ethics of AI and the wider machine learning community, there have recently been more and more calls for a turn to 'critical machine learning' (e.g., [48], with the emphasis on the fairness (bias), transparency and accountability of machine learning systems in general, but specifically of ADM systems powered by the newest deep neural network research and technology. See for instance, [4, 14].

[4] An echo chamber in the online environment is a space in which the same views are shared without any opposing voices, which results in confirmation bias. Clearly echo chambers have the potential to lead to extreme views and social and political instability. See, e.g., [10, 43].

moral justice. I have in mind here the power of AI technologies for value change [42]. It is imperative to realize that this tendency of AI technologies to change human values is a layered issue [9].

Firstly, closest to the technical core of AI systems, is the fact that the very values that drive performance and efficiency are significantly defined by economic - and even political - values. Birhane et al. [9] ask why performance should necessarily be defined in terms of metrics of correctness rather than metrics of social justice, as success might easily also be defined in terms of benefit to a given community rather than exclusively in terms of accuracy or improvement or progress, and yet this does not happen.

Secondly, technical values impact on moral values in concrete ways when we think of structural justice discussed above, think for instance of the power of Large Language Models to contribute to deception (e.g., [8]). More concretely in the philosophy of human-technology relations, writers such as Verbeek [57] and Kudina [32,33] point to the co-shaping role that technology in general plays in determining the sociocultural spaces humans inhabit. In the context of AI technologies, for instance, Kudina [32] writes that Digital Voice Activators such as Alexa "...recontextualize the meaning of the world to the users that in turn, helps to shape them as specific subjects. This may change not only how we understand ourselves but also what we expect from those around us". Clearly this kind of "value-infused-ness" [46] holds significant implications for how the world in which future generations will find themselves, is shaped.

My argument is not that AI technology holds only bad news for future generations, but rather that framing responses to threats to social justice and human flourishing potentially posed by AI technology in intergenerational terms can contribute significantly to establishing concrete measures to underpin AI ethics and governance discourses. To better make this point, let us consider the issues raised in this section from the perspective of data activism in the next section, adding a fourth lens to our discussion in terms of data justice. This discussion, and the discussion in Sect. 4, consider the other part of the non-identity problem in intergenerational ethics, namely, to what extent present generations are in fact able to take interests of future generations into account in their thinking and actions.

3   Data Activism and Social Justice

The focus on future benefit of diverse data communities in data activist and data justice contexts links naturally to intergenerational debates. In this section we will unpack this link in more concrete terms in order to strengthen the call for ascribing (current and) future social justice issues a more central role in AI ethics debates through intergenerational discourse.

In a recent article in 'Communications of the ACM', Vardi [56] points to a curious disconnect between the Association for Computing Machinery (ACM) Code of Conduct [1] and the business model of Big Tech. The ACM code of Conduct states clearly that "computing professionals" can be said to be acting responsibly if they reflect on the "wider impacts of their work, consistently supporting the greater public good". On the other hand, Vardi [56] identifies the business model of Big Tech as that of surveillance capitalism [60] which is a model based on the collection and commodification of personal data for profit or personal gain. And, surely, as he concludes, there is a contradiction between the ACM code of conduct and this model as this model can certainly not outright be said to be aimed at the greater public good.

Keep in mind too that 'Big Tech' consists of a handful of companies that are monopolising the AI scene [18,31]. There are many voices warning of the danger of the hype around recent generative AI advances [7]. This hype is fuelled by prominent figures such as Sam Altman, CEO of Open AI [15,25] and many others. The problem with this hype is that if most of the public believe that AGI is about to be realised, the resulting hype not only financially benefits companies like Google and OpenAI, but also helps them avoid taking responsibility for the bias and harm that result from those systems [21].

Given the above, combined with the threats that AI technology holds for current and future human flourishing and well-being (some of which were discussed in the previous section), there are many voices - such as Emily Bender, Timnut Gebru, Lauren Goodlad, Stefania Milan, Lonneke van der Velden, Becky Kazansky, Alex Hannah, Linnet Taylor, Sasha Costanza-Chock and Lina Dencik - canvassing from various perspectives for what basically would be a profound upending of the current objectives and trajectory of AI business narratives from profit-based to future good- and sustainability-based narratives. Part of such changed narratives is the data activist call for a direct link between the use of human data in AI generated processes and the current and future benefits the predictions and decisions generated by such processes bring for the community whose data drives these processes. Milan and van der Velden [37] write that by "[p]ostulating a critical/active engagement with data, its forms, dynamics, and infrastructure, data activists function as producers of counter-expertise and alternative epistemologies, making sense of data as a way of knowing the world and turning it into a point of intervention" for future benefit, rather than creating a means for manipulation, human and environmental harm, or perpetuation of stereotyping and exclusion.

To provide an alternative to the way in which big tech companies or state agencies engage with data, data activists focus on mobilising data sets for social causes. To provide protection from what Big Tech - and state agencies - do with data, their focus is on developing and encouraging employment of technologies that frustrate massive data collection. Kazansky et al. [29] explain that "[p]roblematizing the mainstream connotations of big data, ... 'data activism' take[s] a critical stance to-wards massive data collection and represent[s] the new frontier of citizens' engagement with information and technological innovation".

In its turn, data activist research is therefore focused on more than doing no harm, more than lessening the moral and social impact of data-driven practices. In this sense, data activist research focuses on actively facilitating change for the better in communities in which data driven practices are applied, which means that data has to be representative of "the needs and [future] interests" of those the generated decisions and predictions are intended to support [29]. Thus, when formulating the notion of 'data activist research', Kazansky et al. [29] suggest a new approach to knowledge production combining "embeddedness in the social world with the research methods typical of academia and the innovative repertoires of data activists". As such, rather than presenting the world with another tick-box AI ethics, this kind of research is an active part of the research, design, development, deployment, and use of AI technologies, generating ethical values and their embeddedness in regulation from the bottom-up with an eye to the future flourishing of data communities in a dynamic, concrete, and interactive manner.

These debates play out in the bigger context of data justice which is a "key framework for engaging with ... datafication and society in a way that privileges an explicit concern with social justice" and denotes "a shift in understanding of what is at stake with datafication beyond digital rights" [13]. In this context, projects such as the Global Partnership on AI and Alan Turing Institute's 'Advancing Data Justice' Project [20], fill a gap in current data justice research by engaging with "individuals with diverse and contextually specific lived experiences of injustice or marginalization and those actively combatting these, in this case, as related to data" [52].

Data activism and data activist research offer a concrete and natural way to achieve engagement with intergenerational concerns, given the focus on future benefit for diverse data communities. The sensitivity to data justice and difference presents an opportunity for fluid bottom-up futuristic reflection. Furthermore, it is from within this broad context that I see intergenerational ethics playing a positive role in concretizing AI governance regulation. Intergenerational thinking can emphasize the urgency of the need for the world waking up to the call for not only data justice, but also environmental, structural, and moral justice, given how these forms of justice are linked to intergenerational justice concerns. In other words, it might very well be that the extent to which present generations are in fact able to take interests of future generations into account in their thinking and actions in the era of AI depends on whether such engagement with future societies happen from within social justice contexts.

4   Intergenerational Ethics as Driver for Actionable Responsible AI

The main challenge AI ethics regulation faces is to hook onto the world, to be rele-vant and dynamic and concrete and actionable. I suggest that considering intergenerational justice as a core AI ethics value is an effective way to ensure responsible AI debates are hooked to the real world. The very data-driven nature of AI algorithms that enables algorithms to predict future behavior based on past behavior in fact points to the need for intergenerational thinking in AI ethics discourses. This is so as any effective AI ethics impact assessment will consider what would happen to a future world if current tendencies were perpetuated or amplified (structural bias), or if AI algorithms would change the world and society in perceptible ways (environmental damage, value change).

I call for a more central role for social justice issues in AI ethics debates in general, and in particular, for AI ethics assessments that frame AI ethics principles in terms of social justice (here including at least environmental, structural, moral and data justice) concerns that relate to intergenerational justice, as this is a valid way in which to consider the conditions for the flourishing and well-being of future societies and the sustainability of future AI technologies in such societies. Endeavouring to limit and effectively address structural bias in training data, and to ensure both stable environmental contexts and value systems that humans can justify in terms of their membership of a global humanity, combined with data justice calls for protecting individual communities against Big Tech commodification of their data with no real benefit to these communities, will contribute significantly to the stability of future societies. This calls for paying attention to the identity of AI ethicists, to who decides what an ethical concern is, who will be impacted on by this concern, how it should be addressed, and when it would be effectively addressed. Thus, a bottom-up perspective guided by the intersection of intergenerational and social justice concerns will refine the formulation of AI ethics principles because such a perspective will ensure that these principles remain relevant in the face of a changing world.

The well-known international and minority studies scholar, Satya Mohanty [39], points out that "interpreting the world accurately requires knowing what it would take to change it" this can, according to him, only be done through "identifying relationships of power and privilege that sustains injustice". It is in response to this claim that I argue here for a social justice activist intersection with intergenerational thinking as the necessary driver for a new narrative for engaging with future societies in the age of AI. One can only become aware of what truly needs to be captured, represented, and protected in the AI domain if one is sensitive to linkages between social power and current contexts in which data is made available and predictions and decisions generated. For just and flourishing future societies and sustainable, just AI practices this sensitivity implies at its core understanding and engaging with the social context within which predictions will be interpreted and acted on now and in the future.

No one can claim that a business model aimed at profiting a handful of companies, and skewing human value systems, and pushing a technology that might exacerbate and entrench existing stereotyping, structural bias, and discrimination of various kinds and potentially harming the environment, will make for socially just and economically stable future societies. It is in this context that this paper is a call to realize the combined force of intergenerational thinking and social justice activism (through all the lenses discussed here, but also including other essential ones such as epistemic justice), and thus a call to change the current AI business model to one based on benefit for the communities whose data fuels AI algorithms and for future generations built on the resultant well-being and flourishing of current communities.

References

1. ACM Code of Ethics and Professional Conduct. https://www.acm.org/code-of-ethics. Accessed 27 August 2023
2. AI, Algorithmic, and Automation Incidents and Controversies (AIAAIC) Repository. https://www.aiaaic.org/aiaaic-repository. Accessed 26 Aug 2023
3. AI Ethics Global Inventory. https://inventory.algorithmwatch.org/. Accessed 26 Aug 2023
4. Ananny, M., Crawford, K.: Seeing without knowing: limitations of the transparency ideal and its application to algorithmic accountability. New Media Soc. **20**(3), 973–989 (2018). https://doi.org/10.1177/1461444816676645
5. Asay, M.: Is tech getting older or less ageist? The answer is complicated. TechRepublic (2022). https://www.techrepublic.com/article/is-tech-getting-older-or-less-ageist-the-answer-is-complicated/. Accessed 26 Aug 2023
6. Beck, U.R.S.: Towards a New Modernity. Sage, London (1992). [1986]
7. Bender, E., Hannah, A.: AI causes real harm. Let's focus on that over the end-of-humanity hype. Scientific American (2023), https://www.scientificamerican.com/article/we-need-to-focus-on-ais-real-harms-not-imaginary-existential-risks/
8. Bender, E.M., Gebru, T., McMillan-Major, A., Shmitchell, S.: On the dangers of stochastic parrots: can language models be too big? In: Proceedings of the 2021 ACM Conference on Fairness, Accountability, and Transparency, pp. 610–623 (2021). https://doi.org/10.1145/3442188.3445922
9. Birhane, A., Kalluri, P., Card, D., Agnew, W., Dotan, R., Bao, M.: The values encoded in machine learning research. In: Proceedings of the 2022 ACM Conference on Fairness, Accountability, and Transparency, pp. 173–184 (2022)
10. Brugnoli, E., Cinelli, M., Quattrociocchi, W., Scala, A.: Recursive patterns in online echo chambers. Sci. Rep. **9**(1), 20118 (2019)
11. Crawford, K.: The trouble with bias (2017). https://www.youtube.com/watch?v=fMym_BKWQzk
12. Danish, M.S.S.: AI in energy: overcoming unforeseen obstacles. AI **4**(2), 406–425 (2023). https://doi.org/10.14763/2022.1.1615
13. Dencik, L.: Data justice. Internet Policy Rev. **11**(1) (2022). https://doi.org/10.14763/2022.1.1615
14. Diakopoulos, N.: Algorithmic accountability: journalistic investigation of computational power structures. Digit. Journal. **3**(3), 398–415 (2015). https://doi.org/10.1080/21670811.2014.976411
15. DW Business Special: Max Tegmark Interview: Six Months to Save Humanity from AI? (2023). https://www.youtube.com/watch?v=ewvpaXOQJoU
16. EU Parliament News: EU AI Act. https://www.europarl.europa.eu/news/en/headlines/society/20230601STO93804/eu-ai-act-first-regulation-on-artificial-intelligence. Accessed 26 Aug 2023
17. European Commission EU Science Hub: Examining artificial intelligence technologies through the lens of children's rights. https://joint-research-centre.ec.europa.eu/jrc-news-and-updates/examining-artificial-intelligence-technologies-through-lens-childrens-rights-2022-06-22_en. Accessed 26 Aug 2023
18. Fernandez, R., Klinge, T.J., Hendrikse, R., Adriaans, I.: How big tech is becoming the government (2021). https://tribunemag.co.uk/2021/02/how-big-tech-became-the-government
19. Galaz, V., et al.: Artificial intelligence, systemic risks, and sustainability. Technol. Soc. **67**, 101741 (2021). https://doi.org/10.1016/j.techsoc.2021.101741

20. Global Partnership on AI (GPAI) and Alan Turing Institute: Advancing Data Justice. https://advancingdatajustice.org/. Accessed 26 Aug 2023
21. Goodlad, L.M., Baker, S.: Now the humanities can disrupt AI (2023). https://www.publicbooks.org/now-the-humanities-can-disrupt-ai/. Accessed 26 Aug 2023
22. Gosseries, A.: What do we owe the next generation(s)? Loyola Los Angeles Law Rev. **35**, 293–354 (2001)
23. Hagendorff, T.: The ethics of AI ethics: an evaluation of guidelines. Mind. Mach. **30**(1), 99–120 (2020). https://doi.org/10.1007/s11023-020-09517-8
24. Halsband, A.: Sustainable AI and intergenerational justice. Sustainability **14**(7), 3922 (2022). https://doi.org/10.3390/su14073922
25. Helmore, E.: 'We are a little scared'. OpenAI CEO warns of Risks of AI (2023). https://www.theguardian.com/technology/2023/mar/17/openai-sam-altman-artificial-intelligence-warning-gpt4. Accessed 26 Aug 2023
26. Heyward, C.: Ethics and climate adaptation. In: Gardiner, S.M., Thompson, A. (eds.) Oxford Handbook of Environmental Ethics. Oxford University Press, Oxford (2017)
27. IEEE Standards Association: Planet positive 2030. https://sagroups.ieee.org/planetpositive2030/. Accessed 26 Aug 2023
28. Jobin, A., Ienca, M., Vayena, E.: The global landscape of AI ethics guidelines. Nat. Mach. Intell. **1**, 389–399 (2019). https://doi.org/10.1038/s42256-019-0088-2
29. Kazansky, B., Guillén, T., Van der Velden, L., Wissenback, K., Milan, S.: Data for the social good: toward a data activist research agenda. In: Daly, A., Devitt, S., Mann, M. (eds.) Good Data, Theory on Demand #29, pp. 244–259. Institute of Network Cultures, Amsterdam (2020)
30. Klockman, V., von Schenk, A., Villeval, M.: Artificial intelligence, ethics, and intergenerational responsibility. J. Econ. Behav. Organ. **203**, 284–317 (2022). https://doi.org/10.1016/j.jebo.2022.09.010.hal-03778525
31. Knowledge at Wharton Podcast: A World without 'Mind': Big Tech's Dangerous Influence (2017). https://knowledge.wharton.upenn.edu/podcast/knowledge-at-wharton-podcast/world-without-mind/. Accessed 26 Aug 2023
32. Kudina, O.: "Alexa, who am I?" Voice assistants and hermeneutic lemniscate as the technologically mediated sense-making. Hum. Stud. 1–21 (2021). https://doi.org/10.1007/s10746-021-09572-9
33. Kudina, O., Verbeek, P.P.: Ethics from Within: Google Glass, the Collingridge Dilemma, and the mediated value of privacy. Sci. Technol. Hum. Values **44**(2), 291–314 (2019)
34. Martin, J.: AI's sustainability conundrum (2023). https://www.greenbiz.com/article/addressing-ais-sustainability-conundrum. Accessed 26 Aug 2023
35. McKinnon, C.: Endangering humanity: an international crime? Can. J. Philos. **47**(2–3), 395–415 (2017). https://doi.org/10.1080/00455091.2017.1280381
36. Meyer, L.: Intergenerational justice (2021). https://plato.stanford.edu/archives/sum2021/entries/justice-intergenerational/
37. Milan, S., Van der Velden, L.: The alternative epistemologies of data activism. Digit. Cult. Soc. **2**(2), 57–74 (2016). https://doi.org/10.14361/dcs-2016-0205
38. Mittelstadt, B.: Principles alone cannot guarantee ethical AI. Nat. Mach. Intell. **1**, 501–507 (2019). https://doi.org/10.1038/s42256-019-0114-4
39. Mohanty, S.P.: The epistemic status of cultural identity: on "beloved" and the postcolonial condition. Cult. Critique 41–80 (1993)
40. Morley, J., Floridi, L., Kinsey, L.: From what to how: an initial review of publicly available AI ethics tools, methods and research to translate principles into prac-

tices. Sci. Eng. Ethics **26**, 2141–2168 (2020). https://doi.org/10.1007/s11948-019-00165-5

41. Pellegrini-Masini, G., Corvino, F., Löfquist, L.: Energy justice and intergenerational ethics: theoretical perspectives and institutional designs. In: Bombaerts, G., Jenkins, K., Sanusi, Y., Guoyu, W. (eds.) Energy Justice Across Borders, pp. 253–272. Springer Open (2020)

42. van de Poel, I.: Embedding values in artificial intelligence (AI) systems. Mind. Mach. **30**(3), 385–409 (2020)

43. Ranalli, C., Finlay, M.: What's so bad about echo chambers? Inquiry (2023). https://doi.org/10.1080/0020174X.2023.2174590

44. van Reisen, M., et al.: FAIR practices in Africa. Data Intell. **2**(1–2), 246–256 (2020). https://doi.org/10.1162/dint_a_00047

45. Rittel, H.W., Webber, M.M.: Dilemmas in a general theory of planning. Policy Sci. **4**(2), 155–169 (1973)

46. Ruttkamp-Bloem, E.B.: The very notion of artificial moral agency and the case of artificial moral dispositions. In: Marmodoro, A., Bauer, W.A. (eds.) Artificial Dispositions: Investigating Ethical and Metaphysical Issues, pp. 193–218. Bloomsbury (2023)

47. SAP SE: Environmental Report for 2022 (2022). https://www.sap.com/integrated-reports/2022/en/environmental-performance.html. Accessed 26 Aug 2023

48. Selbst, A.D., Boyd, D., Friedler, S.A., Venkatasubramanian, S., Vertesi, J.: Fairness and abstraction in sociotechnical systems. In: Proceedings of the Conference on Fairness, Accountability, and Transparency, pp. 59–68. Association for Computing Machinery, New York (2019). https://doi.org/10.1145/3287560.3287598

49. Solum, L.B.: To our children's children's children: the problems of intergenerational ethics (2001). https://scholarship.law.georgetown.edu/facpub/873

50. Steele, J.: The need for a futurist mind-set (2021). https://www.britannica.com/topic/The-Need-for-a-Futurist-Mind-Set-2119786. Accessed 26 Aug 2023

51. Strubell, E., Ganesh, A., McCallum, A.: Energy and policy considerations for deep learning in NLP. arXiv:1906.02243 (2019). https://doi.org/10.48550/arXiv.1906.02243

52. Taylor, L.: What is data justice? The case for connecting digital rights and freedoms globally. Big Data Soc. **4**(2) (2017). https://doi.org/10.1177/2053951717736335

53. Thompson, C.: AI is thirsty (2023). https://clivethompson.medium.com/ai-is-thirsty-37f99f24a26e. Accessed 26 Aug 2023

54. UNESCO: Preliminary Report on the First Draft of the Recommendation on the Ethics of Artificial Intelligence (2021). https://unesdoc.unesco.org/ark:/48223/pf0000374266. Accessed 26 Aug 2023

55. UNICEF: Policy Guidance on AI for Children. https://www.unicef.org/globalinsight/reports/policy-guidance-ai-children. Accessed 26 Aug 2023

56. Vardi, M.Y.: ACM, ethics, and corporate behavior. Commun. ACM **65**(3), 5 (2022)

57. Verbeek, P.P.: Moralizing Technology: Understanding and Designing the Morality of Things. University of Chicago Press (2011)

58. World Economic Forum (WEF): AI Regulation through an Intergenerational Lens. https://www3.weforum.org/docs/WEF_AI_Regulation_through_an_Intergenerational_Lens_2021.pdf. Accessed 26 Aug 2023

59. World Economic Forum (WEF): Global Risks. https://www.weforum.org/global-risks. Accessed 26 Aug 2023

60. Zuboff, S.: The age of surveillance capitalism: the fight for a human future at the frontier of power. Public Affairs (2019)

AI Literacy: A Primary Good

Paige Benton[1,2]([⊠]) [iD]

[1] Department of Philosophy, University of Pretoria, Pretoria, South Africa
paige.benton@up.ac.za
[2] Centre for AI Research (CAIR), CSIR, Pretoria, South Africa

Abstract. In this paper, I argue that AI literacy should be added to the list of primary goods developed by political philosopher John Rawls. Primary goods are the necessary resources all citizens need to exercise their two moral powers, namely their sense of justice and their sense of the good. These goods are advantageous for citizens since without them citizens will not be able to fully develop their moral powers. I claim the lack of AI literacy impacts citizens' ability to exercise their sense of justice and their sense of the good. Without citizens having the ability to understand how AI technology works – including being aware of the social and political implications and the limits and possibilities of this technology broadly speaking – this could impact their ability to participate in a free, equal and fair society and their ability to carry out their conception of the good. Thus this paper is a call for AI literacy to be regarded as a basic good in a liberal constitutional democracy in order for citizens to be able to exercise their freedom and equality.

Keywords: Basic Needs · Ethics of AI · John Rawls · Justice as Fairness · Political Liberalism

1 Introduction

The aim of this paper is not to provide a technical solution for the implementation of AI literacy[1] 1 or add to the literature by expanding AI literacy competencies.[2] Instead, I provide a philosophical foundation for the normative claim namely, that AI literacy is a basic need that should be safeguarded and enforced by the state in liberal constitutional democratic societies. The philosophical foundation of this argument rests on demonstrating how and why AI literacy could be regarded as a primary good in John Rawls's political liberalism.

---

[1] In this paper I have focused on addressing why AI literacy as opposed to digital literacy should be regarded as a primary good. Digital literacy has a wider scope than AI literacy, where the latter is generally subsumed as a feature of the former [30]. The aim of focusing on AI literacy specifically and not digital literacy, in general, is to focus on the harms of AI technology on the moral powers of persons specifically given the timely nature of these harms in liberal democracies worldwide.

[2] For research providing technical solutions to the implementation of AI literacy see: [21, 23, 28, 29].

John Rawls is regarded as one of the twentieth century's greatest political philosophers by supporters and critics alike due to his impact on the field [14, 40]. He developed a liberal theory for a constitutional democratic society by developing the normative foundation for a just society in which people were free, equal, and engaged in fair mutual cooperation [35–37]. Rawls develops a set of five primary goods as the essential resources that are rational for all citizens to want regardless of the particular features of their lives or their conceptions of the good [38]. Primary goods are necessary goods insofar as they help citizens to realise their freedom and equality by enabling citizens to exercise their two moral powers, namely a sense of justice and a sense of the good [35].

I wish to extend Rawls's primary goods to account for a key feature of current circumstances of liberal democratic societies namely the infiltration of AI technology into both the public and private spheres of democratic life, which is impacting how citizens engage in these spheres.

AI literacy is an emerging field in which researchers are attempting to set out the competencies individuals need to critically and responsibly engage with and evaluate AI technologies [1, 25, 26]. There is no agreed-upon set of competencies as yet, or a universal framework in which to foster them. The aim of this paper is not to promote one set of skills or frameworks, but to provide a brief discussion of the competencies academics are currently arguing for in the field and suggest why governments should provide AI literacy in these terms as a basic good. I argue that AI literacy should be regarded as a primary good as AI literacy is a necessary good for citizens to fully exercise their sense of justice and their sense of the good.

To illustrate this argument I begin with a brief exposition of AI literacy in Sect. 2. Here I discuss the need for AI literacy and present a selection of the different kinds of competencies academics suggest are necessary to foster for persons to achieve AI literacy. Once I have discussed AI literacy, in Sect. 3, I provide an account of Rawls's justification for primary goods in his theory of justice. Both of these sections set the theoretical foundation to demonstrate why AI literacy should be a primary good in Sect. 4. In this section I first address how and why AI literacy is a need that is rational for all liberal democratic citizens to want regardless of their rational life plans, then I demonstrate how AI literacy helps citizens exercise their sense of justice and their sense of the good.

No current liberal democratic society is a Rawlsian society, yet Rawlsian theory is relevant for assessing the legitimate claims citizens can request from just institutions, since he developed the just conditions for how a well-ordered liberal democratic society should function.[3] The two key circumstances of justice for Rawls's theory of justice and liberal democracies, in general, are moral pluralism and the conception of the person as free and equal. Both circumstances dictate that a liberal democratic government cannot enforce a comprehensive conception of the good on its citizens as this would undermine core democratic values such as liberty of conscience. Primary goods are legitimate claims citizens can request from the government and the government can provide to citizens, as these goods are necessary no matter the other comprehensive conceptions citizens uphold. I argue that, similar to other primary goods, AI literacy is a rational resource for

---

[3] For further research on the intersection of John Rawls's political philosophy and research in AI see: [3, 18, 19, 50].

citizens to want regardless of their moral and social positioning. Before I arrive at this conclusion, it is now pertinent to discuss AI literacy.

2 AI Literacy

People have been engaging with AI-associated ideas in pop culture from the 1920's with Karel Capek introducing the term robot in his play Rossum's Universal Robots, yet public understanding of AI technology is plagued by misconceptions [5, 46]. Since the 1990s, interest in AI has increased in academia, industry, and pop culture with the rise of deep learning, neural networks, machine learning, and most recently large learning language models. OpenAI's ChatGPT and Google's Baard have exploded everyday persons' interest in actively[4] engaging with AI technology. The fact that this technology produces coherent responses to questions via language manipulation, makes it appear 'intelligent' to those who do not know how this technology works, thus perpetuating further potential misconceptions about the power of AI technologies [2, 12].

Fear of AI technology has increased in the public domain, with citizens concerned for future job prospects, and uncertain about how they are impacted by this technology in the background conditions[5] of their lives. For example, some individuals may wonder if an AI algorithm was 'responsible'[6] for them having a bank loan denied to being inundated with recommendations on social media. The lack of education and the complexity of AI system adds to this fear and misunderstanding of AI technology [4, 24, 27].

In the public domain academics, industry experts, policymakers, and civil associations have drawn attention to the ethical and social harms of AI technology [42]. The culmination of this emphasis is the call for the creation of laws, policies, and recommendations (i.e., AI Bill of Rights, EU AI Act, and UNCESO) to regulate the potential harms of AI technology and ensure that human rights and the dignity of individuals are safeguarded [44, 45, 48]. AI technology is becoming more integrated into the lived experiences of citizens around the world thus, there is a growing need for countries to educate their citizens on the potential impact of AI technology. Education aims to remove the misconceptions users have of AI technology due to the opaqueness of this technology to non-expert users and the misinformation citizens encounter in the public domain [7, 12, 43].

AI literacy is an emerging field in which researchers are attempting to set out competencies individuals need to critically and responsibly engage with and evaluate AI technologies [26]. As stated, there is no agreed-upon set of competencies as yet or a global framework in which to foster them, but there is a consensus that the competencies that need to be developed should not require everyday individuals to have expert

---

[4] By 'actively' I mean to suggest that citizens are choosing to want to use AI technology in an intentional manner to achieve a specific outcome.

[5] By 'background conditions' I mean to suggest that AI technology impacts citizens in a subtle manner, insofar as this technology is influencing persons' economic, political, social and cultural experiences in society.

[6] By 'responsible' I am not suggesting that AI technology has agency to be responsible rather I am implying that citizens may attribute blame to the technology and not the persons using the technology.

knowledge or discipline-specific skills [1, 25, 26]. Below I provide a brief outline of three conceptual frameworks for AI literacy to demonstrate the kinds of competencies may be necessary.[7]

The Artificial Intelligence (AI) for K-12 initiative (AI4K12) developed "five big ideas of AI" as the core ideas around which an AI literacy curriculum could be constructed [1]. Firstly, individuals should acquire an understanding of how computers perceive the world. Secondly, they need to understand how AI technology represents intelligence, and how the reasoning processes of technology are distinct from humans. Thirdly, people need to understand how learning takes place in AI technology by being introduced to machine learning and training data sets. The Fourth aspect is unpacking the mechanisms behind how AI technology expresses outputs in terms of human language. Lastly, developing an understanding of the social harms and advantages of AI technology on society [47]. These 5 ideas are overarching competencies that AI4K12 has developed into a framework for curriculum development which aims to foster different skill sets at different levels of education for these 5 ideas to be achieved by the end of grade 12.

A second conceptual framework for AI literacy is that developed by Long and Magerko [26]. They suggest 16 competencies[8] individuals need to develop to be critical knowers of AI technology: 1 The ability to recognise AI technology. 2 The capability to define and distinguish different types of intelligence. 3 The competency to understand the interdisciplinarity aspect of AI technology. 4 The proficiency to identify the difference between narrow and general AI. 5 The know-how to recognise the strengths and weaknesses of AI technology. 6 The aptitude to consider the future impact of AI technology on the world. 7 The ability to understand how AI technology represents knowledge of the world. 8 The capacity to know how AI technology arrives at decisions 9. Understanding the basics of Machine Learning without the need for a computer science degree. 10 Knowing that AI technology requires human decision-making, 11 Having a basic knowledge of data literacy. 12 The comprehension to how AI technology requires data to learn. 13 The capacity to exercise scepticism with data. 14 Developing insight into the physical impact AI technology can have on people and the environment. 15 Acquiring the knowledge of how computers engage with the world via sensors and how this impacts the data that can be gathered by robotic technology. Lastly, 16 being able to critically identify ethical concerns stemming from AI technology [26]. Unlike the AI4K12 framework, the framework by Long and Magerko is not a list of competencies for primary or secondary education, but rather a general list of competencies all citizens would need to acquire.

In contrast to the above frameworks, a third AI literacy framework developed by Kong, Cheung and Zhag [25], is one developed for university students. This framework contains three elements. Firstly, "AI concepts" require educating citizens on the basics of AI concepts (such as machine and deep learning) as set out by the OECD [33]. This

---

[7] The three frameworks I discuss in this section are not a comprehensive representative sample of all the frameworks in the field or how they are being implemented. For further frameworks see: [6, 11, 28, 29].

[8] There are 15 'design considerations' experts in AI need to consider when developing the technology to facilitate the development of these competencies. Due to space constraints, I have not discussed these considerations, see [26] for discussion.

ensures persons acquire the ability to understand how these technologies function in the world. Secondly, "AI concepts for evaluation", necessitates individuals being able to understand how and why AI is valuable and analysing the social impact of AI technology [25]. The third aspect is developing the ability to use AI concepts to solve real-world problems. This ability requires students to construct novel argumentations, and develop self-confidence for persons to become "future producers of technology and not just to enable the consumption of existing technology" [25].

These three frameworks for AI literacy differ in terms of the construction of their competencies, yet there is a significant overlap in terms of the skills all three frameworks want to foster in persons. All three frameworks aim to equip persons with knowledge of how AI technology works (sufficient for a nonexpert) and ensure participants in each framework can critically assess the ethical and social implications of this technology on their lives and the world around them. Thus, AI literacy research aims to remove the opaqueness of this technology to non-expert users and remove misconceptions about AI from public culture.

The aim of this section is not to present a comprehensive analysis of all AI literacy frameworks, but rather to illustrate the need for AI literacy and the kinds of competencies academics are arguing for in this field. In this paper, I wish to provide a philosophical foundation for AI literacy to be regarded as a primary good, in other words, a good that is necessary for all citizens in a liberal constitutional democracy to want regardless of what else they may desire. Now that I have addressed what AI literacy is, let us turn to primary goods.

3   Primary Goods

John Rawls's argument for primary goods rests on his conception of the person. Rawls develops this conception of the person from the historical tradition of liberal democracies. He claims that a just society is one in which persons are considered free, equal, rational, and reasonable [35]. These latter two mental faculties of the person Rawls regards as moral powers. Primary goods are the necessary resources citizens of a constitutional liberal democracy need to exercise their two moral powers [36]. The two moral powers citizens have in a liberal democracy are the sense of justice and the sense of the good. The sense of justice[9] refers to the ability individuals have to exercise, act from, revise, deliberate and uphold principles of justice, legislative principles, or any rule that is to the benefit of mutual cooperation for all citizens. In contrast, individuals exercise their sense of the good[10] when they develop a rational life plan[11] in which they decide on

---

[9] This moral power associated with the faculty of reasonability, see: [35].

[10] This moral power is associated with the faculty of rationality, see: [35].

[11] A rational plan of life refers to a person's chosen life ends/goals. This plan is informed by one's conception of the good, personal desires, affiliations, and loyalties. These aspects inform one's moral duties and obligations they have for persons in their private lives. Rational life plans are subjective, since a one's goals are determined by the social, economic, moral, and political environment an individual exists within [37].

their own conception of the good[12] to uphold and form associational bonds with others who uphold the same good while having the freedom to revise their conception of the good if desired [37].

One of the core circumstances of justice for liberal democracies is moral pluralism. Moral pluralism is the social condition that accounts for the co-existence of diverse moral beliefs. The philosophical underpinnings for moral pluralism are values such as toleration, liberty of conscience, and freedom of thought which form the justification for the separation of church and state [35]. Given moral pluralism, Rawls argues that a just society is one that does not promote comprehensive conceptions of the good. This promotion would undermine the freedom and equality of citizens as it would imply the state enforcing a moral truth claim onto its citizens [35]. To avoid this, Rawls suggests that the content of justice[13] needs to be that which all reasonable persons can agree to uphold along with their own chosen conception of the good [35].

The question Rawls then must solve is: How can one develop a conception of the good that all citizens need regardless of their chosen good? His solution is primary goods. Primary goods are considered objective goods because they are goods all citizens need regardless of their specific comprehensive conception of the good [38]. To put it another way, these are goods that are rational for all citizens to want since these are essential resources that citizens need to carry out a rational life plan [36].

Primary goods form the public justification for distributive justice, in other words, they provide the justification for political institutions to safeguard these resources and ensure citizens receive these resources so that they can develop and carry out a meaningful life plan. Because citizens cannot appeal to the state to help them achieve their conception of the good, citizens can only appeal to primary goods as the appropriate idea of the good the state can protect [35].

The five primary goods that Rawls argues are to the rational advantage of all citizens are: "(a) First, the basic liberties as given by a list, for example, freedom of thought and liberty of conscience; freedom of association; and the freedom defined by the liberty and integrity of the person, as well as by the rule of law; and finally the political liberties; (b) Second, freedom of movement and choice of occupation against a background of diverse opportunities; (c) Third, powers and prerogatives of offices and positions of responsibility, particularly those in the main political and economic institutions;(d) Fourth, income and wealth; and (e) Finally, the social bases of self-respect." [38].

Primary goods help citizens express and realise their freedom and equality as these goods are goods all citizens need to freely advance their own ends and the state should be structured so that it ensures that the distribution of primary goods is to the benefit of

---

[12] A conception of the good, is an umbrella term to refer to the moral values a person or community consider valuable to hold. These moral values can range from, secular belief systems to philosophical, metaphysical or religious doctrines [38].

[13] In addition to primary goods, the main content of justice is the two principles he proposes. Firstly, the 'liberty principle', which safeguards the equal basic rights and liberties of citizens. The second principle has a dual function; on the one hand, it safeguards fair equality of opportunity among citizens, and, on the other hand, it justifies social inequalities *iff* these inequalities are to the benefit of the worst-off members of society. This second aspect of the second principle is referred to as the difference principle [36].

the most disadvantaged citizens [36]. 'Disadvantaged' here would be citizens who have the lowest proportion of primary goods [36]. Primary goods are the "rights, liberties, and opportunities that all citizens need regardless of how these goods can be used as a means to achieve each person's rational life plans" [36]. For example, if person A and person B have reason to want primary good C expressed as the political liberty to be able to vote, then it becomes irrelevant if voting is more useful to person A than to person B in expressing their conception of the good. All that matters is that to a minimum degree, the right to vote is useful to some degree to both persons. Thus, although primary goods are useful to all citizens, there may be some primary goods that are more useful to some persons than others, yet these primary goods are still fundamental needs of all citizens in a liberal constitutional democracy.

Rawls states that what counts as primary goods depends on "social interdependence" [36]. Thus, the list of primary goods could be expanded if the circumstances of justice change, and so long as these new primary goods are for the rational advantage of all citizens regardless of the conception of the good to which they subscribe. Rawls gives examples of leisure time and freedom from physical pain as two possible primary goods if the list were to expand [36]. In the following section, I justify why I suggest the list of primary goods should be expanded to include AI literacy and the relevance of this for current constitutional liberal democracies.

4 The Necessity for AI Literacy to be Considered a Primary Good

AI literacy is a primary good, insofar as it is an objective good that is rational for all citizens in liberal constitutional democracies to want, no matter what their rational life plan is. Allow me to justify. AI literacy is broadly defined as the resources citizens can utilise so that they have an informed understanding of how AI technology works, know the impact this technology has on individuals and society, and have an awareness of the potential limits and possibilities of this technology [1, 25, 26, 30].

AI technology has become integrated into the fabric of economic, social, and political institutions across the world, as such citizens cannot escape the impact of this technology on their lives no matter what they consider their rational life plan or comprehensive conception of the good to be [22, 41]. Allow a toy example to illustrate my argument. Suppose there are two persons, Person A belongs to a social group that bans the personal use of digital and AI technology, while Person B is an online activist. One may argue that for Person A, AI literacy cannot be a basic need as Person A does not actively engage in the use of AI technology and thus has no need to understand how this technology works. In contrast, Person B as an online activist has direct and constant engagement with AI technology, is a consumer of this technology, and can influence others via the use of this technology. Hence, there is a basic need for AI literacy for Person B, so that she can be a responsible consumer. I thus suggest that Person A and Person B both need access to AI literacy regardless of their chosen rational life plans. The degree of the need may increase for Person B, but it would be a misconception to regard Person A as not having a need.

AI literacy is a basic resource for all citizens since having access to this resource would be advantageous for all citizens including both Person A and B. No matter the economic or social positions persons occupy, they are impacted by AI technology. This impact can be more obvious in cases when one actively chooses to engage with technology by using AI chatbots, smart watches, and generative or speech recognition technology to name a few, but it applies to all citizens at least at a minimum level. To some degree, citizens have the agency to choose not to wear smartwatches or use ChatGPT. But citizens have less[14] personal agency in terms of how AI technology impacts the background conditions of their lives, such as via personalised recommendations on digital platforms, or how their healthcare professional, financial and legal experts may utilise such technology when treating them as patients or clients [32].[15]

In the case of Person A even if their use of AI technology is limited, they have a responsibility to know how AI technology impacts them in the background conditions of their lives, such as being educated in terms of how their data is collected, stored and used. Although all persons occupy different social positions and have different rational life plans which can impact their reliance on AI technology, no person is exempt from this reliance, hence I argue that AI literacy should be regarded as a basic good provided by the political institutions of a just society.

Rawls uses the original position and veil of ignorance as an intuition pump[16] to model the conditions of fair agreement when thinking about the content of justice. When reasoning in line with this hypothetical position, Rawls states that one knows the general social circumstances of the society that they belong to [37]. These general circumstances can change given the historical conditions of the time [35]. Due to the rise of AI technological advancements in liberal democratic societies across the world today, this feature could form part of the general knowledge conditions a citizen would have in the original position when reasoning about justice today. Behind the veil of ignorance, one would not know how AI technology will impact them or how they would interact with this technology since they are unaware if they would be Person A or B in the above example. However, knowing that they would somehow be impacted, would provide the rational justification for one in this position to want the state to not only provide but safeguard this resource to ensure citizens receive access to AI literacy. Therefore, AI literacy could function like the other primary goods, such as the two primary goods of access to basic rights and liberties and income and wealth, as these goods are a "…practicable public basis of interpersonal comparisons based on objective features of citizens' social circumstances…" [35].

---

[14] By 'less' I mean their agency is reduced in comparison to the agency they have when they actively engage in choosing to use certain AI technology such as a smartwatch. It is reduced as these individuals do not have the ability to not engage with algorithmic recommendations for social media, rather they only have the agency to choose how they will engage, and this choice is constrained by factors external to them.

[15] Given the potential impact of AI technology on people's lives and the types of harms that could stem from this, there is a vast field of research on ethical guidelines for AI technology, see: [8, 17, 13, 20].

[16] Daniel Dennett coined the term 'intuition pump' to describe thought experiments that enable a reader to buy into the moral intuitions the relevant thinker wants the reader to grasp [10]. I refer to the original position as an intuition pump since, it is a device of representation that models the intuitions of fair agreement [35–37]. Rawls wants to justify the relevance of his principle's justice given the constraints of reasoning.

So far, I have demonstrated that AI literacy should be regarded as a primary good because all citizens need this good regardless of their comprehensive conceptions of the good. Now I need to show why AI literacy is necessary for citizens to exercise their moral powers. AI literacy contributes to citizens' full development of their moral powers since it helps provide citizens with the educational resources to understand how this technology can impact, on the one hand, their ability to revise, uphold their moral beliefs, carry out their rational plans of life, while on the other impacting their ability to deliberate about the principle's justice and legislative laws.

Algorithmic biases, epistemic bubbles, echo chambers, disinformation and deepfakes, are some of the current examples of harms arising from AI technology that undermine citizens' capacity to be free and equal [9, 31, 34]. Algorithmic biases can amplify historical inequalities by exacerbating the economic inequalities between citizens of different demographics [16]. This could undermine their sense of the good by citizens having to rearrange their rational life plans based on an unfair distribution of income per se. Algorithmic bias can promote societal prejudices that then become further entrenched in the public political culture of society [49]. This can lead to citizens not being able to exercise their sense of justice fully as they may experience testimonial injustices[17] and thus have been seen as an inferior epistemic knower when expressing their justification for legislative change in the public sphere [15].

Epistemic bubbles and echo chambers can limit individuals' access to some information while overexposing these individuals to like-minded perspectives, creating a coercive and unequal epistemic system [31]. This could undermine citizens' sense of the good as individuals may not have equal exposure to alternative moral belief systems and thus may not have the freedom to revise their beliefs. Deepfakes and disinformation have an impact on individuals' free and equal expression, they can undermine citizens' ability to freely deliberate about political events thus undermining one's sense of justice or manipulating individuals into preferencing certain conceptions of the good over others [39].

AI technology offers benefits to citizens to exercise their sense of the good and their sense of justice. For example, a religious group could use social media platforms to express their moral beliefs to a broader audience or expand membership of their group, they could develop chatbots to provide members with spiritual guidance on their website or use generative technology to develop content for seminars. In contrast, the leader from the same religious group could use the same AI technology to argue for reforming current laws or promoting a political stance by mobilising support on digital platforms, engaging in public debate with persons who hold opposing views, and relying on AI technology to send out automated communication to interests parties.

Without AI literacy the line between using AI technology to exercise one's sense of justice and one's sense of the good or using the technology to undermine these

---

[17] Fricker defines testimonial injustice as "…either the prejudice results in the speaker's receiving more credibility than she otherwise would have—a credibility excess—or it results in her receiving less credibility than she otherwise would have—a credibility deficiency" [15]. Both credibility excess and credibility deficiency pose as epistemic danger, in the discussion of an 'inferior epistemic knower' above, an injustice is present because of the latter, a deficiency of credibility.

capacities by harming other citizens has the potential to become entangled since the misconceptions surrounding AI and the opaqueness of the technology would remain in the public political culture. Thus, AI literacy provides citizens with the basic resources they need to understand and engage with AI technology in an ethical and responsible manner that helps facilitate the freedom and equality of all citizens.

So far I have demonstrated that AI literacy is an objective good insofar as it is a good that is rational for all citizens – in liberal constitutional democracies that employ AI technology – to want. Secondly, AI literacy is necessary for citizens to exercise their two moral powers. Without this literacy citizens face potential harms arising from AI technology and will not be equipped to recognise or respond to these harms. Hence, this paper is a call for the recognition of AI literacy to be regarded as a basic good. However, the argument I have presented in this paper for AI literacy as a primary good, could be extended to other kinds of literacies I have not discussed here but are relevant for citizens' participation in liberal constitutional democracies, such as economic, psychological, and civic literacies. One could argue that all three of these literacies are rational for citizens to want regardless of their conception of the good. Moreover, equipping citizens with non-expert knowledge of economic forces, psychological principles, and constitutional knowledge could help citizens better exercise their moral powers. Citizens equipped with these literacies would be able to make more informed decisions about their financial future, their interpersonal relations and their rights and responsibilities as citizens.

In this respect, AI literacy could be one literacy among many that needs to be recognized as a primary good. It would be interesting to reflect on whether education in general may be a primary good. However, traditional education might be safeguarded under the first primary good of equal basic rights and liberties whereas AI literacy is a good that would not naturally fall under this first good as it is not necessarily associated with traditional education systems in liberal constitutional democracies.

5   Conclusion

The argument I presented in this paper provides a philosophical justification for the importance of adding AI literacy to Rawls's list of primary goods. Just like access to income and wealth, AI literacy is an essential resource for the full development of citizens' moral powers. The lack of access to AI literacy can impede individuals' abilities to engage in free, equal, and fair cooperation. Without AI literacy citizens are at a high risk of having their freedom and equality infringed upon or hindering their fellow citizen's freedom by sharing deepfakes or contributing to misconceptions of AI technology.

If citizens lack the ability to understand how AI technology functions and are unaware of the potential harms and possibilities of this technology they could be at a disadvantage when participating in democratic processes (i.e., exercising their sense of justice) or carrying out their conception of the good (i.e., exercising their sense of the good). AI literacy is a basic need for citizens as having this resource safeguards citizens against the potential harms of AI technology. At the same time, AI literacy empowers citizens to use AI technology responsibly and assess the impacts of this technology critically.

Because AI-literate citizens would be educated to be aware of potential bias in algorithmic systems, they would be able to exercise scepticism when viewing videos that could be deepfakes and not be manipulated by fake news, in one instance.

By regarding AI literacy as a basic need liberal constitutional democracies would be taking a step towards helping ensure freedom and equality for their citizens. The gap between the haves (AI-literate) and have-nots (AI-illiterate) will continue to grow if governments do not actively intervene to provide citizens with this basic need. This growth could perpetuate further inequalities as those who are AI-literate could have a better ability to navigate the technological landscape and thus may have access to economic opportunities that AI-illiterate individuals may not. Thus, those that are AI-illiterate may not have the ability to exercise their autonomy as they would not have the ability to assess the validity of information produced by AI technology, this impacts their ability to freely engage in reasonable public deliberation or the rational pursuit of their good. Therefore, I claim that AI literacy is a good that is rational to want no matter what else a citizen wants, and hence should be regarded as a primary good.

References

1. AI4K12. https://ai4k12.org/. Accessed 20 Sept 2022
2. Bender, E.M., Shah, C.: All-knowing machines are a fantasy IAI TV - changing how the world thinks (2022). https://iai.tv/articles/all-knowing-machines-are-a-fantasy-auid-2334. Accessed 10 Mar 2023
3. Binns, R.: Algorithmic accountability and public reason. Philos. Technol. **31**(4), 543–556 (2017). https://doi.org/10.1007/s13347-017-0263-5
4. Bogen, M.: All the ways hiring algorithms can introduce bias. Harvard Business Review (2019). https://hbr.org/2019/05/all-the-ways-hiring-algorithms-can-introduce-bias. Accessed 10 Mar 2023
5. British Science Association, One in Three Believe that the Rise of Artificial Intelligence is a Threat to Humanity (2016). https://www.britishscienceassociation.org/news/rise-of-artificial-intelligence-is-a-threat-to-humanity. Accessed 20 Mar 2023
6. Burgsteiner, H., Kandlhofer, M., Steinbauer, G.: IRobot: teaching the basics of artificial intelligence in high schools. Proc. AAAI Conf. Artif. Intell. **30**(1), 4126–4127 (2016)
7. Burrell, J.: How the machine 'thinks': understanding opacity in machine learning algorithms. Big Data Soc. **3**(1), 1–12 (2016). https://doi.org/10.1177/2053951715622512
8. Cowls, J., King, T., Taddeo, M., Floridi, L.: Designing AI for social good: seven essential factors (2019). https://doi.org/10.2139/ssrn.3388669
9. Crawford, K.: The Trouble with Bias. NIPS Keynote (2017). https://www.youtube.com/watch?v=fMym_BKWQzk. Accessed 20 Sept 2021
10. Dennett, D.C.: Intuition Pumps and Other Tools for Thinking. W.W. Norton & Company, New York (2014)
11. Druga, S., Vu, S.T., Likhith, E., Qiu, T.: Inclusive AI literacy for kids around the world. In: Proceedings of FabLearn ACM, pp. 104–111 (2019)
12. Fast, E., Horvitz, E.: Long-term trends in the public perception of artificial intelligence. In: Thirty-First AAAI Conference on Artificial Intelligence, pp. 963–969 (2017)
13. Floridi, L., Cowls, J.: A unified framework of five principles for AI in society. Harvard Data Sci. Rev. **1**(1), 1–15 (2019). https://doi.org/10.1162/99608f92.8cd550d1
14. Forrester, K.: In the Shadow of Justice: Postwar Liberalism and the Remaking of Political Philosophy. Princeton University Press, New Jersey (2019)

15. Fricker, M.: Epistemic Injustice Power and the Ethics of Knowing. N.Y. Oxford University Press, New York (2007)
16. Friedman, B., Brok, E., Roth, K.S., et al.: Minimizing bias in computer systems. ACM SIGCHI Bull. **28**(1), 48–51 (1996). https://doi.org/10.1145/249170.249184
17. Hagendorff, T.: The ethics of AI ethics: an evaluation of guidelines. Minds Mach. 99–120 (2020).https://doi.org/10.1007/s11023-020-09517-8
18. Iason, G.: Toward a theory of justice for artificial intelligence. Daedalus **151**(2), 218–231 (2022)
19. Hoffmann, A. L.: Rawls, information technology, and the sociotechnical bases of self-respect. In: Shannon, V. (eds.) The Oxford Handbook of Philosophy of Technology, Oxford University Press, pp. 230–49 (2022)
20. Jobin, A., Ienca, M., Vayena, E.: The global landscape of AI ethics guidelines. Nat. Mach. Intell. **1**, 389–399 (2019). https://doi.org/10.1038/s42256-019-0088-2
21. Julie, H., Alyson, H., Anne-Sophie, C.: Designing digital literacy activities: an interdisciplinary and collaborative approach. In 2020 IEEE Frontiers in Education Conference (FIE), pp. 1–5 (2020)
22. Jungherr, A.: Artificial intelligence and democracy: a conceptual framework. Soc. Med. + Soc. **9**(3) (2023). https://doi.org/10.1177/20563051231186353
23. Kaspersen, M.H., Bilstrup, K.E.K., Petersen, M.G.: The machine learning machine: a tangible user interface for teaching machine learning. In: Proceedings of the Fifteenth International Conference on Tangible, Embedded, and Embodied Interaction, pp. 1–12 (2021)
24. Köchling, A., Wehner, M.C.: Discriminated by an algorithm: a systematic review of discrimination and fairness by algorithmic decision-making in the context of HR recruitment and HR development. Bus. Res. **13**, 795–848 (2020). https://doi.org/10.1007/s40685-020-00134-w
25. Kong, S-C., Cheung, W.M-Y, Zhang, G.: Evaluation of an artificial intelligence literacy course for university students with diverse study backgrounds. Comput. Educ.: Artif. Intell., 1–12 (2021).https://doi.org/10.1016/j.caeai.2021.100026
26. Long, D., Magerko, B.: What Is AI literacy? competencies and design considerations. In: Proceedings of the 2020 CHI Conference on Human Factors in Computing Systems, pp. 1–16 (2020). https://doi.org/10.1145/3313831.3376727
27. Menczer, F.: Here's Exactly How Social Media Algorithms Can Manipulate You. Big Think (2021).https://bigthink.com/the-present/social-media-algorithms-manipulate-you/. Accessed 20 May 2023
28. Ng, T.K.: New interpretation of extracurricular activities via social networking sites: a case study of artificial intelligence learning at a secondary school in Hong Kong. J. Educ. Train. Stud. **9**(1), 49–60 (2021)
29. Ng, T.K., Chu, K.W.: Motivating students to learn AI through social networking sites: a case study in Hong Kong. Online Learning **25**(1), 195–208 (2021)
30. Ng, D., et al.: Conceptualizing AI literacy: an exploratory review. Comput. Educ.: Artif. Intell. **2**, 100041 (2021). https://doi.org/10.1016/j.caeai.2021.100041
31. Nguyen, C.T.: Echo chambers and epistemic bubbles. Episteme **17**(2), 141–161 (2020). https://doi.org/10.1017/epi.2018.32
32. Naik, N., et al.: Legal and ethical consideration in artificial intelligence in healthcare: who takes responsibility? Front. Surg. **9**(862322), 1–6 (2022). https://doi.org/10.3389/fsurg.2022.862322
33. OECD.: Artificial Intelligence (2022). www.oecd.org. https://www.oecd.org/digital/artificial-intelligence/. Accessed 20 Aug 2023
34. O'Neil, C.: Weapons of Math Destruction: How Big Data Increases Inequality and Threatens Democracy, 1st edn. Crown, New York (2016)
35. Rawls, J.: Political Liberalism, revised Columbia University Press, New York (2005)

36. Rawls, J.: Justice as Fairness: A Restatement. Harvard University Press, Cambridge Massachusetts (2001)
37. Rawls, J.: A Theory of Justice, revised Harvard University Press, Cambridge, Massachusetts (1999)
38. Rawls, J.: Social Unity and Primary Goods. In: Freeman, S. (ed.) Collected Papers, pp. 359–387. Harvard University Press, Cambridge, MA (1999)
39. Ruiter, A.: The distinct wrong of deepfakes. Philos. Technol. **34**, 1311–1332 (2021). https://doi.org/10.1007/s13347-021-00459-2
40. Sandel, M.J.: Review of political liberalism. J. Rawls. Harvard Law Rev. **107**(7), 1765–1794 (1994). https://doi.org/10.2307/1341828
41. Smith, L., Fay, N.: What social media facilitates, social media should regulate: duties in the new public sphere. Polit. Q. **92**(4), 613–620 (2021). https://doi.org/10.1111/1467-923X.13011
42. Stahl, B.C.: Ethical issues of AI. In: Stahl, B.C., (Eds.) Artificial Intelligence for a Better Future An Ecosystem Perspective on the Ethics of AI and Emerging Digital Technologies. SpringerBriefs in Research and Innovation Governance SRIG, pp. 35–53. Springer, Cham (2021). https://doi.org/10.1007/978-3-030-69978-9_4
43. Stone, P., Brooks, R., Brynjolfsson, E., et al.: Artificial intelligence and life in 2030: one hundred year study on artificial intelligence. Report of the 2015 Study Panel, Technical report, pp. 1–52 (2016)
44. The Council of the European Union. Proposal for a Regulation of the European Parliament and of the Council Laying Down Harmonised Rules on Artificial Intelligence (Artificial Intelligence Act) and Amending Certain Union Legislative Acts (2021).https://data.consilium.europa.eu/doc/document/ST-8115-2021-INIT/en/pdf. Accessed 20 Aug 2022
45. The White House. Blueprint for an AI Bill of Rights. The White House (2022). https://www.whitehouse.gov/ostp/ai-bill-of-rights/. Accessed 20 Nov 2022
46. Thetechnopulse.: AI in Popular Culture: Shaping Perceptions & Inspiring Innovations. The Techno Pulse (2023).https://thetechnopulse.com/ai-popular-culture-impact/. Accessed 20 Aug 2023
47. Touretzky, D., Gardner-McCune, C., Martin, F., Seehorn, D.: Envisioning AI for K-12: what should every child know about AI? In: Proceedings of the 2019 Conference on Artificial Intelligence, pp. 9795–9799 (2019)
48. UNESCO.: Recommendation on the Ethics of Artificial Intelligence (2021). https://www.unesco.org/en/legal-affairs/recommendation-ethics-artificial-intelligence. Accessed 22 Nov 2022
49. Vlasceanu, M., Amodio, D.: Propagation of societal gender inequality by internet search algorithms. Proc. Natl. Acad. Sci. **119**, 1–8 (2022). https://doi.org/10.1073/pnas.2204529119
50. Weidinger, L., et al.: Using the veil of ignorance to align AI systems with principles of justice. Proc. Natl. Acad. Sci. **120**(18), 1–9 (2023). https://doi.org/10.1073/pnas.2213709120

Exploring the Ethical and Societal Concerns of Generative AI in Internet of Things (IoT) Environments

Bridget Chimbga[1,2]([✉]) [iD]

[1] Department of Philosophy, University of Pretoria, Pretoria, South Africa
`bridget.chipungu@gmail.com`
[2] Centre for AI Research (CAIR), Pretoria, South Africa

Abstract. Generative Artificial Intelligence (Generative AI) and the Internet of Things (IoT) are two rapidly evolving technologies that have the potential to revolutionise many aspects of our lives today. Generative AI, a subset of AI technologies, has a unique ability to create and manipulate content based on patterns learned from vast datasets, and when combined with the interconnected nature of IoT devices, they hold great promise across industries. Leveraging deep learning capabilities, Generative AI can intelligently analyze and apply the vast amounts of data generated from the internet to optimize operations and drive innovation. However, the potential integration of Generative AI and IoT systems in the future could raise notable ethical and societal concerns. These may encompass worries about privacy, security risks, biases and fairness challenges, lack of autonomy and human control, and emerging issues like the lack of ethical content creation. Given the relatively new nature of Generative AI technologies, this paper explores the ethical and societal concerns that may arise should Generative AI and the Internet of Things (IoT) environments integrate. By shedding light on these ethical and societal concerns, this research seeks to stimulate a nuanced understanding of the critical balance between technological advancement and responsible deployment, guiding stakeholders towards an ethically sound integration of Generative AI within IoT ecosystems.

Keywords: Generative AI · Artificial Intelligence · Internet of Things

1 Introduction

The Internet of Things (IoT) is a network of physical devices such as vehicles, appliances, and other gadgets that are embedded with sensors, software and network connectivity that allows them to collect and exchange data. The IoT allows these smart devices to communicate with each other and with other devices that are connected to the internet like smartphones and gateways to create a large network of interconnected devices that can exchange data and execute tasks independently [26]. The potential applications of the IoT are enormous and vast, and its impact is felt across a wide range of industries including manufacturing, transportation, healthcare, and agriculture. As the number of devices connected to the internet continues to grow, the IoT is likely to play an increasingly significant role in influencing the way we interact with each other [27]. With over 31 billion IoT devices expected to be in use by the year 2025, the need for processing and analyzing the substantial amount of data they generate is increasing [33]. As IoT devices become more prevalent, ethical, and societal concerns related to their use are being raised. The concerns of IoT use stem from the vast amount of data that IoT devices collect, process, and share. The data can be sensitive and personal, and its collection and processing without appropriate consent or control can lead to security and privacy breaches, discrimination, and authorized surveillance.

Artificial Intelligence (AI) is a technology designed to imitate human intelligence and its goal is to create systems that can imitate human intelligence and decision-making, sometimes even surpassing human abilities. AI plays a vital role in supporting economies, educational institutions, and societies, as well as improving production processes in areas such as healthcare, agriculture, manufacturing, banking, and finance [19]. With the advancement of AI, there is a growing need to consider its ethical and societal implications.

Ethical AI is about ensuring responsible AI usage, with a focus on safety, security, people's well-being, and environmental considerations [31]. This emphasizes the importance of reflecting deeply on the ethics surrounding its impact and how society can effectively address current and future challenges. Several countries and organizations now have their own guidelines regarding how AI should be used, even though the specifics may differ. The UNESCO Recommendation on the Ethics of AI, backed by 193 countries, advocates for AI to be accountable, transparent, and compliant with established rules. The aim of these recommendations is to harness the benefits of AI while mitigating potential risks. They also call for specific actions in significant areas such as education, culture, communication, gender, data governance, and the labor market.

The widespread integration and utilization of AI technologies have sparked numerous ethical and societal concerns, as evidenced by the following case studies. Within healthcare, AI algorithms play a critical role in guiding decisions about patient care. However, [34] exposed racial biases that were present in algorithms that were used in certain US hospitals. Despite Black patients facing more severe health conditions, the algorithm consistently assigned them the same risk levels as healthier white patients. This bias is attributed to the algorithm's reliance on health cost rather than the severity of the illness, leading to an inaccurate perception that Black patients are healthier for the same condition, subsequently reducing the identification of those requiring additional care.

The influence of AI extends into the financial sector, shedding light on gender bias. [18] highlighted significant disparities in interest rates and credit limits within the USA's Apple Card system, with men consistently receiving substantially higher credit limits, even when possessing comparable or superior credit scores to women. Additionally, with AI having been integrated into predictive policing mechanisms like the Correctional Offender Management Profiling for Alternative Sanctions (COMPAS), it was revealed that racial bias was observed in in a case where it was found that black defendants with similar criminal backgrounds were often categorized as having a higher risk of reoffending, underscoring the necessity to reassess the implementation of AI in the criminal justice system [25].

These case studies highlight the critical importance of addressing biases and ensuring fairness in AI systems. If left unaddressed, these biases can perpetuate social inequalities. Therefore, implementing mandatory actions and regulations is crucial to tackle the ethical and societal challenges posed by AI. These measures will help build an ethical framework for AI, promoting responsible use and mitigating the potential adverse impacts on individuals and society.

Advancements in Generative AI and the Internet of Things (IoT) have revolutionized various business sectors such as sales and marketing, human resources (HR), advertising industry, supply chain, automotive industry, and healthcare, promising unprecedented levels of automation, efficiency, and convenience [5]. Amongst the numerous AI applications, Generative AI stands out as a technology capable of creating new content, such as images, text, and audio, based on patterns learned from existing data [35]. The public introduction and rapid adoption of generative AI applications like ChatGPT and Dall-E raise additional ethical and societal concerns regarding AI. They introduce risks such as spreading misinformation, plagiarizing content, violating copyright laws, and generating harmful material. Hence, having a clear ethical framework for AI is highly important. It enables us to understand both the risks and benefits of AI tools and sets guidelines for their sensible application. Creating a clear set of moral principles and strategies for the responsible use of AI involves closely examining major societal concerns and, most importantly, contemplating what defines our humanity.

The UNESCO recommendations on Generative AI in Education address several concerns that include concerns about the potential widening of the digital divide, where access and benefits of AI may not reach all, exacerbating existing inequalities [12]. Another issue is the use of content without consent, emphasizing the importance of respecting intellectual property rights and ensuring that generated content is used ethically and with proper permissions. Furthermore, the ethical use of AI models, particularly those that are unexplainable or used without informed consent, raises important questions about accountability and transparency. The proliferation of AI-generated content on the internet, potentially contributing to misinformation and pollution of digital spaces, is a critical concern. Additionally, the recommendations highlight the risk of AI-generated "deepfakes" that can distort reality and underscore the need for vigilance and ethical guidelines to combat misinformation and deceptive practices in the digital realm [12]. It's important to address these controversies and challenges to ensure responsible and equitable use of Generative AI in and whilst UNESCO's recommendation focus on education, this paper takes a broader look into the concerns that may arise in the future if Generative AI technologies are integrated in IoT environments.

The expected ethical and societal impacts of deploying Generative AI in IoT ecosystems arise from the intricate relationship between intelligent Internet of Things (IoT) devices and AI-powered content generation systems and platforms. Although there is optimism regarding their convergence, some perspectives suggest that the integration of the two technologies might not be extensive. By examining the interplay between

Generative AI and the Internet of Things, the paper seeks to foster a deeper understanding of the potential ethical and societal concerns inherent in this evolving technological landscape.

2 Understanding Generative AI and IoT

2.1 The Intersection of Generative AI and IoT Technologies

Generative AI is a powerful computational innovation that comes up with new and meaningful content, such as the creation of textual narratives, images, and audio, derived from the internet [13]. It is built from foundation models that are driven by training codes from both labeled and unlabeled data. In the domain of Generative AI, foundation models are compelling computer algorithms that have learnt a lot of knowledge from extensive text analysis. They possess the ability to come up with imaginative tales, poetical verses, and even engage in dialogues [9]. These models excel in interpreting human languages and can assist in various tasks involving written and spoken words. They apply their acquired knowledge to come up with novel information that closely mirrors human expression, making them highly effective for tasks such as composition and conversation [20].

Generative AI goes beyond some machine learning algorithms by not only analyzing existing data but also generating new data that imitates and generalizes the patterns and characteristics of the original dataset. Its widespread use, exemplified by Dall-E 2, GPT-4, and Copilot, is revolutionizing todays' work and communication methods [13]. Different Generative AI models can produce content based on various inputs or "prompts," such as generating images from text (e.g., MidJourney, Stable Diffusion, DALL-E) or creating videos (e.g., Gen2 or Meta's Make-A-Video) [13]. According to industry reports, Generative AI has the potential to boost the global Gross Domestic Product (GDP) in the US by 7% and could potentially replace 300 million jobs held by knowledge workers [15].

The anticipated integration of Generative AI with the Internet of Things (IoT) has the potential to revolutionize various aspects of our lives, from smart homes to autonomous vehicles (Parikh, 2023). This possible convergence has sparked innovation in several industries leading to smarter solutions and increased market share for IoT products and services. Wienbot, an AI-powered chatbot in Vienna is using Facebook Messenger to provide answers to an array of user questions about city services to users [2] and the automation of customer service provisioning by city planners such as spotting traffic congestion using images from street cameras [11]. These use cases are capacitated by Generative AI's capacity to understand and generate content that empowers it to handle large volumes of data efficiently and effectively [17]. Another example is in smart homes where users can use Generative AI to detect suspicious activities to monitor conversations in the house with the goal of finding out potential security breaches by analyzing voice patterns recorded live and matching them against those of the homeowners, which would be stored on the cloud [11].

Schneider Electric, General Motors, the BMW Group, and Nokia are utilizing Generative AI and the Industrial Internet of Things (IIoT) to enhance their operations. In

Finland, Nokia is using a video application to alert assembly operators of any irregularities during the production process in their factories. Schneider Electric also came up with a predictive IoT analytics solution based on Microsoft Azure Machine Learning service and Azure IoT Edge to enhance the safety of workers, minimize expenditure, and achieve sustainability goals. General Motors has a system that analyzes images from cameras mounted on assembly robots that allows it to give indications of failing robotic components. Additionally, BMW uses Generative AI and the Industrial Internet of Things (IIoT) to improve production by simulating and improving processes, optimizing efficiency and cost effectiveness. [7]. These examples are in line with a report by Marketreseach.biz which states that globally, the Generative AI in the IoT market is expected to increase exponentially, from USD 947.8 million in 2022 to an estimated worth of USD 8,952.6 million by 2032, with a Compound Annual Growth Rate (CAGR) of 25.9% during the forecast period from 2022 to 2032 [14].

2.2 Ethical and Societal Concerns on the Intersection of Generative AI and the IoT

The rise of wireless technologies and smart cities has led to the widespread adoption of IoT devices such as smart cameras, sensors, smart phones, wearable devices, and home automation gadgets. These devices, interconnected wirelessly, can monitor conversations, locations, actions, and behaviors, raising privacy and ethical and societal concerns. Insufficient security measures in IoT devices can jeopardize personal data, potentially enabling malicious information usage. Furthermore, the collection, storage, and control of data by these devices give rise to worries about unauthorized access. These issues encompass security, confidentiality, informed consent, integrity, and more. The usage of IoT devices also brings up concerns related to discrimination and equality, potentially deepening the digital divide and societal disparities [27].

The rapid development and deployment of AI technologies present ethical and societal concerns, particularly when AI systems impact human lives through decision making. Concerns span from privacy, security, accountability, bias, discrimination to transparency in algorithmic processes. AI can perpetuate societal biases and treat certain groups unfairly [26]. Generative AI introduces its own set of ethical societal challenges such as biased training data that can lead AI models to generate disrespectful content or spread false information. This technology's ability to fabricate realistic yet untrue content raises doubts about authenticity and trustworthiness. Additionally, ownership and copyright of AI-generated content pose uncertain challenges. This could lead to workforce changes and displacement, sparking worries about its impact on affected individuals and communities [26].

3 Ethical and Societal Challenges of Generative AI in IoT

Ethics involves principles for organising, arguing, and recommending right and wrong behavior [26]. In the context of the IoT and Generative AI technologies, ethical and societal considerations encompass the moral responsibilities of the applications, users, and its developers [26]. The following section outlines some ethical and societal dilemmas and concerns expected at the convergence of Generative AI and the Internet of Things.

3.1 Ethical and Societal Concerns and Risks of Generated Content

The possible integration of Generative AI and the Internet of Things (IoT) will facilitate for the creation of AI-generated content such as text, images, or audio. However, the ethical and societal considerations surrounding the use of the generated content within the IoT ecosystem are significant. One of the key ethical and societal concerns is the potential for misinformation and deception. AI-generated content can be used to create deceptive or misleading information that can be disseminated through IoT devices and networks.

An example is the emergence of AI-powered influencers that are making waves online, disrupting social media with content that often rivals human influencers in terms of likes and engagement. These AI influencers create seemingly authentic posts, spanning from AI-generated travel photos to fictional travel vlogs as well as news articles. This trend started with influencer Lia (@_Lia27), who conducted an early experiment on platform X (Twitter) in 2021 [30]. The increase of AI-generated content on social media platforms distorts the boundary between truth and fabrication, potentially undermining, transparency, and authenticity in influencer interactions. The idea of mimicking human behavior by AI influencers by producing seemingly genuine content raises issues of deception, eroding trust, and honesty in online interactions. Moreover, the prominence of AI influencers could trigger worries about human content creators losing opportunities or jobs, particularly if AI-generated content gains more popularity, affecting their income and roles on social media [30].

The extensive use of AI-generated content in several sectors such as Literature and Humanities research may decrease the unique voice of human authors. This issue raises a lot of concern about content plagiarism, intellectual property and copyright infringement as well as undermining the validity of original material produced [1]. AI generated content that is not properly labeled can potentially deceive readers regarding its origin and this raises questions around transparency and authorship. This concern is exemplified by the case in which authors Sarah Silverman, Richard Kadrey, and Christopher Golden have jointly sued Meta, claiming copyright infringement for using pirated copies of their 'Books3' dataset of nearly 200,000 books in training LLaMA, Meta's Generative AI model [29]. This case has raised concerns over copyright, patent, and trademark infringement in AI creations [1], particularly after MediaTek announced plans to integrate Meta's LLaMA with IoT systems such as smartphones, smart vehicles, and smart home gadgets for home automation [23].

Another ethical and societal concern is the lack of protection of intellectual property rights in a case where AI-generated content may rely on the use of copyrighted materials, trademarks, or other proprietary information. For instance, BMW is using Generative AI and Industrial Internet of Things (IIoT) to optimize its manufacturing processes [7]. The use of copyrighted materials to train Generative AI models could result in allegations of copyright infringement. Therefore, organizations and individuals need to ensure that the Generative AI systems they adopt for their IoT applications do not infringe upon the intellectual property rights of others.

The proliferation of internet technology and internet-enabled smart devices has created a platform for several people to generate disinformation narratives easily and cheaply such that it is now increasingly difficult to detect. With the prevalence of Generative AI, deepfakes have been on the increase leading to potential reputation damage, privacy violations and targeted harassment. In May 2023, a notable case of misinformation received global attention when a fake image showing an explosion near the Pentagon in the USA circulated on Twitter (now X). The realistic image prompted the USA Department of Defense and local authorities to counter the false information with official statements. The impact extended to the stock market in the USA, which experienced a brief downturn before rebounding. Analysis by experts later suggested that the image was likely generated by Artificial Intelligence (AI) [6]. Given IoT systems' capacity to analyze data used to provide insights and inform decision making, if such AI-generated fake data infiltrates the dataset, it could lead to inaccurate analytics, misleading predictions based on false information as well as raising concerns that could emanate from the incorrect data processing.

3.2 Fairness and Bias Concerns and Risks

Generative AI models that undergo training using biased datasets have the potential to amplify existing societal biases. For example, when biased training data is used in healthcare IoT systems, it could lead to disparities in diagnosing conditions or recommending treatments for various demographic groups. For instance, a study conducted by the University of East Anglia in the UK suggests evidence of a left-wing bias in ChatGPT, an OpenAI-developed Generative AI model. This bias reportedly manifests as a preference for the UK's Labour Party and the US Democratic Party. The study suggests that given the growing public use of and reliance on Artificial Intelligence platforms such as ChatGPT, concerns about fairness and bias may increase particularly as the use of ChatGPT and its integration in commercial and consumer platforms rises [24].

Beyond that, complex algorithms in Generative AI models and IoT systems can introduce unintended bias, potentially yielding discriminatory outcomes or reinforcing societal biases. For example, a biased placement of IoT sensors in specific neighborhoods can result in unequal data collection, while biased algorithms in IoT systems might lead to an unjust distribution of resources for specific groups through mass surveillance and predictive policing. In a case involving AI and IoT bias, particularly affecting people of color, Artificial Intelligence-driven facial recognition system wrongly identified a pregnant woman, Porcha Woodruff, as being involved in a violent carjacking after an outdated photo of her was matched with surveillance footage online [10]. When viewed in the context of the possible integration Generative AI and IoT, the Porcha Woodruff's case provides insights into potential biases and challenges that can arise in models that mirror biases embedded in their training data. Such concerns are being raised about a company named Clearview AI that has amassed around 30 billion images scraped from platforms like Facebook and other social media sites obtained without users' consent [32]. Critics argue that the utilization of Clearview AI by the police effectively places everyone in an ongoing police lineup. Clearview AI's system empowers law enforcement agencies to upload a facial photo and identify matches in a database containing billions of collected images. The system then provides links to where those matching images appear online. Whilst Clearview AI is recognized by some as one of the most potent and accurate facial recognition companies worldwide [6], its use has been challenged by activists because it perpetuates racial inequalities in policing. It is therefore imperative for companies to meticulously curate and assess training data to ensure equity and prevent the reinforcement of societal biases.

3.3 Privacy and Security Risks and Concerns

Another risk linked to the use Generative AI tools is the potential for data breaches or leaks and this is because these tools are designed to generate content automatically, they often collect and store large amounts of data. Additionally, the complexity and vast interconnectedness of IoT devices present privacy and security challenges because each device acts as a potential entry point for attackers. Therefore, the incorporation of Generative AI in IoT environments has potential to introduce an additional layer of complexity, given the resource-intensive nature of these models and the potential for introducing new avenues for attacks.

In their study, [8] investigate the security risks associated with Generative AI models focusing on ChatGPT as a prime example. The study raised concerns over a range of risks, such as malicious text and code generation, disclosure of personal information, fraudulent services, and the production of unethical content. Despite ongoing efforts to develop ethical and secure AI systems, their analysis highlights that ChatGPT still possesses the ability to generate inappropriate content, showcasing the far-reaching nature of these risks. These concerns are the reason why some financial institutions like JP Morgan Chase and Deutsche Bank have restricted or banned the use of ChatGPT in their workplaces [36].

Privacy and security make up a significant concern for individuals, organizations, and society but unfortunately the risks involved aren't obvious to many [28]. For instance, Tinder is currently testing a Generative AI tool aimed at enhancing user profiles by scanning users' photo albums and selecting the five most appealing photos for their profiles [21]. Whilst there may be potential advantages of integrating Generative AI into dating applications, there remain concerns such as the utilization of Generative AI for photo analysis as well access to personal images and data. The concerns require those responsible to consider increasing security measures to prevent the potential of unauthorized access or breaches. Additionally, the incorporation of Generative AI in profile selection could steer the dating experience towards a more mechanized approach, potentially overshadowing human intuition. Furthermore, should Generative AI-driven photo selection become standard, users might feel compelled to conform to AI-optimized beauty standards, possibly impacting their self-esteem and self-worth [21].

The future convergence of Generative AI and IoT may raise concerns about security and privacy infringement in cases where sensitive information might be inadvertently exposed or misused without proper consent. To mitigate these privacy and security challenges, it is imperative to adopt a multi-layered security approach. This idea includes secure Generative AI model design, robust authentication and access control mechanisms, and encryption techniques to protect data collection, transmission and storage. Regular security updates and patches will need to be implemented to address any identified vulnerabilities in both the Generative AI models and the underlying IoT infrastructure.

3.4 Autonomy and Human Control Challenges

As Generative AI and IoT systems advance in sophistication, there arises potential implications about diminishing human agency and decision-making authority in the ecosystem. The possible intersection of Generative AI, the Internet of Things and digital identity ownership raises concerns especially with regards to transparency, hence striking the appropriate balance between AI autonomy and human control is imperative to uphold human values, preferences, ethical and societal considerations at the core of the integration. A case in point is the emergence of companies like Hyperreal, that specialize in crafting digital identities or AI digital twins known as Hypermodels [3]. These Hypermodels are meant to be used in perpetual productions, even after the death of the individual. The increased use of digital twins and trailers generated by AI tools raises a lot of ethical and societal concerns in the entertainment industry raising fears of potential consequences [16].

Access to the same AI algorithms and tools raises the risk of sidelining originality and creativity in favor of automated imitation. This situation could possibly stifle the emergence of new genres in film, art, or music, as we find ourselves trapped in a cycle of recycled ideas and replicated content. Such a restraint on innovation is a frightening prospect, raising concerns for both artists and their audiences. The possible influence of Generative AI and IoT on the entertainment industry goes beyond creative concerns, it may carry significant consequences for labor and the livelihoods of industry professionals. This technological shift may lead to job losses among actors, writers, and other creative talents. Furthermore, the possibility for studios to create entire characters without the need for human actors may threaten the demand for human talent, putting the livelihoods of countless individuals at risk. This struggle will not only impact the rights and dignity of actors but also threatens the preservation of artistic integrity.

Autonomy has held a central role in moral and political philosophy for centuries, establishing itself as a pivotal element of both justice and well-being [4]. Therefore, it is important to safeguard autonomy and human control from the onset in terms of systems design all the way to consent of data collection and processing. Users must possess a comprehensive understanding of how their data is utilized, and they should have the liberty to opt out if they so desire. Additionally, the content derived from these models must exhibit accuracy and respect towards the individuals they represent as well as refraining from misrepresentation or manipulation.

3.5 Truthfulness and Accuracy

The future of Generative AI technologies on the Internet of Things environment has potential to produce deceptive, damaging, or inappropriate content in sectors such as marketing and education. In the field of marketing, Generative AI can be used for unethical business practices, such as manipulating online reviews for marketing purposes or for creating thousands of accounts with false identities. These concerns can be exacerbated when companies use IoT devices such as smart TV (advertising) and smart phones (social media) to push marketing content to end users [22]. The potential integration of Generative AI technology into smart IoT devices, may enable businesses to have the ability analyze extensive data and identify patterns in consumer behavior enabling the creation of well-structured marketing strategies. However, if the generated content is not adequately reviewed and edited, it may result in inaccuracies or misleading information, potentially causing harm to a brand's reputation.

In education, technological advancements and the internet had the most profound impact in the way we teach and learn. The Internet of Things (IoT) has allowed education and schooling to be an environment of technology learning and learning. Features on the Internet of Things technologies make learning more accessible, affordable, and convenient for students thus improving the quality of teaching. From online courses to integrated mobile technology, the IoT-enabled connectivity continues to replace pencils, paper, and chalkboards as instructional approaches as learning opportunities evolve.

However, since its initial release, Generative AI tools like ChatGPT have been subjects of hot debate in education. The concerns of Generative AI such as being misused for generating false or misleading information that is presented as fact, have been topical. Owing to the increase in online learning during the COVID (19) shutdown, schools are incorporating IoT-enabled learning into their educational platforms to encourage interactive learning. However, the emergence of Generative AI systems such as ChatGPT in the IoT educational platforms may cause students to be misled [9]. Generative AI models can generate fabricated information, a phenomenon researchers describe as "hallucination." According to [9], ChatGPT has been positioned as an alternative to search engines like Google for seeking out information and students can become particularly vulnerable to misinformation thus diminishing literacy skills. Moreover, ChatGPT can be used to create material that is not only factually incorrect but also ideologically biased and can peddle conspiracy theories or cite fake scientific studies.

4 Conclusion

The possible convergence of Generative AI with IoT systems represents a significant step in harnessing the immense potential of data generated by smart devices. By leveraging the abilities of Generative AI, there is a possibility of unearthing valuable insights, promote innovation, reinforce security, and establish adaptable systems, pushing the IoT landscape towards heightened efficiency and productivity. This transformation ushers in a state of limitless possibilities for the IoT sector, paving the way for a more intelligent and interconnected world.

However, amidst these promising discussions, this paper offers an anticipatory glimpse into the ethical and societal concerns that demand attention regarding the possible integration of Generative AI and the IoT. Ensuring the ethical and responsible integration of Generative AI within IoT environments involves dealing with various legal and ethical considerations, including intellectual property rights, privacy, data protection, and human autonomy.

To navigate this anticipated landscape, staying informed and up to date with the current and future developments in Generative AI and the Internet of Things is essential. Its critical to be familiar with efforts led by Organisations and institutions researching on Generative AI and the Internet of things such as, The International Telecommunications Union (ITU) that is exploring the risks of Generative AI in smart cities as well as UNESCO that has recently published a document on the guidance for Generative AI in

Education and Research [12]. Cultivating awareness and understanding the future impact of Generative AI and the Internet of Things in businesses, individuals, and society hinges on how we address the risks it presents.

Furthermore, in the realm of regulations and policies, the development and subsequent enforcement of regulations that mandate ethical and societal practices for Generative AI within IoT environments need to be considered. These regulations will need to set standards for how Generative AI systems interact with the IoT devices, ensuring data usage and content generation adherence to ethical and societal guidelines. This includes rules for data collection, storage, and processing, helping mitigate potential privacy breaches and security vulnerabilities. Additionally, guidelines for responsible model deployment that prioritize user well-being and prevent negative consequences need to be developed. Additionally, balancing automation with human involvement is also important to leverage the full potential of Generative AI while mitigating potential negative consequences.

In the context of education and awareness, raising awareness and educating users about AI-generated content and the IoT will be crucial for promoting responsible Generative AI interactions in IoT environments. Users will need to be informed to be able to make decisions about their interactions with systems based on transparency, solid understanding of the capabilities, benefits, and potential risks of AI-generated content in the IoT environments. Providing training and resources to Generative AI developers will equip them with the knowledge and tools to make ethical and societal decisions throughout the development process. Equally vital will be importance raising public awareness through campaigns and initiatives that emphasize the importance of data privacy, security, ethical and societal considerations in Generative AI-driven IoT interactions.

Finally, it is important to note that the extent and implications of this convergence are subjects of ongoing discourse. Whilst complete convergence may not materialize or could be restricted, the exciting prospects of such convergence persist due to the associated benefits.

References

1. Appel, G., Neelbauer, J., Schweidel, D.A.: Generative AI has an intellectual property problem. Harvard Business Review (2023). https://hbr.org/2023/04/generative-ai-has-anintellectual-property-problem
2. Arkaraprasertkul, N.: SmartCityGPT: How generative AI creates smart and sustainable cities (2023). https://doi.org/10.13140/RG.2.2.14090.03522
3. Black, J.: 'don't put your head in the sand': stars are quietly inking deals to license their AI doubles. The Information (2023). https://www.theinformation.com/articles/dont-putyour-head-in-the-sand-stars-are-quietly-inking-deals-to-license-their-aidoubles?utm_content=art iclen11101&utm_source=sg&utm_medium=email&utm_campaign=article_email
4. Calvo, R.A., Peters, D., Vold, K., Ryan, R.M.: Supporting human autonomy in AI systems: a framework for ethical enquiry. In: Burr, C., Floridi, L. (eds.) Ethics of Digital Well-Being. PSS, vol. 140, pp. 31–54. Springer, Cham (2020). https://doi.org/10.1007/978-3-030-50585-1_2
5. Chui, M., Manyika, J., Miremadi, M.: What AI can and can't do (yet) for your business (2018). https://www.mckinsey.com/capabilities/quantumblack/our-insights/what-ai-can-and-cant-doyet-for-your-business

6. Clayton, A.: Fake AI-generated image of explosion near Pentagon spreads on social media (2023). https://www.theguardian.com/technology/2023/may/22/pentagon-ai-genera ted-image-explosion

7. Columbus, L.: 10 ways AI is improving manufacturing in 2020. Forbes (2023). https://www.forbes.com/sites/louiscolumbus/2020/05/18/10-ways-ai-is-improvingmanufacturing-in-2020/?sh=295d9ba31e85

8. Derner, E., Batistič, K.: Beyond the safeguards: exploring the security risks of chatGPT. arXiv.org. https://arxiv.org/abs/2305.08005. Accessed 10 Aug 2023

9. Dilmegani, C.: Generative AI ethics: Top 6 concerns. AI Multiple (2013). https://research.aimultiple.com/generative-ai-ethics/#easy-footnote-bottom-2-59046

10. Eberhart, C.: AI facial recognition led to 8-month pregnant woman's wrongful carjacking arrest in front of kids: Lawsuit, Fox News (2023). https://www.foxnews.com/us/ai-facialrec ognition-led-8-month-pregnant-womans-wrongful-carjacking-arrest-front-kids-lawsuit

11. Farina, M.: Chat GPT in smart home systems: prospects, risks, and benefits. J. Smart Environ. Green Comput. **3**, 37–43 (2023). https://doi.org/10.20517/jsegc.2023.11

12. Fengchun, M., Wayne, H.: Guidance for generative AI in education and research. https://une sdoc.unesco.org/ark:/48223/pf0000386693

13. Feuerriegel, S., et al.: https://www.researchgate.net/publication/370653602_Generative_AI (2023)

14. Generative AI in IOT market size, share, trends and forecast 2032 (2023) MarketResearch.biz (2023). https://marketresearch.biz/report/generative-ai-in-iot-market/

15. Goldman Sachs.: Generative AI could raise global GDP by 7% (2023). https://www.goldma nsachs.com/intelligence/pages/generative-ai-could-raise-global-gdp-by-7-percent.html

16. Grimes, C.: Hollywood actors seek new deal over use of AI 'Digital Doubles'. Financial Times (2023). https://www.ft.com/content/ffa4e333-b691-4964-ada1-20bf6ce60396

17. İbrahim., İ., Okay, F.Y., Özdemir, S.: FogAI: an AI-supported fog controller for next generation IOT. Internet of Things (2022). https://www.sciencedirect.com/science/article/abs/pii/S25426 60522000646

18. Knight, W.: The Apple Card didn't "see" gender-and that's the problem. https://www.wired.com/story/the-apple-card-didnt-see-genderand-thats-the-problem/

19. Lawton, G., Wigmore, I.: What are AI ethics: definition from What is. https://www.techta rget.com/whatis/definition/AI-code-of-ethics

20. Li, F.-F., et al.: Generative AI: perspectives from Stanford Hai. Genera-tive_AI_HAI_Perspectives (2023). https://hai.stanford.edu/sites/default/files/202303/Gen erative_AI_HAI_Perspectives.pdf

21. de Luna, E.: Tinder will soon let Ai pick your dating profile photos for you. Mashable (2023). https://mashable.com/article/tinder-ai-profile-photos

22. Macpherson, L.: Lies, damn lies, and generative artificial intelligence: how GAI Automates Disinformation and what we should do about it. Public Knowledge (2023). https://public knowledge.org/lies-damn-lies-and-generative-artificial-intelligence-how-gaiautomates-dis information-and-what-we-should-do-about-it/

23. Mathur, C.: Mediatek teams up with meta for smarter generative AI on your phone (2020). https://www.androidpolice.com/mediatek-meta-smarter-generative-ai-android/#:~:text= Chip%20maker%20MediaTek%20announced%20plans,a%20connection%20to%20cloud% 20infrastructure.

24. Motoki, F., Neto, V.P., Rodrigues, V.: More human than human: measuring chatGPT political bias. Public Choice (2023).https://doi.org/10.1007/s11127-023-01097-2

25. O'Brien, T.: Compounding injustice: the cascading effect of ... - georgetown law. https://www.law.georgetown.edu/mcrp-journal/wp-content/uploads/sites/22/2021/05/GT-GCRP21 0003.pdf

26. Parikh, N.: The software product management framework is not the software product manager's framework a systematic literature review. https://doi.org/10.2139/ssrn.4450114. (2023)
27. Ramasamy, L., Seifedine, K.: Internet of things (IoT) (2021). https://doi.org/10.1088/978-0-75033663-5ch1
28. Rawat, B., et al.: AI based drones for security concerns in Smart Cities, APTISI Transactions on Management (ATM) (2023). https://doi.org/10.33050/atm.v7i2.1834
29. Reisner, A.: Revealed: the authors whose pirated books are powering generative AI, The Atlantic (2023). https://www.theatlantic.com/technology/archive/2023/08/books3-ai-metallama-pirated-books/675063/
30. Sweat, Z.: The rise Of AI-generated influencers: a new era in social media? MSN (2023). https://www.msn.com/en-us/news/technology/the-rise-of-ai-generated-influencers-anew-era-in-sociamedia/arAA1fiYhN?ocid=msedgdhp&pc=U531&cvid=b623f6e3751546eb8c6da0515b38f311&ei=56
31. Talagala, N.: AI ethics: what it is and why it matters. https://www.forbes.com/sites/nishatalagala/2022/05/31/ai-ethics-what-it-is-and-why-it-matters/?sh=d0341003537a
32. Tangalakis-Lippert, K.: Clearview Ai scraped 30 billion images from Facebook and other social media sites and gave them to cops: It puts everyone into a "perpetual police line-up (2023)." https://www.businessinsider.com/clearview-scraped-30-billion-images-facebook-police-facialrecogntion-database
33. The impact of generative AI on smart, connected devices: the impact of generative AI on smart, connected devices (2023). https://www.neubloc.com/ai-devices.html
34. Vartan, S.: Racial bias found in a major health care risk algorithm. https://www.scientificamerican.com/article/racial-bias-found-in-a-major-health-care-risk-algorithm/
35. Wieboldt, E.: Beyond the black box: unpacking the impacts of generative AI in academia, ICAI (2023). https://academicintegrity.org/resources/blog/113-2023/may2023/435-beyond-the-black-box-unpacking-the-impacts-of-generative-ai-in-academia
36. Zhang, J.: The data security risks of using third-party generative AI tools - Wiz Ai, WIZ AI - A.I Conversational Talkbot Platform (2023). https://www.wiz.ai/the-data-security-risksof-using-third-party-generative-ai-tools

Warfare in the Age of AI: A Critical Evaluation of Arkin's Case for Ethical Autonomy in Unmanned Systems

Maxine Styber[1,2,3]([✉]) [iD]

[1] Department of Philosophy, University of Pretoria, Pretoria 012, Gauteng, South Africa
U20428376@tuks.co.za
[2] The Centre for Artificial Intelligence Research, Rondebosch, Cape Town 7701, South Africa
[3] Council for Scientific and Industrial Research, Pretoria 001, Gauteng, South Africa

Abstract. The threat posed by Lethal Autonomous Weapon Systems (LAWS) is among the most severe challenges facing humanity today. Climate change, geopolitical instability and economic recessions all form part of the current 'permacrisis'. I would like to add 'threats from AI weaponry' to this list. This paper importantly explores in-depth the ethical issues and possible negative consequences of allowing the development and deployment of AI weaponry. In this paper, I demonstrate that the use of LAWS in warfare is unethical and that it violates not only human rights and International Humanitarian Law (IHL) but human dignity as well. I do this by first examining Roland C. Arkin's article. I then explore the debate on LAWS from a consequentialist ethical perspective. I outline the arguments both in opposition and in support of the deployment of LAWS to determine the outcome of the moral calculus that much of the debate has been centred around. I then explore the principled arguments against the use of LAWS. These arguments focus on possible violations of human rights, IHL and human dignity. The final section offers a way forward in terms of how a ban on LAWS might be implemented and the role of governments, academics, and the public.

Keywords: Lethal Autonomous Weapon Systems · Artificial Intelligence · Human Rights · International Humanitarian Law

1 Introduction

The Fourth Industrial Revolution and its incredible technologies have the potential to be both beneficial but also detrimental to humanity. This dichotomy is dangerous. Recent developments in Artificial Intelligence (AI), especially in the defence industry, have presented humanity with a profound ethical dilemma. If left unregulated, Lethal Autonomous Weapon Systems (LAWS) could undermine fundamental values of democratic societies and geopolitical stability [5].

This paper will engage with Roland C. Arkin's article *The Case for Ethical Autonomy in Unmanned Systems* as a starting point to assess the potential risks, rewards and ethical issues LAWS pose to humanity [2]. Ultimately, I will utilize the literature to determine

A. Pillay et al. (Eds.): SACAIR 2023, CCIS 1976, pp. 57–68, 2023.
https://doi.org/10.1007/978-3-031-49002-6_5

whether the deployment of LAWS is ethical or, if not, how they ought to be regulated. The first section is a summary of Arkin's article. Next, I explore the potential moral risks and rewards of deploying LAWS and, thereafter, the principled arguments. These arguments are concerned with issues of human rights, dignity, and legal principles such as International Humanitarian Law (IHL). The last section considers a way forward in determining how humanity can navigate this imminent, or even present, crisis.

Before continuing, I wish to clarify some technical terms. It is imperative to clearly distinguish between Lethal Autonomous Weapon Systems (LAWS), Autonomous Weapon Systems (AWS) and Unmanned Systems. Though defining LAWS is somewhat contentious, the working definition I will be using comes from the International Committee of the Red Cross (ICRC). This definition states that a lethal autonomous weapon system is.

> "any weapon system with autonomy in its critical functions. That is, a weapon system which can select (search for or detect, identify, track, select) and attack (use force against, neutralize, damage or destroy) targets without human intervention" [14].

AWS are similar to LAWS but differ in terms of the intended purpose [4]. LAWS have "a specific use in terms of deploying lethal force while AWS have a wider set of functions including anti-material, damage, and destruction" [4]. For this essay, I will focus primarily on LAWS as AWS do not necessarily pose the same ethical issues in terms of lethal force. Another important difference is that of autonomous and unmanned systems. Unmanned systems such as unmanned aerial vehicles (UAV) are not autonomous since, though the system is being piloted remotely, a human is still in control and these systems have no degree of autonomy [18]. There is some level of disagreement in the literature regarding the definition of autonomy and issues of human supervision and 'degrees of meaningful control' [4]. My understanding is that an autonomous machine can operate without direct human involvement in its critical functions [8]. So though humans are involved in the programming and initiating the operation of an autonomous system, no direct human input is involved in the particular actions of an autonomous system [26]. Degrees of autonomy in these systems do vary, but this essay will focus on systems known as human-out-of-the-loop weapons which refers to weapons that can act and deliver force without human input [8].

2 Arkin's Argument

Arkin claims that the further development and deployment of LAWS in conflict situations is not only ethical but is a moral imperative as it will reduce human suffering and instances of war crimes. For Arkin, there are certain inherent human failings which lead to unethical behaviour from soldiers and these failings are not present in artificially intelligent systems. The main failing is the emotional stress human soldiers face which negatively impacts their decision-making capabilities. LAWS, according to Arkin, are immune to these emotional influences and he suggests that AI systems will be able to make more rational, ethical, and better-informed decisions than humans. Therefore, LAWS have the potential to act more ethically than human soldiers and adhere more strictly to IHL[1]. However, there are some key issues that Arkin fails to account for in his argument. These include issues of technical feasibility, risks of hacking and cyber warfare, destabilizing geopolitical relations, the proliferation of these weapons and possible violations of human rights. The following sections use Arkin's article as a basis for the positive views on LAWS, after which is an exploration of some of the shortcomings of LAWS that Arkin neglects.

3 The Risks Versus the Rewards

This section of the essay explores the consequentialist arguments both in favour of and in opposition to the deployment of LAWS in warfare. Consequentialism states that an action can be deemed right or wrong based on the outcomes that that action produces. Hence, an action is morally good or ethical if the outcome of that action maximises benefit and reduces harm [24].

3.1 The Rewards

The Moral Advantages. Those who advocate for LAWS feel that their cause is just because innovative AI technology will reduce suffering and the human cost of war. LAWS reduce the number of personnel on the battlefield, thus reducing the risk of harm and death to one's own forces [25]. Additionally, using autonomous systems rather than unmanned systems is preferred since it would prevent both lethal and psychological harm to soldiers. Automating the act of killing is preferential since the greater the separation between the system and the operator, the greater the reduction in harm [12].

The gist of this argument is that it is morally good to deploy LAWS if technology can exclusively reduce harm to soldiers (physical, psychological, and otherwise) [12]. Galliott and McFarland [12] even make a legal case for this based on IHL since the principle of unnecessary risk requires states to use autonomous systems over human soldiers where possible. Riesen [20] goes further and states that human soldiers should only be in the loop if there is some moral reason in favour of having a human involved that outweighs the physical, moral, and psychological risk to the soldier. This is a major theme in Arkin's [2] article and the sentiment has been echoed by others, including US military officials [6]. Systems could outperform human soldiers and have greater operational effectiveness. LAWS do not get tired or hungry, they do not forget orders or have feelings of fear, revenge, or anger [12]. Even with current technology, autonomous weapons are faster and more accurate than humans, they can function without rest and in environments where human remote control is not possible [18]. Many states, including the US and Great Britain, contend that LAWS can reduce risks of harm to civilians by automating target identification and advancements in speed, precision, and accuracy [22].

---

[1] International Humanitarian Law is the law which regulates the conduct of war and limits the effects of armed conflicts [9].

Strategic Necessity. The proponents of LAWS argue that the outcomes of having these weapons result in a net gain in terms of not only reducing fatalities and suffering in war [21], but also in terms of state security, geopolitical stability, and tactical advantages. Many military strategists, politicians and those involved in the realm of international relations feel that it is better to stay ahead of adversaries and invest in AI weaponry, as not doing so will threaten a state's national security. The essence of this argument is that it is better for a state to have LAWS to maintain a balance of power[2] and protect its citizens than to not have them and be vulnerable to attacks from adversaries. The British and American position on the issue is that there is a "utilitarian need to maintain military superiority over any potential enemies who would develop AI weapons" [22]. An arms race for AI weapons is already underway as weapons systems that are capable of full autonomy and use of lethal force already exist [22]. Proponents of LAWS argue that a ban would be futile and that it is better to have LAWS for protection than to not have them and be vulnerable.

Another prominent argument for LAWS is one based on tactical advantage. The belief is that because LAWS have access to more information that can be processed faster, this will lead to greater accuracy and operational effectiveness [25]. Thus, arguments in favour of increased autonomy and keeping humans out of the loop are based on the tactical advantages gained from the sheer speed that LAWS can operate at [18]. With continuing advances in military technology, humans will have less time to respond, and the battlefield may become too complex for human operators to navigate, therefore creating autonomous weapons is imperative to ensure tactical advantages on the battlefield [25].

Riesen [20] argues that we should take seriously the possibility that technical issues can be solved in the near future and that the very nature of warfare will change [20]. In short, the tactical advantages of autonomous weapons are exceptionally beneficial and are thus morally good since they increase a state's ability to protect its citizens. Moreover, given that AI systems will provide better situational awareness than humans, wars can be less lethal as enemy combatants can more easily be incapacitated rather than killed. Enhanced accuracy will lead to greater proportionality[3] in engaging with the enemy [21].

To summarise, proponents of LAWS argue that deploying AI weapons in warfare will lead to an overall net gain for humanity. The main points proponents cite are that 1) LAWS will lead to a reduction in atrocities and fatalities in war; 2) due to the arms race for AI weapons that is already underway, states must keep up with their adversaries to ensure state security; 3) increasing autonomy in weapons systems leads to faster and more accurate responses which will increase operational effectiveness and could again make wars less deadly. For many proponents of LAWS, opposing the deployment of LAWS is analogous to holding back on effective life-saving strategies and tactics [21].

---

[2] A balance of power refers to the idea that states try to maintain an unequal distribution of power to avoid dominance by one [1].

[3] Proportionality here refers to one of the three principles of *Jus in Bello* that must be met for a military action to be considered legal. "A military action is proportional if the harms are proportionate to the military advantage" [16].

3.2 The Risks

Technical Feasibility. One of the main objections raised by opponents to LAWS is based on issues of technical feasibility. For an autonomous system to be able to function in complex battlefield environments, these systems would need to be exceptionally sophisticated [3]. The *International Committee of the Red Cross* (ICRC) argue that the limits of technology inhibit LAWS from functioning within international legal constraints and ethical norms [3]. Those who advocate for a ban on LAWS list several technical issues with these systems which may produce negative outcomes.

With current technology, AI systems cannot accurately predict how actions affect the environment and cannot weigh alternatives from an ethical or legal perspective [25]. An autonomous system would need to be able to accurately perceive its environment to make tactical decisions [22]. This is where many doubt the ability of LAWS to operate effectively. These systems would require "artificial general intelligence (AGI) with human-level perception" [25]. Estimates for this kind of technology range between 10 years and never [25]. Thus, perception is seen as a major limitation and the risks of poor perception include failing to distinguish combatants from civilians [25]. Issues with perception are so persistent and remedying this seems to be exceptionally challenging. An AI weapon mistaking a child with a toy gun for an enemy combatant exemplifies the dangers of these technical limitations [22]. Also, distinguishing between combatants and non-combatants is exceptionally challenging even for human soldiers since in today's world of irregular warfare it is not simply a question of what uniform a soldier is wearing [21].

Another major problem is that algorithms don't show their working [7]. And we cannot understand how decisions are made. Thus, they do not necessarily meet the ethical requirement of explainability. The decisions made in war are subject to legal review, so to adhere to legal requirements, a chain of evidence must be provided [25]. Because the internal workings of an AI system are opaque to users [7], what actions a system takes are unpredictable and not traceable [21]. Additionally, if an algorithm makes a mistake, engineers and operators will struggle to correct that mistake and due to the nature of machine learning, mistakes may be replicated over time [7]. Algorithms are also not immune to biases [11], bad input data can lead to discriminatory biases [7]. The prospect of algorithmic errors leading to the deaths of innocent people and civilians is not just a worrying future possibility but is happening right now [7]. Jeremy Davis [7] cites the *Persistent Ground Surveillance System* (PGSS) as an example, where an AI system mistakenly identified an innocent farmer for an enemy target and if it weren't for human intervention an innocent man would have been killed [7]. The last and possibly the most severe problem with deploying LAWS on the battlefield is the possibility of hacking by adversaries. The possibility of hacking poses unimaginable risks not only to civilians and non-combatants but to one's own forces and allies as well [25]. Though advances are being made in cyber security and defense, the possibility of malicious cyber-attacks can never be ruled out [21]. The more sophisticated an AI system is, the more brittle it becomes. These systems are open to unpredictable endogenous shocks that occur through the implementation of complex technologies that humans cannot fully control [17].

Risks to Geopolitical Stability. The use of LAWS will fundamentally change the way wars are fought and therefore destabilise international relations globally [22]. A worrying side effect of using machines instead of human fighters is that it may be easier for political leaders to resort to war since their own troops are not at risk [8]. This would lower the perceived risk of waging war so those states with a greater capacity to deploy LAWS will view war as a viable policy alternative [21]. Unlike what some proponents claim, wars would not become completely bloodless. Winning a war would depend on the vulnerability of one's civilian population [21] and the burden of war[4] would thus shift from soldiers to civilians [8]. These weapons could also invite attacks on civilians since killing civilians would be the only way to end the conflict [12]. Furthermore, low-cost and low-tech barriers to these types of weapons could lead to mass proliferation [25] and the emergence of a new arms race [22]. This new arms race means that more ruthless actors will prevail over those who show restraint [21]. This is already an issue since some engineers in US and European tech firms are refusing to contribute to the creation of systems with lethal applications, while Chinese defence contractors have not expressed the same reservations [21]. As the arms race continues, some may become negligent with safety standards. LAWS would not increase a state's security but decrease it [22]. Actors might deploy unsafe AI systems in a bid to win the arms race and thus endanger themselves as much as their adversaries [21].

Lastly, since LAWS require minimal human supervision and a massive attack could be launched at the push of a button, humanity has a lot to lose if these weapons proliferate and fall into the hands of hostile regimes and non-state actors [22]. Russel [22] argues that these weapons could reduce the national security of all states. The destructive capabilities of LAWS may be on par with nuclear bombs since small quadcopters could all be deployed simultaneously and launch a massive attack on any population group [22]. The endpoint for LAWS is that these weapons would become "cheap, selective and scalable weapons of mass destruction" [22]. This would be devastating for international security; it is in the self-interest of all governments to ban these weapons.

3.3 Conclusion: The Moral Calculus

After all this, which side comes out on top in terms of the moral calculus? If we were to look purely at the consequentialist arguments, I would argue that the negative outcomes do in fact outweigh the potential strategic and supposed moral benefits. Yet, this is all very confusing because so much of the consequentialist argument rests on speculation and unknowable future outcomes. I believe that principled arguments offer a stronger foundation for determining the ethicality of deploying LAWS and that rights-based concerns can better guide us in regulating the use of LAWS for humanity's benefit.

---

[4] The burden of war refers to those most directly impacted by the negative consequences of war. Though human soldiers are usually the direct victims of military operations this may shift with the introduction of LAWS on the battlefield [19].

4 Beyond Consequentialism

I aim to show in this section that there is more to the debate and that rights-based considerations are important and should not be dismissed. I believe these principled arguments in combination with the consequentialist concerns I explored earlier make a strong case in favour of a ban on LAWS.

4.1 International Humanitarian Law

IHL refers to a set of rules that seeks to limit the effects of armed conflict. There are three key principles that must be satisfied in the conduct of war: distinction, proportionality, and necessity [16]. The principle of distinction states that the targeting of non-combatants is not permissible, the principle of proportionality states that harming non-combatants is permissible only if those harms are proportional to military advantages, the principle of necessity states that the least harmful means of achieving a military objective must be chosen [16]. Many authors have argued that LAWS lack the required judgement to ensure that targeting decisions are consistent with IHL [13]. *Human Rights Watch* argues that LAWS cannot adhere to IHL and that these weapons would increase the risk of harm or death to civilians [8]. As discussed previously, part of the problem lies with technical feasibility and problems with perception as LAWS cannot adequately discriminate between civilians and combatants [25]. Also, there are no codified or programmable definitions of what a civilian is [23]. Thus, LAWS will inevitably fail to adhere to the principle of distinction [23].

In terms of the principle of proportionality, weighing up military advantage and risk to civilians requires human judgement. AI systems cannot weigh up alternatives from an ethical or legal perspective because they cannot predict how actions affect the environment [25]. The principle of proportionality requires "responsible accountable human commanders, who can weigh the options based on experience and situational awareness" [23]. The principle of necessity, often thought of in terms of the use of minimal force, also cannot be adhered to by LAWS argues Blanchard and Taddeo [4]. LAWS are faster than humans and can achieve military objectives with greater efficiency, but this is a common error whereby necessity is confused for expediency [4]. LAWS would likely not abide by the requirements of the principle of necessity even if they did save lives, money and time [4]. Necessity permits measures that achieve legitimate military objectives. However, that does not mean that whatever is necessary is permissible, only that everything that is permissible must also be necessary [4]. I would agree that the use of LAWS is not strictly necessary for achieving military objectives and for a variety of reasons already considered are not morally permissible due to the additional risks they pose to civilians, combatants and geopolitical relations. Therefore, it is unlikely that current or future LAWS would be able to adhere to the three principles of IHL.

Furthermore, Heyns insists that even if they could adhere to IHL, these systems should still not be used, since errors made would lead to issues of accountability and the lack of human decision-making would make any killing arbitrary[5] [23]. Peter Asaro

---

[5] The arbitrary deprivation of life violates IHL the argument here is that the taking of life by an autonomous system would automatically violate this clause [28].

also argues that a fundamental requirement of IHL is that lethal force only be used as a result of intentional human decision-making. For him, IHL requires human judgement [23]. The decision to use force cannot be delegated to an autonomous machine [13]. This would automatically contravene IHL, deaths that result from LAWS are arbitrary, and would be considered unlawful deprivations of life [13]. IHL thus requires humans to be in-the-loop[6] when lethal force is used [6]. IHL requires common sense, interpretation, understanding of social norms and situational awareness which cannot be programmed into an algorithm. The very nature of IHL presumes that combatants are human agents. Hence, human reflection is essential for justice, morality, and law [23].

4.2   Human Rights and Human Dignity

The arguments outlined below contend that deploying autonomous weapons would breach the international standard framework of human rights and that allowing machines to decide whether a person lives or dies is morally wrong in and of itself [6]. As the use of LAWS in warfare infringes on the right to live a dignified life, which is foundational to the development of both human rights and IHL [13]. Heyns, Birnbacher and Asaro make several arguments as to why LAWS infringe on human rights and the right to live a dignified life in ways that other weapons do not.

The first argument examines how deploying LAWS could lead to severe levels of stress and psychological harm for civilians. Birnbacher argues that LAWS cause extreme mental stress for civilians because they are unpredictable and inscrutable, have a limited capacity for discrimination and adherence to the principle of proportionality, and the nature of their deployment may be disproportionally asymmetrical [23]. Furthermore, Birnbacher [23] asserts that LAWS uniquely violate four central capacities essential for living a dignified life; living a life of standard length, having freedom of movement, not having one's emotional development impaired by fear and anxiety, and having control over one's environment [23]. Allowing algorithms to decide who lives and who dies, reduces humans to objects that need to be destroyed [13]. A central tenant of human rights is that humans should not be treated as objects with instrumental value or no value at all[7] [13].

The use of LAWS threatens the quality of life, and it is essential to acknowledge that what is at stake is not only the continuation of simple biological life but the preservation of a dignified life [13]. Heyns asserts that LAWS should be banned even if they could save lives through more accurate targeting because their use is not acceptable on principled grounds [13]. The analogy Birnbacher and Heyns use is that of torture. Torture is wrong in principle, and it defies a central value of our society [13]. So even if torturing a person for information could save lives, the international community considers it impermissible under all circumstances since it violates both human rights and human dignity.

Another argument looks at the relationship between human emotions and dignity. Opponents of LAWS do not accept the premise that emotions impede sound judgement

---

[6] "Humans-in-the-loop are systems that select targets and deliver force with a human command, human-on-the-loop are systems that select targets and deliver force with oversight from a human operator and human-out-of-the-loop are systems that can select targets and deliver force without any human input or interaction" [8].

[7] As is the case of rape, slavery, or genocide [13].

[21]. Emotions are linked to moral agency and without emotion, judgement on the battlefield will be profoundly unmoral [21]. Emotions provide human soldiers with the capacity to show compassion, make exceptions, and refuse unethical orders [6]. Furthermore, Purves, Jenson and Strawser argue that proper moral decision-making requires moral judgement. LAWS cannot replicate human moral judgement and thus they could never be trusted to make sound moral decisions [20].

There is also an argument that for battlefield killing to be considered consistent with human dignity, combatants would need to engage with one another as moral equals [21]. Due to the absence of an equal relationship, LAWS have no moral right to lethal force [21]. Accordingly, LAWS or even UAV systems with humans in-the-loop assume no risk unto themselves, and there is no moral justification for fighting a risk-free war [21]. To kill human beings based on decisions made by algorithms would degrade warfare into mere slaughter, like killing animals on a conveyor belt [21]. Though this is not a comprehensive analysis of the principled arguments, I believe it shows that LAWS present a unique threat to the rights and dignity of combatants and civilians that other weapons do not.

5 A Call for a Ban on LAWS

This section presents recommendations for policymakers, international regulatory bodies, and governments to implement to mitigate the potential harms the deployment of LAWS could create. Having considered the moral, legal and political ramifications of developing and deploying LAWS I assert that all weapons systems without sufficient levels of meaningful human control be banned following the guidelines and norms of a binding international treaty. I propose that there are three crucial elements needed for a successful ban on LAWS. First, is the creation of a well-defined international treaty, appealing to state interests to ensure major powers become signatories and implementing verification protocols managed by the United Nations to guarantee compliance. The creation of a legally binding international treaty is possible, and The *Stop Killer Robots* campaign has already developed a well-defined treaty structure, and the recommendations that they propose provide a solid foundation for the creation of a sufficiently comprehensive treaty.

The most challenging part of implementing a successful ban on autonomous weapons is encouraging major powers to comply. Appealing to state interests is key to encouraging states with advanced militaries to adhere to a ban on LAWS. Since individuals are more susceptible to humanitarian concerns than governments and governments are obligated, through the democratic process, to carry out the will of the people. I believe that through civil society campaigns, the public can be made aware of the harms LAWS can create. Thus, citizens should hold their governments accountable for prohibiting the development and use of LAWS. Accordingly, lobbyists and AI ethicists need to appeal to the public's concerns about how LAWS may infringe on human rights and contravene IHL. Additionally, with growing anxiety about the state of the environment, it may be helpful to educate people on the environmental impacts of AI systems. AI technology across a range of sectors has a high carbon footprint and emissions created by data centres have a significant environmental impact [15]. It is estimated that by 2030 AI data centres will account for 8% of global energy consumption and be responsible for 1% of global greenhouse gas emissions [10].

Thus, following the suggestions from the UNESCO Recommendation on the Ethics of Artificial Intelligence [27], the public must be made aware of the impacts of AI through accessible education, civic engagement, and literacy training [27]. Research and the production of AI literature must continue so that the public can be properly informed about the adverse impacts AI systems and LAWS may have on human rights and fundamental freedoms, as well as their impact on the environment and fragile ecosystems [27].

6 Conclusion

This paper sought to engage with the literature to determine whether the deployment of LAWS in warfare would reduce the brutality of war and have a net positive outcome or if the potential risks of deploying LAWS are greater than the proposed benefits. I first summarised Arkin's [2] article as a starting point as he is a prominent supporter of deploying LAWS and proposed several positive moral benefits of using AI weapons over human soldiers. Yet, it is evident that there are several ethical and technical problems with deploying LAWS which Arkin failed to consider. After weighing up the risks and rewards based on a consequentialist ethical framework, I then argued that though consequentialism is a valid ethical approach it is not sufficient to determine the ethicality of deploying LAWS. This is because one cannot be absolutely certain of what future technologies will be capable of and whether LAWS would indeed have a positive or negative global impact. I thus turned to principled and rights-based arguments in the literature for more clarity. Through the work of Heyns, Birnbacher and Asaro I concluded that LAWS would in fact violate the rights and dignity of both civilians and combatants in ways that other weapons would not. The last section of this essay looked at how LAWS ought to be regulated. Based on the literature I argue for an international legally binding prohibition on the development and deployment of LAWS in warfare. The challenge with such a ban is that major powers are unlikely to adhere to such a prohibition unless it aligns with the national interest of the state. Therefore, those wanting to ban LAWS should focus on educating the public about the potential harms and violations of human rights.

References

1. Anderson, S.M.: Balance of power. Wiley Library (2018). https://onlinelibrary.wiley.com/. Accessed 23 June 2023
2. Arkin, R.: The case for ethical autonomy in unmanned systems. J. Milit. Ethics 9(4), 332–341 (2010)
3. Baker, D.: Should We Ban Killer Robots? Polity Press, Cambridge (2022)
4. Blanchard, A., Taddeo, M.: Jus in bello necessity, the requirement of minimal force, and autonomous weapons systems. J. Milit. Ethics 21(3–4), 286–303 (2022). https://doi.org/10.1080/15027570.2022.2157952

5. Blanchard, A., Edgar, E., McNeish, D., Taddeo, M.: Ethical principles for ai in national defence. Philos. Technol. **34**(1), 1707–1729 (2021). https://doi.org/10.1007/s13347-021-004 82-3
6. Cernea, M.: The ethical troubles of future warfare. on the prohibition of autonomous weapon systems. ANNALS Univ. Bucharest Philos. Ser. **66**(2), 67–89 (2018)
7. Davis, J.: Ethics and insights: the ethics of AI in warfare. Naval Postgraduate School (2021). https://nps.edu/documents/110773463/135759179/Ethics+and+Insights+The+Ethics+of+ AI+in+Warfare.pdf/dfa0271f-1b93-9495-69a3-fa160ebb2f77?t=1652136179368. Accessed 10 July 2023
8. Docherty, B.: Losing Humanity: The Case Against Killer Robots. Human Rights Watch (2012). https://www.hrw.org/report/2012/11/19/losing-humanity/case-against-killer-robots. Accessed 11 May 2023
9. European Commission. European Civil Protection and Humanitarian Aid Operations (2019). https://civil-protection-humanitarian-aid.ec.europa.eu/what/humanitarian-aid/intern ational-humanitarian-law_en. Accessed 26 Apr 2023
10. Ezra, A.: Renewable energy alone can't address data centre's adverse environmental impact. Forbes Technology Council (2021). https://www.forbes.com/sites/forbestechcouncil/2021/ 05/03/renewable-energy-alone-cant-address-data-centers-adverse-environmental-impact/? sh=384bdd5b5ddc. Accessed 07 Aug 2023
11. Fallmann, D.: Human cognitive bias and its role in AI. Forbes (2021). https://www.forbes. com/sites/forbestechcouncil/2021/06/14/human-cognitive-bias-and-its-role-in-ai/?sh=43f a1f6527b9. Accessed 23 May 2023
12. Galliott, J., McFarland, T.: Autonomous systems in a military context (part 2): a survey of the ethical issues. Int. J. Robot. Appl. Technol. **4**(2), 53–68 (2016). https://doi.org/10.4018/ IJRAT.2016070104
13. Heyns, C.: Autonomous weapons in armed conflict and the right to a dignified life: an African perspective. South Afr. J. Hum. Rights **33**(1), 46–71 (2017). https://doi.org/10.1080/ 02587203.2017.1303903
14. ICRC. What You Need to Know About Autonomous Weapons Systems (2022). https://www. icrc.org/en/document/what-you-need-know-about-autonomous-weapons. Accessed 24 May 2023
15. IEA. Data Centers and Data Transmission Networks (2023). https://www.iea.org/energy-sys tem/buildings/data-centres-and-data-transmission-networks. Accessed 07 Aug 2023
16. Lazar, S.: War. The Stanford Encyclopedia of Philosophy. Zalta, E.N. (ed.) (2020). https:// plato.stanford.edu/entries/war/. Accessed 28 June 2023
17. Maas, M.: Regulating for 'normal AI accidents': operational lessons for the responsible governance of artificial intelligence deployment. In: Proceedings of the 2018 AAAI/ACM Conference on AI, Ethics, and Society (2018)
18. Müller, V.: Autonomous killer robots are probably good news. In: Di Nucci, E., Santo-nio de Sio, F. (eds.) Drones and Responsibility: Legal, Philosophical and Socio-Technical Perspectives on the Use of Remotely Controlled Weapons, pp. 67–81. Ashgate, London (2016)
19. Plümper, T., Neumayer, E.: The unequal burden of war: the effect of armed conflict on the gender gap in life expectancy. Int. Organ. **60**(3), 723–754 (2006)
20. Riesen, E.: The moral case for the development and use of autonomous weapon systems. J. Milit. Ethics **21**(2), 132–150 (2022). https://doi.org/10.1080/15027570.2022.2124022
21. Reichberg, M.G., Syse, H.: Applying AI on the battlefield: the ethical debates. In: von Braun, J.S., Archer, M., Reichberg, G.M., Sánchez Sorondo, M. (eds.) Robotics, AI, and Humanity, pp. 147–159. Springer, Cham (2021). https://doi.org/10.1007/978-3-030-54173-6_12
22. Russel, S.: Banning lethal autonomous weapons: an education. Issues Sci. Technol. **38**(3) (2022)

23. Sharkey, A.: Autonomous weapons systems, killer robots and human dignity. Ethics Inf. Technol. **21**(2), 75–87 (2018). https://doi.org/10.1007/s10676-018-9494-0
24. Sinnott-Armstrong, W.: Consequentialism. The Stanford Encyclopedia of Philosophy. Zalta, N.E., Nodelman, U (eds.) (2022). https://plato.stanford.edu/entries/consequentialism/. Accessed 31 May 2023
25. Swett, B., Hahn, E., Lourens, A.: Designing robots for the battlefield: state of the art. In: von Braun, J.S., Archer, M., Reichberg, G.M., Sánchez Sorondo, M. (eds.) Robotics, AI, and Humanity, pp. 131–146. Springer, Cham (2021). https://doi.org/10.1007/978-3-030-54173-6_11
26. Taddeo, M., Blanchard, A.: A comparative analysis of the definitions of autonomous weapons systems. Sci. Eng. Ethics **28**(37). https://doi.org/10.1007/s11948-022-00392-3
27. UNESCO. Recommendation on the Ethics of Artificial Intelligence. UNESDOC Digital Library (2022). https://unesdoc.unesco.org/ark:/48223/pf0000380455. Accessed 07 Aug 2023
28. UNGA. International Covenant on Civil and Political Rights (2019). https://www.ohchr.org/sites/default/files/Documents/ProfessionalInterest/ccpr.pdf. Accessed 08 Aug 2023

Socio-Technical and Human-Centered AI Track

The Decision Criteria Used by Large Organisations in South Africa for Adopting Artificial Intelligence

Fawwaz Mohammed$^{(\boxtimes)}$ and Lisa F. Seymour

Department of Information Systems, University of Cape Town, Cape Town, South Africa
mhmfaw002@myuct.ac.za, lisa.seymour@uct.ac.za

Abstract. Artificial Intelligence adoption in South Africa is considered low, as experts suggest that large organisations are not AI-ready. However, there is little evidence to prove these sentiments. The research explained the current AI adoption criteria of four large organisations in South Africa from different industries and compared the differences. This explanatory study followed an interpretive philosophy with an inductive approach to explain the AI adoption process of organisations through semi-structured interviews. The findings revealed that organisations are AI-ready, and the low adoption is mainly because AI use cases have not been evident. Organisations view AI as a tool to enable the business to solve business problems and in return create business value. The findings provide an explanation of the decision criteria used by large South African organisations for adopting AI.

Keywords: Artificial · Intelligence · Adoption · Business · Value · Technology · Environmental · Organisational · Decision · South Africa

1 Introduction

Artificial Intelligence (AI) is a disruptive technology which has affected numerous organisations worldwide. The technology has been identified as an essential component to add value and realise future business goals [1]. AI allows organisations to reduce overall costs, increase work efficiency and enhance customer satisfaction [2], enabling organisations to become 40% more efficient in their business processes [3]. Currently, South Africa (SA) is in the infancy of AI adoption, with low maturity and a lack of education [4]. There are many reasons for the low adoption. These vary from Eskom's load-shedding problems to Telkom's ageing infrastructure [5, 6].

Large organisations in SA employ over 13 million citizens, and presently unemployment is as high as 35% [7]. Yet, an Accenture report [8] claims AI could replace 20–30% of jobs in SA. If large organisations adopt AI, they will need to consider employing more skilled workers and reducing unskilled labour. An Accenture [8] report indicated South Africans fear AI as it will cause job loss due to automation, and the fear is justified.

The current literature has explained AI adoption for various organisations in developed countries and remarkably few in the context of SA. There are many differences between SA and developed countries, and assuming the factors influencing AI adoption are the same is problematic. Furthermore, each industry is different and has different reasons for AI adoption. Hence this study explains the decision criteria used by large organisations in SA when considering AI adoption.

2 Literature Review

There are many frameworks developed to explain technology adoption at an organisation level. These frameworks include the TOE (Technology, Organisation, Environment) framework and the CFOIA (Conceptual Framework of Organisational Innovation adoption) [9, 10].

The TOE framework considers the organisation's internal and external requirements for innovation [9, 10]. The technological elements include the advantages, compatibility, and perceived risks for the organisation [9, 10]. The organisational context refers to the organisation's characteristics, such as the size, structure, and employees [9, 10]. Environmental context refers to the industry, vendors, service providers, customers, and competitors in the market or even government regulations. Organisations are inclined to innovate depending on environmental factors surrounding them, which can be opportunities or threats [9, 10].

The CFOIA considers a sequential decision-making process that is evaluated before the initiation phase of technology adoption [9, 10]. The framework evaluates multi-disciplinary factors at an organisational level which influences the decision-making process of innovation adoption such as the direct and indirect factors [9]. Furthermore, the CFOIA extends the TOE framework by including perceived innovation characteristics, supplier marketing activity and social network [9]. The complex nature of technology can influence decisions, and organisations aim to pursue technology that complements their current infrastructure. Furthermore, there are additional characteristics that can be reviewed which are likely specific to an organisation, and these factors are complexity, trialability, observability, and uncertainty.

Yet researchers have identified limitations with these frameworks and with AI, there is no one-fits-all framework as requirements and expectations differ between organisations [10, 11].

2.1 Decision-Criteria for AI Adoption

AI has been adopted by many organisations globally and researchers explain that the business value of AI has not yet been clarified. Experts claim that organisations lack the understanding of how and when AI should be used [2, 12]. Researchers suggest organisations should have relevant use cases or a business problem before adopting AI, and AI should be used as a tool to solve these problems [12–14]. While identifying a business problem is the first step for AI adoption, organisations should review the key characteristics of their business to ensure a successful adoption [15]. In addition, studies have identified technology readiness, organisational factors, and environmental influences as key determinants for AI adoption [14–17].

Technology Readiness. Data and IT infrastructure are the core technology characteristics required to adopt AI [14–17]. AI applications utilise large data sets, and organisations need computing power to manage the outputs [18]. Data quality is also essential as it allows AI applications to provide accurate and reliable predictions [17–20]. Furthermore, an organisation's IT infrastructure requires flexibility and scalability to deploy AI applications [14, 17]. Cloud providers such as Amazon, Google and Microsoft provide the necessary infrastructure for a service fee. Hence cloud-ready organisations can adopt AI with less effort [21].

Organisational Factors. Organisational characteristics such as compatibility, culture, skills, employee trust and AI strategy are essential criteria for adopting AI successfully. Employee trust is an important building block regarding AI adoption in alignment with change management and skills availability [13, 15]. AI is known to automate ad-hoc and manual tasks, causing job loss [2, 22]. Organisations can either create new roles or offer employees redeployment opportunities [23]. The necessary training and awareness should be available to employees so trust can be harnessed [13, 15, 23]. AI requires diverse and unique skills, as many tools and applications are used [15, 19]. If the skills are unavailable internally, the necessary upskilling can be provided, alternatively organisations will need to compete in the market for in-demand AI skills [15]. In addition, organisations need to include AI in the existing IT strategy or develop an AI strategy to guide planning, adoption, and implementation [22, 24].

Environmental Influences. Organisations operate in constantly changing environment, and environmental influences such as government regulations, competitors, customers, and partners impact how organisations conduct business [17]. Government regulations have been implemented to ensure compliance for data protection. Many of these regulations are specific to the location in which the business operates [15, 17].

2.2 Current Adoption of AI in Large SA Organisations

SA organisations are investing in the core AI technologies such as chatbots, robotic automation and advanced analytics [25]. Many industries in SA, such as banking, agriculture and retail have adopted AI technologies, while industries, such as mining have not [26, 27].

The mining industry in SA has contributed significantly to the GDP and has employed more than 450,000 people in SA [28]. A PWC report [29] claims that the mining industry will undergo significant transformation during the next decade.

AI has penetrated the agriculture industry. Aerobotics is a corporation in SA that builds and deploys drones that track pest and disease prevention for crops for the agricultural industry in SA. These drones provide data and AI performs predictive analysis to provide farmers with output estimates and harvest timings. Hence, AI has assisted farmers in managing their farms more efficiently and sustainably [30]. In addition, the study by [30] advises that the SA government are among a few African nations considered AI-ready in the agriculture industry [30, 31]. Yet there is limited research on the SA agriculture community.

Banks are using cutting edge technologies to create new offerings to provide customers with a better experience, propelling the industry into innovation adoption [32].

Examples of applications used in banking are fraud detecting, chatbots, customer relationship management, predictive analysis, and credit risk management [31, 33].

Retail organisations operate on small margins of profit, and that has created a need for efficiency. AI technologies were used for automating inventory management, demand forecasting and merchandising, supply chain optimisation, and customer relationship management. Even though the industry has adopted many emerging technologies, AI in retail is still in its infancy [34].

2.3 Literature Summary

Adoption of technology in SA is complex as organisations have different challenges depending on the industry. The banking and agriculture industries are more prevalent to adopting emerging technologies because of external factors such as competitors, customer expectations and vendor relations. Despite the importance of understanding the adoption of AI in large organisations in SA, there has not been extensive literature in this area of study and more research is required [35].

3 Method

This study aimed to explain the current reality of AI adoption in large organisations in SA. The interpretive philosophy was chosen to provide insight into these organisation's views towards AI adoption [36]. In addition, this study used a multiple case study approach and selected four large SA organisations, each from a different industry. These industries were retail (C1), mining (C2), banking (C3) and agriculture (C4). Semi-structured interviews were conducted, as displayed in Table 1. A combination of purposive and snowball sampling was used for case and participant selection [37]. The researcher initially knew one expert on the phenomenon, and then asked the participants to recommend further experts.

Table 1. Participants Interviewed

Unique ID	Case ID	Participant ID	Industry	Role
C1P1	C1	P1	Retail	Head of Architecture and Innovation
C1P2	C1	P2	Retail	IT Group Executive
C2P1	C2	P1	Mining	Group CIO
C2P2	C2	P2	Mining	IT Portfolio Manager
C3P1	C3	P1	Banking	Division CIO
C3P2	C3	P2	Banking	IT/Business Executive: Digital
C4P1	C4	P1	Agriculture	IT Executive: Data and Innovation
C4P2	C4	P2	Agriculture	IT Executive: Digital

Inductive reasoning was chosen to gain new insights into the study phenomenon in the context of SA [38]. Thematic analysis followed Braun and Clarke's method [38].

Participants received individual and organisational consent forms detailing that participation was voluntary. A primary objective was to ensure anonymity and privacy for all participants. Each participant was given a unique ID.

4 Findings and Discussion

The inductive findings presented here explain the perceived innovation process which is the decision-making process that organisations go through before they adopt technology. The study then describes the determinants influencing the adoption. These were classified according to the TOE framework as the additional categories from the CFOIA framework were not found to be significant.

4.1 Perceived Innovation Process

Identifying a Business Problem. All the participants indicated that the starting point for technology adoption is driven by a business problem or an opportunity statement. The business units of each organisation drive the initiative by bringing a business problem to IT. They then use technology to enable their business to solve the relevant problem. Technology is not adopted without a use case and these findings aligns with findings in the study [15]. *"We never implement tech for the sake of tech... it must be linked to a problem" [C4P1] "wouldn't necessarily always review emerging technologies for the sake of reviewing emerging technologies, we always looking at business solutions that will improve our business" [C1P1].*

Collaborating Between IT and Business. All the participants discussed the relationship between IT and business which influenced AI or technology adoption. The banking and agriculture cases have consciously structured their business to ensure IT and business can work harmoniously to ensure the organisation achieves strategic goals. *"But now the culture has changed within the IT environment with or the IT environments very closely living, living close to the to the business units" [C4P2].*

In contrast, participants in the retail and mining cases believe their IT team should not aspire to be more. Their strategy is to ensure IT supports the business. *"In the past as an IT function, I think maybe we've been, we've had a desire to be more than we need to be to support the business." [C1P1].* As [2] discussed, it is essential to have synchronous communication between business units to ensure successful AI adoption. The banking and agriculture cases have benefited from the internal collaboration of teams.

Evaluation by Decision-Making Unit. All four cases follow a sequential process to consider AI adoption. Decisions for AI adoption are proposed to a senior leadership committee. While the name of the decision-making unit (DMU) is unique to each case, the purpose of the DMU is the same role across all cases. The findings align with the CFOIA framework, which refers to a DMU whose perceptions and innovative propensity influence adoption [9]. The starting point is a proposal from the business or IT team to the DMU. The DMU then looks at multiple criteria points before considering AI adoption. Some criteria are value to the business, costs, skills and change management. *"So, we have worked with what we call an investment committee, that word evaluates any* *investment, whether it's technology related, or, or store related" [C1P1]. "the way that we make the recommendation will also be based on the collective expertise we've got in the team deciding on the direction that we're taking" [C2P1].*

Having an AI Strategy. Furthermore, the discussions lead to whether AI is part of the IT roadmap or strategy. Large organisations have structured and rigid roadmaps to ensure the organisation runs efficiently and cost-effectively [22]. All four cases recognise the importance of AI and thus include the technology in their IT strategy and roadmap. *"Strategically, three, four years ago, and we said, there are five to six major emerging technological trends that are coming at us. Based on those six, how well positioned are we from a platform perspective to harness them all? So where are we on Cloud? Where are we on mobile? Where are we on AI and machine learning?" [C1P2] "One of them is AI. Another one is, things like predictive analytics, you know, big data reporting, etc. So, there's a multitude of different themes that are part and parcel of the of the IT strategy and roadmap" [C2P1].*

The retail, agriculture and banking cases have already developed an AI strategy for their long-term IT goals. However, the mining case has not identified the value of an AI strategy. Studies [19] discuss the value of an AI strategy before adoption.

Even though AI is part of the retail and mining cases' IT strategy, participants noted that their organisations only informally discuss AI. One participant from the mining case even stated it does not warrant formal discussions as it has not proved to add value to the industry at this point. However, the retail case has a 3-year plan in which senior leadership is tasked to evaluate how AI can improve the business. The participant believes it will be added to the roadmap, and future use cases will likely occur. *"I would say informally, it's not really a discussion topic. It doesn't warrant dedicated time." [C2P1].*

In contrast, the banking and agriculture cases have innovation teams who are regularly discussing AI solutions during internal forums. The IT and business teams collaboratively discuss solutions on how to improve the business or creating opportunities. *"We have a very good roadmap around what we want to do with our recommenders. And the roadmap is basically we want our recommenders from a corporate perspective to appear anyway?" [C3P1].*

The agriculture industry is AI-ready. As one participant advised that they have many AI tools available. This aligns with a study [30] on the AI readiness of agriculture in SA. *"I'm going to say it exactly like I said, we got a big toolbox behind us. I've got partners, we've got very good AI sitting in that box waiting for an opportunity. Same as the RPA developer, same as the guys with IoT sensors." [C4P1].*

Assessing Costs. The financial benefit of AI was a theme that was well-argued in the study. All participants claimed that there are two related costs, operational and implementation costs. The cases reiterated the importance of considering costs for AI or technology adoption. Furthermore, identifying these costs beforehand is essential to adoption success. A study [2] explains that the costs of infrastructure and skills are critical determinants for AI adoption. *"So, you can't purchase tech that cost that that has a high operational expense, because you can afford it during the good times, but you're not going to be able to afford it during the bad times." [C2P2].*

One participant from each of the banking and mining cases argued their organisation must justify the costs of AI with specific KPIs, such as whether it will bring a return on investment, or if it reduces the costs of current solutions, and whether the operational costs are sustainable. *"Cost is always a factor. Cost is not just what it what the what the outlay is, it's about what's the cost to live with it."* [C2P2]. *"Cost is a big factor, right? Like, how costly is the technology to implement and use sustainably is a big one?"* [C3P1].

Building a Proof of Concept (POC). The banking and agriculture cases have innovation teams that work harmoniously with the business. The teams have the capability available to build proof of concepts for the business and trial the solution to the DMU. The trialability is meaningful as the DMU has a relevant use case demo on how an existing business problem could be solved. The findings align with the CFOIA framework, which refers to a trialability factor, where organisations first trial a solution with a cost-effective proof of concept [9]. *"So, I guess in a way they are, by having set up a team to do this, you know, that like, you know, just be ready for any new technology, understand it POC it? Because we don't just experiment with a tech, we also say okay, what would it take to also train up the organization right to the ability to leverage this technology going forward"* [C3P1].

The agriculture case had proven examples where POCs solved business problems, and the DMU had an easy decision regarding the adoption AI. *"You must give it a very specific small use case, and then componentized and build on top of that. So, I think that's, that's probably our approach, and we keep iterating against that thing and keep, as we prove the POC, or as we complete a POC"* [C4P1].

4.2 Technology Characteristics

Scalable IT Infrastructure. All participants advised their IT infrastructure ready to implement AI. All four cases have either completed a cloud migration project or are currently in process. Therefore, they don't believe scalability would be an issue when adopting AI. This aligns with studies that claim cloud-ready organisations can adopt AI much easier [14, 17]. *"So, if you look at our infrastructure that's sitting in the cloud, if we needed to quadruple the size of us of our infrastructure there, it's very agile, we can just simply, purchase new environments within AWS, spin them up within an hour and we've got kind of capacity. So that's very agile"* [C1P1].

Data Readiness. For organisations to adopt AI effectively, an essential consideration must be the status of their data. Not only do organisations need to have large sets of data, but they also need quality data to produce quality results. While the findings align with the presumptions from various studies [16–18, 20–22], participants gave contrasting views of their current data landscape and approach.

Participants from the banking case first considered if the data was accessible and whether it was legally accepted. Furthermore, POPIA and global regulations were discussed by one participant from the banking case. *"Look at is where's the data? And what are we allowed to use it for?"* [C3P2] *"Or you can access it but not allowed to use it in the way that you need to solve the problem. You're wasting your time. And it's a new and it's a complex world, this consent POPIA who owns it?"* [C3P2].

A participant from the mining case noted the importance of maturity and quality of data. *"It's the data, it's the maturity. So, it's taking into consideration all those aspects to ensure that you realize your benefits and you get the adoption that you need."* [C2P2].

One participant from the retail case discussed how they would build data lakes. Extending that, a participant from the agriculture case discussed their strategy for data clean-up. They are currently running a project to collate all their data tools and construct quality data for AI solutions to leverage. *"When we are making a decision on which technology should we build a data lake, and that's something that we have to obviously take a take a much longer and more kind of collaborative and multifaceted kind of approach on"* [C2P2] *"I would say now, our focus is pretty much on data and data enablement and data segmentation data lakes"* [C4P2].

4.3 Organisational Characteristics

Approach to Innovation. The banking and agriculture cases explained that they have innovation teams dedicated to upskilling on AI and relevant emerging technologies. These teams are searching the market for technology that will improve the business, gain competitive advantage, or improve customer experience. The agriculture case has many innovation projects running concurrently. The innovation teams emphasize the positive culture towards innovation from these two industries. This positive attitude towards innovation supports a study [13]. *"My team was sort of primarily tasked with doing that, which is one to do the technology research, right to understand that, you know, the various technology trends in the market"* [C3P1]. *"Innovation team, right? Anything that's new, I can tell you about three experiments I've currently got running and kind of how we got to where we were."* [C4P1].

With a different approach, participants from the retail and mining cases explained that having an innovation team to research AI is not feasible. Additionally, they advised that it is too expensive to employ people not adding immediate value to the business. They prefer being a fast-follower and have a strategy in which they first observe if the solution works in the market and then evaluate whether it is something the business would benefit from. *"that's a big thing. There's not there's not a lot of time to do research, unfortunately. And I think that goes for, I think every industry, but I think in, in my company specifically, I think that's a bad thing. We don't spend our time on research."* [C2P1]. One participant from the retail case advised that senior leadership has KPIs to stay aware of market trends around AI adoption. *"it's a very good growth component of what we look for from the leaders so that they build markets and partnership awareness of what's around them all the time. And encourage them to stay on top of trends and on top of movements in the market"* [C1P2].

Although all cases have different approaches to AI and technology adoption, the attitude towards innovation is positive. The approach of the mining case aligns with a PWC [29] report which notes the industry's reluctance towards digital transformation.

Verifying Value. A study [15] has noted the need to find a business reason for adopting AI, which aligns with the approach of large organisations in SA. The mining case only considers the value to the business. *"How does this add value to my company? If it's not going to add value? why don't why am I spending time on it?"* [C1P1]. The priority is to ensure that the business gains value from any technology undertaking. It could be a competitive advantage, first to market or solving existing technical problems [9]. *"Business benefit is what has to be the overriding factor, there might be some technical benefits" [C1P1].* A core part of adopting AI is realising the value to the business. Furthermore, retail, banking and agriculture cases have a customer-first strategy to strive to improve customer experience. *"I think it would tend to be very proactive when it comes to front end, or client facing technologies. So, like, things that would impact client experience" [C1P1].*

Planning Change Management. Researchers have advised that an organisation's culture impacts AI adoption [13]. Organisations that have built a culture of innovative thinking find it easier to manage change when AI is adopted [13, 15]. There are two aspects to change management. Business change management and IT change management. All four cases collectively endorsed the importance of both. All participants believed that IT employees have a continuous improvement mindset and do not often find it challenging to adapt to change. However, on the business side, it is slightly different. *But I think we've got it right where all our tech guys have quite a strong, continuous improvement mindset. So, they're all constantly trying to assist in doing something just slightly better [C4P1].* Furthermore, they indicated that change management is a massive part of their current strategy, and they emphasized the importance of ensuring that all employees are aligned with business goals and values. The cases have established programs such as training and workshops. This aligns to a study [13] which encourages promoting change by providing training for improvement and development. *"But you know, change management is never an easy thing. So, we're currently on our roadmap is a big part of change management is embedded within me and C4P1 sort of dynamic environment in our strategic key performance indicators, because we know that that change management is needed." [C4P2].* The banking case has internal certification initiatives to encourage change management within the organisation. *"But one thing that Bank has done very well is make training very pervasively available." [C3P1].*

Although all participants remarked that they focus on change management within their organisations, but that it is a very complex landscape and it's not easy to manage. *"It's a fairly, the landscape is a complex landscape. So, there's largely a resistance to change across the entire landscape. Because change is always difficult." [C1P1].*

Resistance to AI Adoption. An Accenture [8] report indicates that AI is likely to contribute to further job loss in SA. Additionally, a study [22] discusses the reasons employees fear AI adoption. A survey [4] confirmed these barriers to AI adoption. Our findings align with the aforementioned reports. Participants in the retail case reported that workers in stores oppose automation as it causes fear around job loss. However, there has not been a problem with the IT or business teams. *"So, I do think so the combination of AI and robotic process automation, I think probably would drive that degree of fear." [C1P1].* Every participant identified a case where AI has disrupted jobs in their organisations by either automating an administrative task or making a position redundant. *"It would be in more repetitive, manual. You know, kind of less thinking tasks." [C2P1]* Although AI has replaced jobs, industries such as mining, and agriculture have initiatives to redeploy people within the organisations and choose not to reduce headcount. *"Technology introduction, one of our key criteria to look at is redeployment of people."* *[C2P1]*.

Resource Capability for AI. All participants indicated that they do not have the internal skills for all branches of AI. Their skills for AI are predominantly for the RPA and predictive analytics applications. *"One is the confidence in our team to be able to support stuff, because we know, we know it, and we have strong team"* *[C1P2]*. *"I think most times you have to actually go and find the skills and then train them up at the same time."* *[C3P1]*. A Gartner [39] report also indicates the low availability of skilled AI resources and how organisations should invest in training programs.

When the business needs a niche skill for specific AI branches, all organisations reported they are probably don't have the skills available even though banking and agriculture cases have innovation teams. They then partner with vendors. *"Predominantly, it would be vendors and contractors? And, yeah, and to be honest, it's a struggle. Because like I said, most of these new types of technologies, the skill sets are in demand, not just by me by everyone else as well."* *[C2P1]*.

The mining participant advised that they don't have any AI skills available. This is echoed by a PWC [29] report that indicated mining has not embraced AI. *"New stuff we would be what's the flavour of the month? And does it make sense for us? Okay, it makes sense for us, who's the right vendor to work with?"* *[C2P1]*.

All participants suggested that external skills for AI could be challenging as South Africans have become attractive to global markets which makes AI skills expensive to retain. *Just because something is available internationally, but it comes at very high rates, becomes unsustainable for you to run"* *[C3P1]*.

Although the participants believe SA has the skills available, its scarce and expensive as echoed in prior studies [2, 15]. *"Finding good solid tech skills is not easy. And if you get them, you're paying a premium, and it's getting more expensive. And then in the in the South African contexts,"* *[C4P1]*. *"And then to keep up with the way the world has changed around that like Project culture where a lot of the guys don't want to be permanently employed"* *[C1P2]*.

The Covid-19 pandemic has disrupted the IT job market with a considerable shift to remote work. It has also encouraged skilled resources to pursue opportunities abroad as the compensation packages are more appealing. Some participants agreed that many skilled resources are choosing to work remotely for international companies. *"So, the competition for skill has lifted and shifted. And with, you know, what the pandemic has taught us about working remotely, people don't necessarily have to relocate nowadays, either."* *[C1P2]*. *"a lot of good guys actually leaving SA, and they're going to, you know, international shores"* *[C3P1]*.

One participant from the agriculture case argued there are skills and experience in other industries but not agriculture. *"South Africa lacks experience, not skills. We don't have enough people working in this industry."* *[C4P1]*.

The agriculture case speculated that many AI-skilled South Africans are predominately working for start-ups. The reason for choosing start-ups over large organisations is that it allows freedom of innovation. Large organisations have red tape and, in most cases, only support what the business needs. *"So, you'll get start-ups with guys doing crazy stuff. But it won't be the consensus so there's and the guys who have the skill don't work for corporate, so they decide to go on their own and build their own type things. I think there's pockets of excellence" [C4P2].* Accenture [8] discusses the dynamism and entrepreneurial drive of start-ups are essential to AI adoption.

4.4 Environmental Characteristics

Pressure from Competitors. Most participants responded in an apathetic manner to the questions posed about the pressures of adopting AI because of competitors. Although all cases agreed there are pressures from competitors, it's not a significant concern.

The retail and mining cases indicated they would not consider technology adoption because of competitor pressures. The organisation's strategy allows a competitor to invest in AI or innovations, and they would be a fast follower and learn from mistakes made by the competitor and develop a better value proposition. *"We would never follow a competitor to directly, we would always try and understand what that means for us" [C1P1]* and *"being first to market, and by being first to market didn't necessarily guarantee that you'd always lead the market." [C1P1].*

The banking case accepted that there is competitor pressure, as they motivate their innovation teams to be first to market for innovation adoption. As stated in [26] and [30]. The banking industry has competition from various start-ups and competitors, and the pressure drives innovation in the industry [26]. *"Is there pressure when competitors put something out? Absolutely. It's a bit of an arms race up there" [C3P2].* The banking case accepted that there is competitor pressure, as they motivate their innovation teams to be first to market for innovation adoption. As stated in [26] and [30]. The banking industry has competition from various start-ups and competitors, and the pressure is known to drives innovation in the industry [26]. *"Is there pressure when competitors put something out? Absolutely. It's a bit of an arms race up there" [C3P2].*

The agriculture case indicated they have very little competition from competitors. *"Not really no, not to my understanding, I think there isn't real pressure from the external macro" [C4P2].* However, the industry is built on trust amongst the farmers and the existing businesses. *"Generally, first to market. It's not always coordinated because we've got three big segments, right consumer business and corporate Banking. So, it's not always coordinated but for example, like we've we champion the adoption of Salesforce in Africa" [C4P2].*

Pressure from Government. All participants explained that government does not interfere with AI or technology adoption. Other than with regulatory governance and compliance. From a technology and operations perspective, there is autonomy. *"So, you have a lot of autonomy and freedom in how you run the front. And to some extent the back." [C3P1].* Globally, European countries have been sanctioned with many regulations for AI and data privacy [2]. Other than POPIA, participants advised that there is no government interference in their AI adoption.

Constraints of Local Infrastructure. Participants from retail and agriculture cases explained the problems encountered with Eskom and Telkom. With the current national constraints of power shortages and Telkom's ageing infrastructure, the retail industry considers these issues major barriers to innovation and AI adoption. These considerations are essential for the technical efficiency required for AI solutions. These concerns were briefly discussed in a study [5] and a report [6] which outlined the National Infrastructure Plan. *"And so those are those are, that's a component. I think Eskom is a big factor is a, you know, big issue, which is outside our control to a degree."* *[C4P1]*.

Pressure from Vendors, Suppliers, or Partners. All participants explained that vendors constantly reach out regarding different technology solutions. However, when it relates to AI, IoT or blockchain, these technologies are instead proposed by local start-ups. *"If I can't connect what you're trying to sell me to an actual solution or proper return on investment, it's not going to happen."* *[C4P1]*. Agriculture has significant pressure from farmers, always aspiring towards new technology for efficiency and sustainability. *"Vendors not so much. I would say our biggest pressure comes from our customer centric focus in our customers being Farmers and they want to embrace technology."* *[C4P2]*.

One participant from the retail case advised that their organisation is keen to partner with start-ups as they have the skills and applications like AI available. *"We've come across a very interesting local start-up and then decided to work with him to build that capability"* *[C2P1]*. Additionally, the participant advised that they rely on start-ups and vendors to propose emerging technology. *"So, what we tend to rely on is our partner ecosystem, applying the emerging technologies to solutions that they can then propose to us."* *[C2P1]*. The banking and agriculture cases have partnered with start-ups for innovative technology, this aligns with studies [26, 30]. The findings indicate that retail industry is keen to partner with start-ups for innovative solutions, however it was not supported by literature. The banking and agriculture cases have partnered with start-ups for innovative technology, this aligns with prior studies [26, 30]. The findings indicate that the retail industry is keen to partner with start-ups for innovative solutions, however it was not supported by literature.

4.5 Conceptual Model

The conceptual model provides a theoretical framework for explaining the decision criteria used by large SA organisations for adopting AI. The perceived innovation process goes through different steps which are detailed. While the order of these steps presented is not necessarily followed, all sectors started with identifying the business problem and built a proof of concept prior to further adoption. While organisations considered environmental and technology criteria in making their decisions, the characteristics of the individual organisations played a dominant role (Fig. 1).

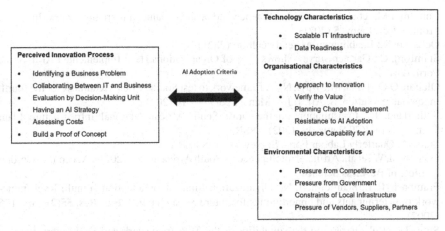

Fig. 1. Decision Criteria for AI Adoption In SA

5 Conclusion

This study explains the decision criteria large SA organisations use for adopting AI. SA and global organisations use similar decision criteria for adopting AI. Slight nuances in characteristics and processes across industry sectors were presented.

While an Accenture report [8] indicates that AI adoption in SA is low, this is an incomplete picture. This study found that SA organisations are currently AI-ready. However, in comparison to global leaders the adoption is low. In agriculture, SA is leading the way. In mining, retail and banking, SA is not lagging far behind. The four cases studied indicated that AI use cases are not necessarily feasible and therefore the adoption is low. They further indicated they are AI-ready from a technological point of view, and they are actively focusing on change management to reduce AI adoption disruption. They are optimistic about innovation and believe employees and management are well-equipped for change. The biggest concern is the financial costs of skills for AI.

Yet the study has limitations. Semi-structured interviewees can create biased opinions. The banking and agriculture cases were market leaders, while the retail and mining cases were not. Hence the organisation's market position could influence the results. Additionally, a case per industry isn't necessary apparent to the entire industry.

Data readiness is paramount to leverage AI benefits, and large SA organisations have large data sets that can be utilised. There is an opportunity to investigate the current data readiness strategies. In addition, the findings uncovered that organisations are AI-ready, and the low adoption is mainly because AI use cases are not evident. However, there is an opportunity to further investigate the reasons for the low adoption.

References

1. Grover, P., Kar, A.K., Dwivedi, Y.K.: Understanding artificial intelligence adoption in operations management: insights from the review of academic literature and social media discussions. Ann. Oper. Res. **308**(1–2), 177–213 (2020)

2. Enholm, I.M., et al.: Artificial intelligence and business value: a literature review. Inf. Syst. Front. J. Res. Innov. **5** (2021)
3. Oclarino, R.: Enabling an AI-ready culture (2021)
4. Stamford, C.: Gartner Survey Shows 37% of Organizations Have Implemented AI in Some Form (2019)
5. Olaitan, O.O., Issah, M., Wayi, N.: A framework to test South Africa's readiness for the fourth industrial revolution. S. Afr. J. Inf. Manag. **23**(1), 1–10 (2021)
6. Sutherland, E.: Consultation on the Draft South African National Infrastructure Plan: Comments on ICT Networks. (2021). SSRN
7. StatsSA, Quarterly Labour Force Survey. 2021, StatsSA
8. Schoeman, W., et al.: Artificial intelligence: is South Africa ready? (2021). Accenture: Gordon Institute of Business Science
9. Frambach, R.T., Schillewaert, N.: Organizational innovation adoption: a multi-level framework of determinants and opportunities for future research.pdf. J. Bus. Res. **55**(2), 163–176 (2002)
10. Smit, D., et al.: Exploring the suitability of the TOE framework and DOI theory towards understanding AI adoption as part of sociotechnical systems. In: Gerber, A., Coetzee, M. (eds.) South African Institute of Computer Scientists and Information Technologists. SAICSIT 2023. CCIS, vol. 1878, pp. 228–240. Springer, Cham (2023). https://doi.org/10.1007/978-3-031-39652-6_15
11. Jöhnk, J., Weißert, M., Wyrtki, K.: Ready or not, AI comes— an interview study of organizational AI readiness factors. Bus. Inf. Syst. Eng. **63**(1), 5–20 (2020)
12. Mikalef, P., Gupta, M.: Artificial intelligence capability: conceptualization, measurement calibration, and empirical study on its impact on organizational creativity and firm performance. Inf. Manag. **58**(3) (2021)
13. Lee, J., et al.: Emerging technology and business model innovation: the case of artificial intelligence. J. Open Innov. Technol. Mark. Complex. **5**(3) (2019)
14. Wamba-Taguimdje, S.-L., et al.: Influence of artificial intelligence (AI) on firm performance: the business value of AI-based transformation projects. Bus. Process. Manag. J. **26**(7), 1893–1924 (2020)
15. Pumplun, L., Tauchert, C., Heidt, M.: A new organizational chassis for artificial intelligence - exploring organizational readiness factors. In: Twenty-Seventh European Conference on Information Systems (ECIS2019). Stockholm, Sweden (2019)
16. Afiouni, R.: Organizational learning in the rise of machine learning. In: Fortieth International Conference on Information Systems. Munich, Germany (2019)
17. Baier, L., Jöhren, F., Seebacher, S.: Challenges in the deployment and operation of machine learning in practice. In: Twenty-Seventh European Conference on Information Systems (ECIS2019). Stockholm, Sweden (2019)
18. Gregory, R.W., et al.: The role of artificial intelligence and data network effects for creating user value. Acad. Manag. Rev. **46**(3), 534–551 (2021)
19. Alsheibani, S., Cheung, Y., Messom, C.: Artificial intelligence adoption: AI-readiness at firm-level. In: Twenty-Second Pacific Asia Conference on Information Systems 2018, Japan (2018)
20. Demlehner, Q., Laumer, S.: Shall we use it or not? Explaining the adoption of artificial intelligence OFR car manufacturing purposes. In: Proceedings of the 28th European Conference on Information Systems (ECIS) (2020). An Online AIS Conference
21. Schmidt, R., et al.: Value creation in connectionist artificial intelligence a research agenda. In: Americas Conference on Information Systems (2020). Online
22. Keding, C.: Understanding the interplay of artificial intelligence and strategic management: four decades of research in review. Manag. Rev. Q. **71**(1), 91–134 (2020). https://doi.org/10.1007/s11301-020-00181-x

23. Makarius, E.E., et al.: Rising with the machines: a sociotechnical framework for bringing artificial intelligence into the organization. J. Bus. Res. **120**, 262–273 (2020)
24. Finch, G., Goehring, B., Marshall, A.: The enticing promise of cognitive computing: high-value functional efficiencies and innovative enterprise capabilities. Strategy Leadersh. **45**(6), 26–33 (2017)
25. Ogbemhe, J., Mpofu, K., Tlale, N.S.: Achieving sustainability in manufacturing using robotic methodologies. In: Conference on Sustainable Manufacturing. 2017. Stellenbosch, South Africa (2017)
26. Matsepe, N.T., Lingen, E.V.D.: Determinants of emerging technologies adoption in the South African financial sector. S. Afr. J. Bus. Manag. **53**(1), 2078–5585 (2022)
27. Rapanyane, M.B., Sethole, F.R.: The rise of artificial intelligence and robots in the 4th industrial revolution: implications for future South African job creation. Contemp. Soc. Sci. **15**(4), 489–501 (2020)
28. Leeuw, P., Mtegha, H.: The significance of mining backward and forward linkages in reskilling redundant mine workers in South Africa. Resour. Policy **56**, 31–37 (2018)
29. AMine2020, Essential and Resilient, PWC, Editor. 2020, PWC
30. Sampene, A.K., et al.: Artificial intelligence as a path way to Africa's transformations. Artif. Intell. **9**(1) (2022)
31. Gwagwa, A., et al.: Artificial Intelligence (AI) deployments in Africa: benefits, challenges and policy dimensions. Afr. J. Inf. Commun. **26**(26), 1 (2020)
32. Mhlanga, D.: Industry 4.0 in finance: the impact of artificial intelligence (AI) on digital financial inclusion. Int. J. Financ. Stud. **8**(3) (2020)
33. Ndoro, H., Johnston, K., Seymour, L.: Artificial intelligence uses, benefits and challenges: a study in the western cape of South Africa financial services industry. In: SACAIR 2020 Proceedings: AI in Information Systems, AI for Development and Social Good (2020). online
34. Oosthuizen, K., et al.: Artificial intelligence in retail: the AI-enabled value chain. Australas. Mark. J. **29**(3), 264–273 (2020)
35. Pichlak, M.: The innovation adoption process: a multidimensional approach. J. Manag. Organ. **22**(4), 476–494 (2015)
36. Walsham, G.: Doing interpretive research. Eur. J. Inf. Syst. **15**(3), 320–330 (2017)
37. Saunders, M.N.K., Lewis, P., Thornhill, A.: Research Methods For Business Students. Pearson Education Limited, London (2019)
38. Clarke, V., Braun, V.: Thematic analysis. J. Posit. Psychol. **12**(3), 297–298 (2016)
39. Goasduff, L.: While AI adoption is increasing, challenges remain. Organizations must understand what AI can and cannot do (2019)

Let's Play Games: Using No-Code AI to Reduce Human Cognitive Load During AI Solution Development

Armand Graaff[1]([✉])(iD), Danie Smit[1](iD), and Sunet Eybers[2](iD)

[1] University of Pretoria, Pretoria, South Africa
armandgraaff@gmail.com

[2] University of South Africa (UNISA), Science Campus, Johannesburg, South Africa

Abstract. Understanding and developing cutting-edge technologies like artificial intelligence are widely seen as complex tasks and significantly strain human cognitive capacity. Cognitive fit theory is an established theory that proposes that task completion performance is enhanced when there is a congruent relationship between the problem statement and task execution. Despite efforts to simplify the tasks, the tasks may still pose a challenge when it comes to understanding and execution. This paper argues that artificial intelligence, particularly no-code artificial intelligence, can reduce human cognitive burden. The paper aims to illustrate how an artificial intelligence artefact can be used to assist humans in transforming a task with a high human cognitive load into a task with a low human cognitive load. A simple game of Tic-Tac-Toe was developed, which is easy to play and comprehend, therefore representing a low human cognitive load. This was followed by an isomorph Scrabble card game, which is more challenging to play, introducing a higher human cognitive load. Winning the latter served as the problem representation in this paper. Two design science research cycles were used during solution development. During the first cycle, an artificial intelligence agent was developed to play and win both games on behalf of the human. The coding required to develop the agent, however, introduced a high human cognitive load. Subsequently, in the second design cycle, an artificial intelligence agent that could win both games was developed using the no-code artificial intelligence platform DataRobot. Overall, this resulted in a low cognitive load in both solving the problem (winning the Scrabble card game) and developing the problem solution (artificial intelligence agent). On a theoretical level, this research contributes to information systems research by demonstrating the value of cognitive fit theory in the context of developing artificial intelligence solutions.

Keywords: Artificial intelligence · cognitive processing · no-code · design science artefact · DataRobot · cognitive fit theory · information systems

A. Pillay et al. (Eds.): SACAIR 2023, CCIS 1976, pp. 86–99, 2023.
https://doi.org/10.1007/978-3-031-49002-6_7

1 Introduction

Cognitive processing is a key characteristic of humans. It refers to the attainment of information from the environment, making sense of concepts, and the subsequent utilisation thereof for decision-making, knowledge creation or problem-solving [13]. For many years, the primary distinction between humans and computers was the ability to reason, make decisions and solve problems. The ability of computers to perform cognitive processing was predominantly depicted in science fiction movies and academic concept papers, such as the "Computing Machinery and Intelligence" published in 1950 by Alan Turing [18]. Subsequently, the concept of artificial intelligence (AI) was born. Since then, giant technological advancements have not only introduced AI into our everyday lives but embedded it as part of our "normal" daily tasks. Deep learning, for example, involves a class of AI, that learns without human supervision, drawing from data that is both unstructured and unlabeled data and is used in applications such as Siri and ChatGPT [4].

As technology evolves, the requirements for AI improvement increase, not only in the number and extent of the requirements but also in the task complexity and reasoning ability. It comes as no surprise that the continuous evolution and improvement of AI's cognitive processing ability is a popular research topic in the fields of information systems (IS) and computer science (CS). Superior processing capability can lead to more rapid and effective decision-making, which is crucial in larger-scale product development endeavours, such as self-driving cars [15].

The development and improvement of AI solutions is challenging. Challenges can include data issues and bias or variance in model outputs [2,3,12]. Usually, these challenges can be addressed if the problem space is well understood and articulated [14]. For a human, this is often a difficult task. However, good no-code AI platforms can assist developers by providing measures to safeguard against various types of errors that can be overlooked by a data scientist while modelling, including over-fitting, under-fitting and target leakage [8,9]. These no-code AI platforms can also assist humans with problem representation and articulation. Furthermore, it can help in the democratisation of AI [17]. An understanding of the problem and subsequent problem statement articulation and representation is essential since the level of complexity and the relevance of the topic to the individual can often prompt a human to prioritise a more simple and relevant solution over another [6]. The efficient representation of a problem is crucial and should be assessed in relation to the specific task of interest. Cognitive fit theory (CFT) is an established theory postulating superior task completion when a matching relationship between the problem representation and task could be established [20–22]. Therefore, the theory seems appropriate for examining the congruence between problem representation and problem solution in an attempt to make it easier for humans to comprehend. This is further explained in Sect. 3.1.

Given this background, the research question, "How can no-code AI be used to reduce human cognitive load when solving challenging problems?" is addressed in this research. To answer the research question, the game of Tic-Tac-Toe (low cognitive load) [11] and an isomorph[1] of Tic-Tac-Toe in the form of a Scrabble card game (high cognitive load) [13] were used to illustrate how AI can reduce human cognitive load by assisting with AI solution development *i.e.* we use AI to build AI. We explain how no-code AI can be used to help humans solve difficult problems, i.e. enabling the human to transform problems that would generally require a "high cognitive load" to solve, into problems that only need a "low cognitive load" to solve.

The paper starts with a background section explaining no-code AI, the design science research (DSR) approach and methodology. The CFT is described in the context of an AI problem space and is followed by two DSR cycles. The paper concludes with a summary and conclusion section.

2   Background

Artificial intelligence (AI) has become deeply embedded in society as we know it (who can live without Siri and Grammarly?). The term encapsulates many kinds of algorithms, including, but not limited to, computer vision, natural language processing, forecasting, predictions and speech recognition, producing many different types of outputs [4]. Research shows that in the short term, the use of AI in organisations will tend towards a human-machine partnership [10,16]. Hence, neither humans nor AI will operate independently. The human-machine partnership is the focal point of this article, which demonstrates a model for humans and machines (AI) to work together in a more harmonious and productive manner due to reduced human cognitive load. Knowledge in AI artefacts is either obtained through theory (rules) or from experimental observations (data) [5]. Many early AI systems were rule-based, meaning they were created based on human knowledge and hard-coded as rules. Machine learning, on the other hand, is a process that uses data to train a model, where the model captures the relationships between inputs (different features) and outputs (the target variable) [23].

No-code AI is a tool that assists data scientists and business practitioners in exploring problems using machine learning by writing little to no code [8]. In this instance, AI tools can be used to build AI artefacts. These tools introduce and enable non-technical practitioners to use machine learning and assist data scientists in exploring patterns in data that may have otherwise been missed. These tools also help to speed up the machine learning lifecycle process by assisting data scientists with pre-processing data, although this functionality may be limited. One of the major challenges posed to organisations aiming to increase the

---

[1] Isomorphism refers to a one-to-one correspondence that preserves binary relationships between elements of two sets. For instance, the set of natural numbers can be mapped to even natural numbers by multiplying each by 2. This maintains the addition operation, making the sets isomorphic for addition. If both sets are reduced to their base form, they would be identical. [website], https://www.britannica.com/science/isomorphism-mathematics, (accessed 25 August 2023).

use of AI, is the gap in data literacy between technical and business practitioners. No-code AI aims to close this gap by leveraging machine learning operations (MLOps) to enable faster iterations between ML solutions.

In this research, we introduce a novel, emerging technology in AI software [17], namely DataRobot, and illustrate its' use through the development of an AI agent during a second cycle of our DSR approach. This is important as organisations are progressively trying to derive valuable insights from business data and are looking for easy-to-use AI tools to assist with this. More specifically, companies are looking into machine learning systems to extract value from operational data [17].

3 Research Approach and Methodology

Design Science Research (DSR) is an approach that lends itself towards developing artefacts in a quest to enhance knowledge and understanding of a set problem [7]. Given the understanding of the problem and subsequent problem statement articulation and representation complexity, as explained in the introduction, DSR seems suitable to the present inquiry. Furthermore, CFT is used as a theoretical lens to not only frame the AI problem space (awareness of the problem), but also propose, develop and evaluate a possible solution. This is done to explore how a simple Tic-Tac-Toe game and a more complex isomorph Scrabble card game can be used to illustrate how no-code AI can reduce cognitive load when developing AI solutions. The approach allowed for two DSR cycles: (1) during the first cycle a model was developed that can play the game of Tic-Tac-Toe and the Scrabble card game using Python code, subsequently producing a low human cognitive load to play the games (solve the problem), but producing a high cognitive load in terms of solution development (creating the model using code); (2) a second design cycle was triggered with the objective to reduce the high human cognitive load in terms of solution development *i.e.* to create an AI agent (as indicated by the red block in Fig. 1 below). The solution to the problem of high cognitive load when developing an AI agent was to use a no-code AI platform to develop the AI agent without using code (therefore reducing the cognitive load).

3.1 CFT in an AI Problem Space

The CFT determines the effectiveness of a given problem-solving process. On the highest level, CFT sees the problem solution as the outcome of the relation between the external problem representation and the problem-solving task, the combination of which is seen as the mental representation of the problem [1]. In this study, the mental representation is the relationship between the understanding of an AI problem (problem representation) and the ability to solve the problem (problem solution) using AI. Therefore, in most cases, the problem-solving task would be to build an AI artefact. Depending on the problem-solving task, the mental representation could therefore have either a "low cognitive load"

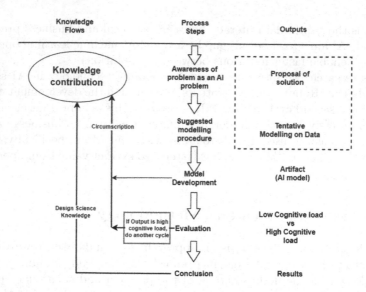

Fig. 1. Design Science Research (DSR) steps followed adapted from the diagram used by Vaishnavi et al. [19].

(for example our Tic-Tac-Toe game) in instances where the task representation could be simplistically broken down or a "high cognitive load" in more complex problems (represented by our isomorph Scrabble card game).

Figure 2 graphically depicts the representation of the CFT in terms of an AI problem space, as adapted from the article "The effectiveness of innovation", by Abou-Zeid et al. [1].

3.2   The Scrabble Card Game as an Isomorph of Tic-Tac-Toe

In the book "The Science of the Artificial" by Herbert Simon (2019) [13], Simon explains a card game called Scrabble and refers to it as an isomorph of the traditional Tic-Tac-Toe game. The book elaborates on the rules of the Scrabble card game as follows: "The game is played by two people with nine cards - let us say the ace through to the nine of hearts. The cards are placed in a row, face-up, between the two players. The players draw alternately, one at a time, selecting any one of the cards remaining in the centre. The aim of the game is for a player to make up a 'book', that is, a set of exactly three cards whose spots add to 15 before the opponent can do so. The first player who makes a book wins; if all nine cards have been drawn without either player making a book, the game is drawn" [13].

Furthermore, the representation of the design is explained: "The magic here is made up of numerals from 1 through 9. Each row, column, or diagonal adds to 15, and every triple of these numerals that add to 15 is a row, column, or diagonal or the magic square. From this, it is obvious that making a book in

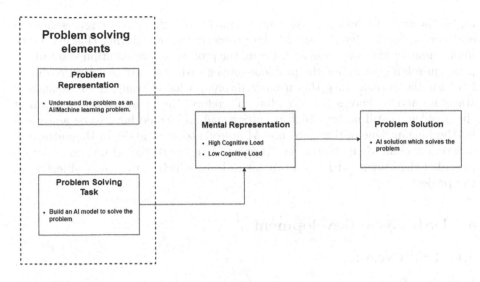

Fig. 2. Graphical representation of CFT in terms of an AI Problem Space [1]

number scrabble is equivalent to getting three in a row in the game of Tic-Tac-Toe" [13].

A Tic-Tac-Toe board can be represented as a perfect square where every cell is represented by a unique integer in the range 1 to 9. Notice, that every row, column, and diagonal sums to 15, and there is no way to get a sum of 15 from three cells unless they share a row, column or diagonal [13]. When playing a more complex representation of a game, as with the isomorph Scrabble card game example explained by Herbert Simon, it is understandable that playing the Scrabble card game compared to playing Tic-Tac-Toe requires more cognitive effort. Because the AI agent was trained on data and not programmed according to the rules of Tic-Tac-Toe, the AI agent had no issue being able to play the Scrabble card game. This shows the ability of AI to pick up complex relationships in data, which is much more difficult for humans to complete.

3.3 A Practical Example of DSR Cycles in AI Solution Development with Reference to the CFT

In DSR, the problem awareness stage marks the start of the research process and represents the problem solution according to the CFT as defined in the problem-solving task (see Fig. 2). This is now the first attempt of a human to remove the high cognitive load from a given problem. The human succeeds in turning the high cognitive load of a complex card game (isomorph Scrabble card game) into a low cognitive load problem, by creating an AI agent that can play the game on his behalf. The problem of high cognitive load is, however, not solved since the human still has to solve the high cognitive load problem of building an AI agent model. As per the CFT, seen in Fig. 2, the high or low cognitive load is described under the mental representation, which is made up of the problem representation and the problem-solving task. In the proposed solution during this cycle, the high cognitive load was transferred from the problem representation state to the problem-solving task, but the problem-solving task still carries a high cognitive load for the human since this usually involves a lot of complex mathematical statistics and/or knowledge of coding. Therefore, the mental representation of the problem is still under a high cognitive load and would have to be addressed by the human. One option is to use AI no-code tools to assist in the automatic creation of code when creating an AI agent. As a result, the AI no-code solution provided the human with a low cognitive load, which is the main objective of the project.

4   DSR Cycle Development

4.1   DSR Cycle 1

In the first DSR cycle a neural network was trained to play a game of Tic-Tac-Toe, as well as a more advanced Scrabble card game (only more advanced from the human perspective), by utilising experimental observations. During this cycle "traditional" AI tools were used (*i.e.* coding). Playing the Tic-Tac-Toe game[2] introduced a low human cognitive load whilst playing the advanced Scrabble card game[3] introduced a high human cognitive load. The problem awareness of this cycle therefore is to lower the high human cognitive load of playing the card game.

Artefact Development in DSR Cycle 1: The technical approach during this cycle was inspired by image classification. On a conceptual level, we would like an AI agent to look at a Tic-Tac-Toe board, observe the moves being played and subsequently establish if those moves lead to a win. This is like a human watching someone else play. However, the AI agent does not know the rules of Tic-Tac-Toe, nor was it explained by anyone to the AI agent. The AI agent had to learn by experimenting and observing. The two main development processes were: (1) creating training data through self-play; and (2) training the neural network using game-play data generated by the first process. This machine-learning approach is known as reinforcement learning. The game must be played on a pre-determined board. For the AI agent (the neural network) to see the board and understand the game state, the board and the data depicting the moves were reformatted using one-hot encoding. The 'x' columns represent the board state, and the 'y' columns represent the move played. The eighteen fields in the 'x' columns represent the board state. The first 9 represent the 'x' positions and the second 9 represent the 'o' positions. The moves in the 'y' columns are represented by nine fields, indicating which move was played. The board-state

---

[2] You can play Tic-Tac-Toe against the AI agent: http://3.20.219.21/tictactoe.

[3] You can play Scrabble Card against the AI agent: http://3.20.219.21/tictactoecard.

was the input into the neural network and the moves played were the labels that the neural network was trained on.

To generate training data for our AI agent (referred to as game-play data), an untrained agent (Agent U) played against another untrained agent (Agent S). Random moves were made by the agents. During the game of Tic-Tac-Toe, g training data was generated and stored in a Gameplay file. The Gameplay file contributed to subsequent training cycles whereby data about the board state, the moves that were made and who won the game was stored (Fig. 3).

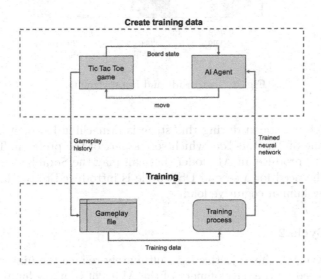

Fig. 3. Training Process

Before training, in terms of the number of wins, the number of games played was equal to the outcome. Therefore, if ten games were played, five were won and five were lost on average for each agent. This was to be expected as both of the agents were playing random moves. The statistics are displayed in Fig. 4 with the label "before training". The moves that contributed to a win were used to train the neural network. The neural network was created by a trained agent, using the following logic to generate a move while keeping randomness. The trained agent (Agent S) and the untrained agent (Agent U) played against each other, where the trained Agent S won more games than the untrained agent (labelled as "after training"). In the final round of training, the trained agent (Agent S) played against itself. This generated more training data and was used to train Agent L (labelled as "Agent L" in Fig. 4). It should be noted, that even in a simple AI example, an iterative approach is required [23].

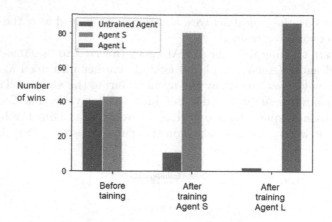

Fig. 4. Test result and improvement

The artifact developed during this stage is a machine learning model, that can play a game of Tic-Tac-Toe, which solves part of the problem. The process was repeated to produce an AI model that can play the Scrabble card game as well. Hence, the need for a second DSR cycle is introduced to produce a better result with low human cognitive load.

4.2   DSR Cycle 2

During DSR cycle 2, the objective is to ultimately reduce the high human cognitive load caused by the development of the AI agent to a low human cognitive load. In other words, an AI agent should be able to play a complex isomorph Scrabble card game on behalf of the human. As demonstrated in Design Cycle 1, this is no easy task as creating an AI agent using code is a high human cognitive load-producing task. One solution to this challenge is the utilisation of no-code AI. Therefore, in this cycle of the DSR, a no-code AI platform was used as an "AI assistant" to build the AI agent that will play the Scrabble card game on the human's behalf. The human used the no-code AI platform since it can understand the problem representation, irrespective of whether it is Tic-Tac-Toe or the Scrabble card game. The no-code AI can create an AI agent, without requiring the human to write any code. The no-code AI assistant, therefore, turns the high cognitive load problem of building a model, into a low cognitive load problem for the human.

Artefact Development in DSR Cycle 2: The tool used to develop our AI agent was DataRobot - a no-code AI platform that assists data scientists with pre-processing and machine learning projects. Figure 5 shows the model training process of one of the many models created by DataRobot. The only requirement for using DataRobot is that the user has to upload the data and select the "y" variable (the moves that were made on the board) as the target.

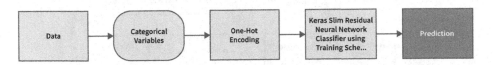

Fig. 5. A blueprint of the modelling process, provided by DataRobot in order to show the human exactly what was done to train the model.

In the Python code for the Tic-Tac-Toe game in DSR Cycle 1, the neural network used for the model training was set by the human. In instances where the model was not specifically adapted to the data it was given, the model may not have worked well. In the example of using the no-code AI tool, no model was specified beforehand. The system therefore looked at the data and used a repository of thousands of models to select and train models most suited to the data. After the models were trained, the tool created blueprints to explain to the user exactly how each of the models was trained, see Fig. 6 for the blueprint of the neural network with the best results.

This is another area where no-code AI is clearly useful in terms of the CFT, namely that the representation of the models as seen in Figs. 5 and 6, makes it easily understandable to users. DataRobot therefore aids in bridging the gap between machine code and solution representation which is understandable to humans. Therefore lowering the cognitive load both in terms of solution development (creating an AI model) and solution representation (allowing a human to understand the results created).

The final artefact developed in this DSR cycle was a machine learning model, or AI agent, developed and trained by a no-code AI tool that was able to play both the Tic-Tac-Toe game and the Scrabble card game. The AI agent was able to learn how to play Tic-Tac-Toe by only self-play and this ability improved over time. Therefore, the problem of high human cognitive load was addressed by using AI. As a result, the final human cognitive load was low which marked the end of the DSR cycles.

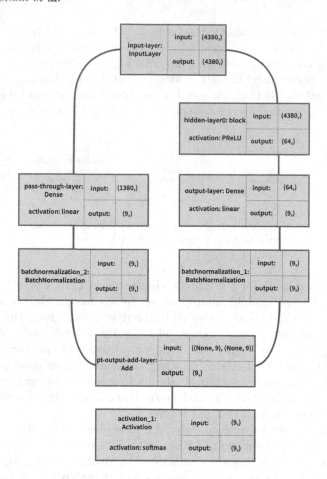

Fig. 6. Blueprint of the neural network created by DataRobot

In Fig. 7, the cognitive load and the source thereof are depicted for each of the different states. In the first block, the human is in the problem representation state, where it is clear that playing the card game caused a high cognitive load on the human. After kicking off the DSR cycles, it is clear from Cycle 1 that playing the card game is now a low cognitive load since the human has an AI agent that can play the game on his behalf. Creating this agent, however, is also a high cognitive load activity. After transitioning to DSR Cycle 2, it can be seen that, as a result of no-code AI, building the model, and playing the card game are both low cognitive load activities, hence, the goal is reached and another DSR cycle is not necessary.

It should be noted that there is a risk involved with using no code AI, namely that no-code methods can lead to low-quality solutions, and a non-expert might not have the competence to evaluate the results.

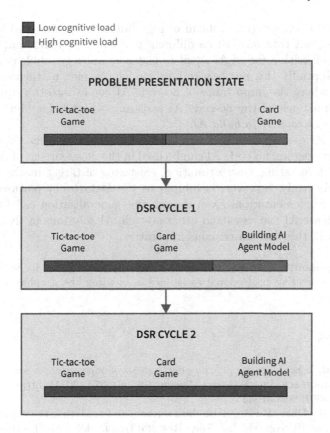

Fig. 7. DSR cycle cognitive load transition transition

5   Conclusion

CFT is a psychological theory that seeks to explain how individuals process information to solve problems [22]. According to CFT, if the presentation format of the information aligns well with the mental representation an individual uses for problem-solving, the person is likely to experience a cognitive fit. No-code AI allowed data scientists to make use of artefact blueprints, automatically generated by a tool such as DataRobot, to understand the problem solution (for example how the AI agent was created). In organisations, the problems that practitioners face are much more complex than the example of Tic-Tac-Toe. Today's practitioners need decision support tools to make sense of the huge amount of data in a timely fashion [16]. AI can detect complex underlying patterns in data, not always apparent and visible to humans.

As illustrated in this paper, no-code AI tools might be the answer to addressing the cognitive fit problem, by reducing the complexity introduced by a high cognitive load problem. In this example, the high cognitive load of solving a difficult problem (the Scrabble card game) is transferred onto an AI agent that can win the game. However, the problem of high human cognitive load is not completely solved, but transferred to a different problem for the human, namely, to create an AI algorithm for an AI agent to play a complex Scrabble card game on his behalf. Normally this involves coding and/or complex mathematical statistics. This is where the importance of no-code AI comes into the equation. The human can make use of the no-code AI assistant, which helps him build other AI models, i.e. *using AI to build AI*.

No-Code AI can be used in a similar fashion in many other disciplines, for example in companies, no-code AI can be used in the development of AI solutions such as sales forecasting, cost estimation, clustering of target markets, etc. The results in this study, however, is limited to the DataRobot platform and the Tic-Tac-Toe implementation. Even though this generalisation can be extended to more intricate AI use cases and other no-code AI solutions in theory, further investigation in this domain remains imperative.

Acknowledgements. The authors wish to express their gratitude towards Renee Lightfoot for providing professional assistance in creating the graphics of Figs. 5, 6, and 7.

References

1. Abou-Zeid, E.S., Cheng, Q.: The effectiveness of innovation: a knowledge management approach. Int. J. Innov. Manage. **08**, 261–274 (2011). https://doi.org/10.1142/S1363919604001052
2. Agarwal, R., Dhar, V.: Big data, data science, and analytics: the opportunity and challenge for IS research. Inf. Syst. Res. **25**(3), 443–448 (2014). https://doi.org/10.1287/isre.2014.0546
3. Baesens, B., Bapna, R., Marsden, J.R., Vanthienen, J., Zhao, J.L.: Transformational issues of big-data and analytics in networked business. MIS Q. **40**(4), 807–818 (2016)
4. Benbya, H., Pachidi, S., Jarvenpaa, S.L.: Special issue editorial: artificial intelligence in organizations: implications for information systems research. J. Assoc. Inf. Syst. **22**(2), 281–303 (2021). https://doi.org/10.17705/1jais.00662
5. Dubois, D., Hájek, P., Prade, H.: Knowledge-driven versus data-driven logics. J. Logic Lang. Inform. **9**, 65–89 (2000)
6. Günther, W.A., Rezazade Mehrizi, M.H., Huysman, M., Feldberg, F.: Debating big data: a literature review on realizing value from big data. J. Strat. Inf. Syst. **26**(3), 191–209 (2017). https://doi.org/10.1016/j.jsis.2017.07.003
7. Hevner, A., Chatterjee, S.: Design science research in information systems. In: Hevner, A., Chatterjee, S. (eds.) Design Research in Information Systems. Integrated Series in Information Systems, vol. 22, pp. 9–22. Springer, Boston (2010). https://doi.org/10.1007/978-1-4419-5653-8_2
8. Hurlburt, G.F.: Low-code, no-code, what's under the hood? IT Prof. **23**(6), 4–7 (2021)
9. Iyer, C.K., et al.: Trinity: a no-code AI platform for complex spatial datasets. In: Proceedings of the 4th ACM SIGSPATIAL International Workshop on AI for Geographic Knowledge Discovery, pp. 33–42 (2021)

10. Keding, C.: Understanding the interplay of artificial intelligence and strategic management: four decades of research in review. Manage. Rev. Q. **71**(1), 91–134 (2021)
11. Pilgrim, R.A.: TIC-TAC-TOE: introducing expert systems to middle school students. ACM SIGCSE Bull. **27**(1), 340–344 (1995)
12. Reis, L., Maier, C., Mattke, J., Creutzenberg, M., Weitzel, T.: Addressing user resistance would have prevented a healthcare AI project failure. MIS Q. Exec. **19**(4), 279–296 (2020)
13. Simon, H.A.: The Sciences of the Artificial, Reissue of the Third Edition with a New Introduction by John Laird. MIT Press (2019)
14. Simon, H.A.: The Science of the Artificial, 3rd edn. The MIT Press, Cambridge (2019)
15. Simoudis, E.: The Big Data Opportunity in Our Driverless Future. Corporate Innovators, Menlo Park (2017)
16. Smit, D., Eybers, S., De Waal, A.: A data analytics organisation's perspective on the technical enabling factors for organisational AI adoption. In: AMCIS 2022 Proceedings, p. 11 (2022)
17. Sundberg, L., Holmström, J.: Democratizing artificial intelligence: How no-code AI can leverage machine learning operations. Bus. Horizons (2023). https://doi.org/10.1016/j.bushor.2023.04.003
18. Turing, A.M.: Computing machinery and intelligence (1950). The Essential Turing: the Ideas That Gave Birth to the Computer Age, pp. 433–464 (2012)
19. Vaishnavi, V., Kuechler, B., Petter, S.: Design science research in information systems (2004)
20. Vessey, I.: Cognitive fit: a theory-based analysis of the graphs versus tables literature. Decis. Sci. **2**, 219–240 (1991)
21. Vessey, I.: The theory of cognitive fit: one aspect of a general theory of problem solving? In: Human-computer interaction and management information systems: Foundations, pp. 155–197. Routledge (2015)
22. Vessey, I., Glass, R.L.: Applications-based methodologies development by application domain. Inf. Syst. Manag. **11**(4), 53–57 (1994)
23. Zhang, Z., Nandhakumar, J., Hummel, J.T., Waardenburg, L.: Addressing the key challenges of developing machine learning AI systems for knowledge-intensive work. MIS Q. Exec. **19**(4), 221–238 (2020). https://doi.org/10.17705/2msqe.00035

Algorithmic, Data Driven and Symbolic AI

Unit-Based Genetic Algorithmic Approach for Optimal Multipurpose Batch Plant Scheduling

Terence L. van Zyl[✉][iD] and Matthew Woolway[iD]

University of Johannesburg, Johannesburg 2092, GT, South Africa
{tvanzyl,mjwoolway}@uj.ac.za

Abstract. The paper studies the scheduling challenge in multipurpose batch chemical production plants. The study focuses on advancing the efficiency of genetic algorithms in solving schedules for medium-to-long-term time horizon production. Current global-based approaches have excessively high dimensional chromosome representations for these scheduling problems. The paper proposes shifting from a global-based to a unit-based event point chromosome representation. The new unit-based representation exploits further characteristics of batch scheduling problems to reduce the intrinsic dimensionality of the problem. The investigation aims to reduce the problem's dimensionality and, in so doing, further streamline the scheduling process. The analysis compares state-of-the-art (SOTA) genetic algorithm unit-based approaches against two new global-based approaches. The study compares these algorithms using three well-known literature examples through extensive experimentation. The new models are tested in profit maximisation scenarios, showcasing several advantages, including reduced dimensionality, faster computation times, and improved accuracy. Results indicate a significant improvement in computational efficiency compared to established methods. This paper contributes to the ongoing research in this field by proposing a more effective scheduling tool for multipurpose batch plants, suggesting unit-based approaches as promising avenues for future investigations.

Keywords: Metaheuristics · Genetic Algorithms · Batch Process Scheduling

1 Introduction

In the last three decades, chemical production scheduling has seen a surge of interest, particularly in multipurpose batch plants. The flexibility and adaptability of these multipurpose plants provide unique opportunities for manufacturers, but they also present substantial challenges. Batch plants that can produce multiple products within a single facility introduce additional complexity in their scheduling. The literature focuses on developing optimisation methods to solve specific schedules within an acceptable computational timeframe.

© The Author(s), under exclusive license to Springer Nature Switzerland AG 2023
A. Pillay et al. (Eds.): SACAIR 2023, CCIS 1976, pp. 103–119, 2023.
https://doi.org/10.1007/978-3-031-49002-6_8

These methods have been extensively reviewed by Mendez *et al.* [15], Shaik *et al.* [17], Li and Ierapetritou [11], Maravelias *et al.* [13,14], Sundaramoorthy and Maravelias [19], Harjunkoski *et al.* [5] and Georgiadis *et al.* [4]. Mixed-Integer Linear Programming (MILP) models have been widely used due to their success and computational efficiency in short-term scheduling. However, MILP models become computationally intractable as the time horizons extend.

Despite their potential to address the computational intractability of batch process scheduling, we noticed a scarcity of research on metaheuristic approaches. Notable efforts by He and Hui [6–8] applied metaheuristic approaches to single-stage batch scheduling and incorporated a genetic algorithm for multipurpose batch scheduling. Despite these advancements, these implementations did not propose a universal framework. A recent series of publications by Woolway and Majozi [20,21], Bowditch *et al.* [1] and van Zyl and Woolway [22] have introduced a generalisable metaheuristic framework that can generate near-optimal solutions with significantly reduced computational times. Compared to leading mathematical programming approaches, this framework reported computational time reductions of up to 98% in some cases, emphasising the potential of metaheuristic approaches for scheduling over medium-to-long time horizons.

This paper builds upon and enhances the foundational framework introduced by Woolway and Majozi [21]. Specifically, the study aims to adjust how time is represented in the existing framework, which currently uses a global-based event point chromosome representation. While this representation simplifies implementation, it results in an overabundance of event points for the optimal schedule solution. This paper proposes shifting from a global-based to a unit-based event point framework, reducing dimensionality and streamlining the scheduling process. The new unit-based event point chromosome representation exploits a novel snapping phenomenon inherent in batch processes to reduce the intrinsic dimensionality of the problem further. The study contributes the following:

1. Transition from a global-based event point representation to a unit-based one, building on prior investigations.
2. Demonstrate the application of the new unit-based event point representation in profit maximisation scenarios, highlighting the applicability and effectiveness of the framework in these operational goals.
3. Emphasise the benefits of the proposed unit-based event point representation, including its reduced dimensionality, faster computation times, and improved accuracy due to the smaller search space.
4. Validate this novel framework using select literature examples to demonstrate its robustness, versatility, and superiority over existing methods.

1.1 Motivation

A general network-represented process is often employed in manufacturing plants where batch processes require merging or splitting, and specific mass balance requirements must be met. Within the scope of continuous-time formulations, there exist two prominent approaches: (i) *Global event-based models*, where event or time points are used for both tasks and units, resulting in event points that are not tied to specific units, and (ii) *Unit-specific event-based models*, where event points are strictly associated with individual units. In unit-specific models, the event points for one unit may differ from those of another, enabling concurrent tasks in different units at a given event point. A theoretical illustration of these models is depicted in Fig. 1(a) for a global-based event model and Fig. 1(b) for a unit-specific event model.

The advantage of the unit-based specific model lies in its potential for reducing the total number of event points compared to global-based event point models. This is due to the model's ability to assign different positions for event points across various units, allowing separate tasks to proceed concurrently at different times. Despite the advantages of the unit-based approach, the metaheuristic framework introduced by Woolway and Majozi [21] relied on a global-based event point representation. This choice was driven primarily by the ease of implementation, and a framework employing a unit-based event point representation was not considered.

With this research, we aim to address this gap by developing a framework based on a unit-specific event point model, hypothesising that such a model could further improve the efficiency of the scheduling process. A successful demonstration of the proposed unit-specific event point model could provide a powerful tool for more efficient scheduling of multipurpose batch plants, opening a new avenue of research in this field.

1.2 Problem Statement

We extend and build upon the work of Woolway and Majozi [21], who analysed three different metaheuristics and found the Genetic Algorithm (GA) using a global-based chromosome representation to be the most effective. We further develop this GA framework in the current investigation to include a unit-based event point chromosome representation. We then compare this new unit-based event point approach with the global-based event point representation employed in the original research.

To ensure the robustness and validity of this comparison, the research examples originally used by Woolway and Majozi [21] are revisited. The first of these (hereafter referred to as Example 1) is a multiproduct batch plant. This example was initially proposed by Ierapetritou and Floudas [9]. It involved the production of a single product via a series of mixing, reaction, and purification stages, as illustrated by the State-Task-Network (STN) in Fig. 2.

The second example (hereafter referred to as Example 2) was introduced by Sundaramoorthy and Karimi [18]. It extends Example 1 by adding complexity, including parallel units feeding intermediate storage states. These are bottlenecked through a single-unit process, which feeds another set of parallel units. The modifications are represented in the STN shown in Fig. 3.

The third example (hereafter referred to as Example 3) is a multipurpose batch plant, first introduced by Kondili *et al.* [10], where three feed materials

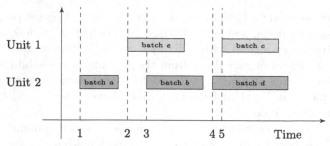

(a) Example of Global-Based Event Point Framework

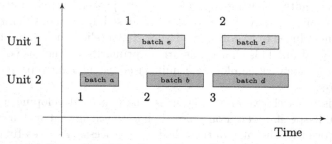

(b) Example of Unit-Based Event Point Framework

Fig. 1. Example of Global-based vs Unit-Based Event Point Framework

undergo reactions, heating, and separation processes to produce two distinct products. The corresponding STN for this example is depicted in Fig. 4.

This study retains the constraints initially outlined in the respective literature across all three examples, comparing the performance of four distinct GAs, each with unique methods of event point representation. These include the global-based event point represented GA (GA-GP) first introduced by Woolway and Majozi [21], the Q-Global-based approach (GA-GQ), which is a new framework equivalent of GA-GP, and the Unit-based event point represented GA (GA-UP). The GA-GQ utilises a two-matrix system and a snapping phenomenon to enhance computational efficiency. The Mixed Q-Global and Unit-based Approach (GA-MP) combines the strategies of GA-GQ and GA-UP, effectively addressing the curse of dimensionality while maintaining the flexibility of a unit-based representation.

For a comprehensive evaluation, the results obtained from these GAs are compared with the solutions yielded by the State-Sequence-Network (SSN) MILP model used by Woolway and Majozi [16,21]. We assess the practicality and effectiveness of the proposed GA-GQ, GA-UP, and GA-MP against the established GA-GP and MILP methods. The goal is to ascertain whether the proposed representations can improve performance regarding computational efficiency and solution quality. The evaluation will consider each method's ability to provide feasible solutions, their speed of convergence, and their ability to identify the global optimum.

Fig. 2. STN for Example 1

Example 1 focuses on a multiproduct batch plant [9]. Figure 2 visually presents the State-Task-Network (STN) for this example, while Table 1 lists the related problem data. In this table, the column τ denotes the average processing time of each unit. We use the coefficients α and β to determine the batch time based on the batch volume (v). We can express the batch time through the linear equation:

$$\alpha + \beta \cdot v. \tag{1}$$

Here, α stands for the "start-up" cost of the unit, and β quantifies the unit's processing rate. The batch volume directly impacts the additional processing time required. This relationship remains consistent across other examples. We use this example as the foundation for examining our proposed unit-based event point GA representations.

Table 1. Problem Data for Example 1

Unit	Capacity	Suitability	τ	(α)	(β)
Unit 1	100	Mixing	4.5	3.0	0.0300
Unit 2	75	Reaction	3.0	2.0	0.0267
Unit 3	50	Purification	1.5	1.0	0.0200
State	Storage Capacity	Initial Amount	Price		
State 1	Unlimited	Unlimited	0.0		
State 2	100	0.0	0.0		
State 3	100	0.0	0.0		
State 4	Unlimited	0.0	1.0		

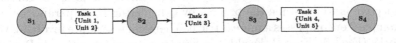

Fig. 3. STN for Example 2

Example 2 introduces a basic chemical processing sequence with three principal tasks: Task 1, Task 2, and Task 3. The STN for this sequence contains four

Table 2. Problem Data for Example 2

Unit	Capacity	Suitability	τ	(α)	(β)
Unit 1	100	Task 1	2.0	1.333	0.01333
Unit 2	150	Task 1	2.0	1.333	0.01333
Unit 3	200	Task 2	1.5	1.000	0.00500
Unit 4	150	Task 3	1.0	0.667	0.00445
Unit 5	150	Task 3	1.0	0.667	0.00445
State	Storage Capacity	Initial Amount	Price		
State 1	Unlimited	Unlimited	0.0		
State 2	200	0.0	0.0		
State 3	250	0.0	0.0		
State 4	Unlimited	0.0	5.0		

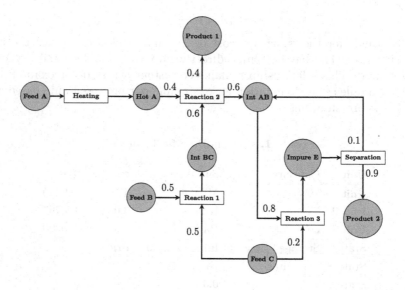

Fig. 4. STN for Example 3

states (s_1, s_2, s_3, and s_4) and five units (Unit 1 to Unit 5). We illustrate this in Fig. 3. Each unit possesses distinct capacity, task suitability, process time, and cost parameters, as outlined in Table 2. When aiming for profit maximisation, our objective is to fine-tune the process by considering these parameters.

Example 3 encompasses a more intricate chemical procedure involving various feedstocks (Feed A, Feed B, Feed C), several reactions (Reaction 1, Reaction 2, Reaction 3), and a distinct separation stage. The STN, showcased in Fig. 4, maps out complex material flows across different stages, with certain streams diverging and others converging. Each participating unit (Heater, Reactor 1, Reactor 2, Still) has capacity, task suitability, process time, and cost parameters, which we

Table 3. Problem Data for Example 3

Unit	Capacity	Suitability	(τ)	(α)	(β)
Heater	100	Heating	1.0	0.6667	0.00667
Reactor 1	50	Reaction 1	2.0	1.3333	0.02664
Reactor 1	50	Reaction 2	2.0	1.3333	0.02664
Reactor 1	50	Reaction 3	1.0	0.6667	0.01332
Reactor 2	80	Reaction 1	2.0	1.3333	0.01665
Reactor 2	80	Reaction 2	2.0	1.3333	0.01665
Reactor 2	80	Reaction 3	1.0	0.6667	0.00833
Still	200	Separation	2.0	1.3342	0.00666

State	Storage Capacity	Initial Amount	Price
Feed A	Unlimited	Unlimited	0.0
Feed B	Unlimited	Unlimited	0.0
Feed C	Unlimited	Unlimited	0.0
Hot A	100	0.0	0.0
Int AB	200	0.0	0.0
Int BC	150	0.0	0.0
Impure E	200	0.0	0.0
Product 1	Unlimited	0.0	10.0
Product 2	Unlimited	0.0	10.0

detail in Table 3. Our overarching goal is to boost the production of the main products, namely Product 1 and Product 2.

2 Methodology

This section introduces the four GA variants we studied. We also discuss the novel process snapping technique, which facilitates a lower-dimensional unit-based event point chromosome representation. Additionally, we discuss the primary performance metrics and the experimental design we adopted for this research. Collectively, these elements underpin our strategy to improve scheduling in multipurpose batch chemical production plants.

2.1 GA-GP: Global-Based Approach

We base the GA-GP method on the strategy Woolway and Majozi presented [21]. However, we made modifications to align it with the specific experimental conditions of our research, including adjustments in termination criteria and population sizes. The GA-GP lays the groundwork for the GA-GQ, offering a reference point for comparisons and deepening our grasp of the impact of the new snapping paradigm, which we detail in Sect. 2.2.

2.2   Batch Process Snapping Approaches

Batch process representations often possess exceedingly high dimensions, posing challenges in tackling specific problems. In addressing this high dimensionality, we note the following characteristics of batch processes: (i) Every batch invariably progresses from its start to its completion without interruption, and (ii) Succeeding batches can only utilise the partial products from preceding batches that have been completed.

While these observations might seem intuitive, they imply that any batch's commencement always coincides with another batch's conclusion. We refer to this phenomenon as "snapping" throughout this article.

GA-GQ: Q-Global-Based Approach. The snapping phenomenon highlights the limitations of global-based event point representations. Batch processes need an embedding space of much lower dimensionality than what these representations offer. In this paper, we introduce an alternative "Q-Global" representation for the global-based approach by utilising the snapping phenomenon. Notably, despite being a global-based representation, it stands distinct from the approach Woolway and Majozi proposed [21].

In this new representation, we employ two matrices: M and V. M functions as the Unit-Action matrix, while V acts as the Unit-Volume or temporal matrix. In both M and V, each row stands for an individual processing Unit. The columns detail the actions planned at every event point. Importantly, for any given row, actions linked to events on the left precede those on the right.

The Unit-Action matrix holds the exact action $a \in A$ that Unit m should perform at event point e, denoted by the location (row, column) $M_{m,e} = a$. Simultaneously, the Unit-Volume matrix captures the volume, $v \in \mathbb{R}$, of action a that Unit m produces at event point e, expressed as $V_{m,e} = v$. In this context, $m \in [1 \ldots |M|]$ ranges from 1 to the total number of Units $|M|$, and $e \in [1 \ldots |E|]$ spans from 1 to the max event points $|E|$, considered a free parameter.

Our novel approach introduces a third matrix, G, to the representation. This matrix, G, lists actions like "global snap m" ($a_{\mathrm{GOP}-m}$), where m denotes the Unit to which every new action tries to align, given it is feasible. This feasibility is crucial since an action should not align with points before the respective Unit is ready. Moreover, given that actions might not occur during an event point, we have incorporated the "no operation" (a_{NOP}) action.

In Fig. 5(a) the diagram for matrices

$$M = \begin{bmatrix} a_1 & a_2 & - \\ a_3 & - & - \\ a_4 & a_5 & - \end{bmatrix}, \quad V = \begin{bmatrix} v_1 & v_2 & - \\ v_3 & - & - \\ v_4 & v_5 & - \end{bmatrix}, \quad \text{and} \quad G = \begin{bmatrix} a_{\mathrm{GOP}-2} & - & - \end{bmatrix}, \tag{2}$$

are shown. Since G_1 contains the action "global snap 2" ($a_{\mathrm{GOP}-2}$) action a_2 and a_5 are snapped to the end of a_3 the current action for Unit 2.

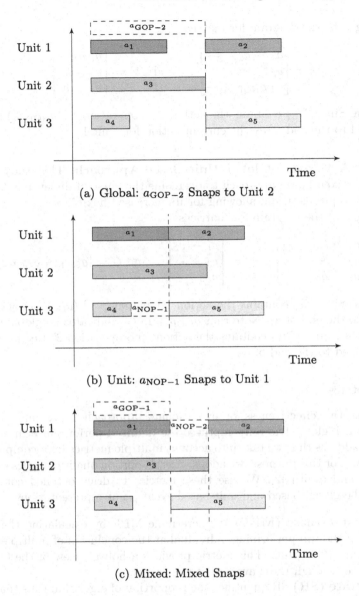

(a) Global: $a_{\text{GOP-2}}$ Snaps to Unit 2

(b) Unit: $a_{\text{NOP-1}}$ Snaps to Unit 1

(c) Mixed: Mixed Snaps

Fig. 5. Examples of Global, Unit and Mixed Snapping Procedures

GA-UP: Unit-Based Approach. The snapping phenomenon enables us to adopt a unit-based approach with a "lower dimensionality" than a simplistic solution would demand. In such a representation, an event point might not require any action from a unit. Our study adds the "no operation m" ($a_{\text{NOP-}m}$) as an additional action to address this. This action signifies that a unit remains inactive, and any subsequent action aligns with unit m.

In Fig. 5(b) the diagram for matrices

$$M = \begin{bmatrix} a_1 & a_2 & - \\ a_3 & - & - \\ a_4 & a_{\text{NOP}-1} & a_5 \end{bmatrix}, \quad \text{and} \quad V = \begin{bmatrix} v_1 & v_2 & - \\ v_3 & - & - \\ v_4 & - & v_5 \end{bmatrix}, \tag{3}$$

are shown. Since M_{32} contains the action "'no operation 1" ($a_{\text{NOP}-1}$) action a_5 is snapped to the end of a_1 the current action for Unit 1.

GA-MP: A Mixed Q-Global/Unit-Based Approach. This study proposes a third or mixed-based approach to overcome the curse of dimensionality in the unit-based representation, allowing for its increased flexibility.

In Fig. 5(c) the diagram for matrices

$$M = \begin{bmatrix} a_1 & a_{\text{NOP}-2} & a_2 \\ a_3 & - & - \\ a_4 & a_5 & - \end{bmatrix}, \quad V = \begin{bmatrix} v_1 & - & v_2 \\ v_3 & - & - \\ v_4 & v_5 & - \end{bmatrix}, \quad \text{and} \quad G = \begin{bmatrix} a_{\text{GOP}-1} & a_{\text{GOP}-0} & - \end{bmatrix}, \tag{4}$$

are shown. Since G_1 contains the action "global snap 1" ($a_{\text{GOP}-1}$) action a_5 is snapped to the end of a_1. Note that $a_{\text{GOP}-0}$ in G_2 indicates no global snapping should occur. Since M_{12} contains the action "'no operation 2" ($a_{\text{NOP}-2}$) action a_2 is snapped to the end of a_3

2.3   Metrics

Evaluating the effectiveness of evolutionary approaches presents challenges because no single metric fully captures the nuanced performance in every scenario. To address this, we take into account multiple metrics for a comprehensive assessment. For this purpose, we adopt several solution quality metrics as defined by Eiben and Smith [2]. We use these metrics to draw detailed comparisons between the global-based and unit-based event point representations.

Mean Best Fitness (MBF) We determine MBF by calculating the average fitness of the top-performing individual at the conclusion of multiple Genetic Algorithm (GA) runs. This metric provides a holistic view of the GA's performance through these average fitness scores.

Success Rate (SR) SR quantifies the proportion of algorithm runs that satisfy a set success benchmark. In this research, we designate a successful outcome as one reaching either 99% or 95% of the optimal value as determined by the SSN model.

Average Evaluations to a Solution (AES) AES represents the mean number of evaluations needed by the GA to achieve a solution with a specified SR, such as an SR of 99%. When computing this metric, we exclude GA runs that fall short of the designated SR.

Efficiency We assess efficiency by averaging the CPU time the GA consumes to achieve a specified SR. This metric enables comparisons of the computational demands of different strategies.

By employing this diverse set of metrics, we comprehensively evaluate the global-based and unit-based event point representations, granting insights into their advantages and challenges across various contexts.

2.4 Experimental Design

We carried out a systematic experimental procedure for each of the GAs studied: GA-GP, GA-GQ, GA-UP, and GA-MP. This procedure involved generating an initial population, setting an adaptive termination criterion, and optimising parameters via a grid search.

2.5 Initial Population, Offspring Generation and Termination

We created an initial population for each GA to guarantee a robust starting point. We set the population size at 100 times the size corresponding to a specific time horizon, which we determined by: $PopSize(H) = 500 \times 2^{((H-8)/2)}$, where the calculation depends on the time horizon H, starting from a base size of 500.

We customised the termination criterion for each GA for our study. While GAs traditionally halt after a set number of iterations, our approach varied. We set the termination threshold at 30 consecutive generations without any improvement but with an upper limit of 5000 generations. By introducing this dynamic termination point, we aimed to minimise redundant calculations, focus computational resources on promising trajectories, and facilitate the computation of quality metrics.

Selecting parents for offspring generation is crucial in GAs as it dictates which individuals pass on their genetic material to the subsequent generation. We use a blend of elitism, rank-based selection, and deterministic crowding [12] for this selection process.

With the rank-based selection, we rank each population at every iteration. We then choose a subset of the top-ranking individuals, known as the elite population, for upcoming crossover and mutation procedures. We also carry this elite population forward to the next generation, a practice termed as elitism. Additionally, our GAs use deterministic crowding, a niche-based selection technique that evaluates phenotypic features using inverse hamming distance. This method ensures that diverse individuals are retained, mitigating premature convergence.

Within GAs, crossover and mutation serve as foundational mechanisms to produce new potential solutions from chosen parents. They emulate natural reproductive and genetic variation processes. Given that the chromosomes M and G are categorical and V is a real value, we pick suitable operators.

For the categorical attributes like actions in M and G, we use uniform flip mutations. For the continuous variables in V, we apply Gaussian mutations, adding a slight Gaussian-random value. We also employ intermediate recombination with BLX-α crossover [3] for these continuous variables and use single-point crossover for categorical ones.

2.6   Grid Search for Parameter Optimisation

After generating the initial population and setting the termination criterion, we conducted a grid search for each GA to fine-tune the elite, crowding, crossover, and mutation parameters. This grid search employed the revised termination criterion to pinpoint the optimal solution and average of the elite population. We computed the score for every parameter set using the following: score $= \frac{1}{2}$ (best $+$ eliteaverage). We selected parameters yielding the peak score for application in the respective GA. Our goal with this method was to identify the parameter combination that best balances exploration and exploitation. We applied this experimental framework to all four GAs to evaluate their strengths, aiming to prevent stagnation and computational inefficiencies.

3   Results

Results for Examples 1–3 are provided in the corresponding sections. All computations were performed on an AMD Ryzen^{TM} 3800x @ 3.9GHz, equipped with 32 GB RAM @ 3200 MHz, running Ubuntu 20.04. The GA-GP and GA-UP algorithms were implemented in Julia 1.4.1, while the MILP models were built on GAMS 24.8.5 r61358 with CPEX 12.7.1.0.

Example 1. The study proposes a "snapping phenomenon" as an alternative global-based approach to overcome the high dimensionality in batch process optimisation. The proposed method is tested for feasibility in Example 1 (Table 4). The results demonstrate the similar performance of GA-GQ in metrics Success Rate at 99% (SR@99), Average Evaluations to SR@99 (AES@99), and Efficiency@99. These results are achieved by using fewer even points than the GA-GP. The results also highlight the challenges posed by high dimensionality on GA-UP, which fails to converge to within 95% of the MILP at 60 h. GA-MP obtains the best SR@99 and Mean Best Fit (MBF) at 60 h.

The Relative Gap (RG) defines the variance between the ongoing optimal solution and the relaxed MILP's maximum upper bound. A value of 0 signals a convergence to the global solution, with the convergence time provided in brackets. Particularly in Example 3, RG values draw attention to the potential for exponential growth.

The results in Table 4 indicate that the snapping phenomenon is an exploitable characteristic of batch processes. The smaller number of event points in GA-GQ indicates that this approach can effectively reduce dimensionality, improve SR@99, reduce AES@99, and enhance Efficiency@99. While the global-based GA (GA-GP) has been previously demonstrated to solve the batch process problem effectively, these results affirm its effectiveness and highlight additional avenues for optimising global representations. The results underscore the necessity of a mixed approach over a unit-based approach in high-dimensional spaces.

While these findings are promising, the simplicity of the setup might inadvertently favour global-based representations. To disentangle the contributions of

Table 4. Results for Example 1 (best has grey backdrop; top two in bold; red not solved).

Model	Hours	EP	BOV	MBF	SR 95	SR 99	AES 95	AES 99	Efficiency 95	Efficiency 99	RG
MILP	–	5	71.5	–	–	–	–	–	–	–	0 (0.75)
GA-GP	12	5	71.5	**71.5**	100.00	**100.00**	162.63	**162.63**	4.16	**4.16**	–
GA-GQ	–	4	71.5	**71.5**	100.00	**100.00**	19.8	**31.46**	1.93	**2.90**	–
GA-UP	–	5	71.5	65.6	6.67	6.67	143.00	173.00	12.55	13.58	–
GA-MP	–	5	71.5	67.3	33.33	33.33	70.30	187.10	7.13	14.32	–
MILP	–	9	249.9	–	–	–	–	–	–	–	0 (1.38)
GA-GP	24	10	249.8	**245.8**	100.00	6.67	287.56	377.00	13.02	**17.82**	–
GA-GQ	–	10	249.9	241.6	80.00	**10.00**	62.87	**91.00**	17.52	26.92	–
GA-UP	–	9	249.9	241.0	90.00	3.33	91.51	**106.00**	6.26	**4.47**	–
GA-MP	–	9	249.9	**243.5**	100.00	**36.67**	82.70	310.63	38.09	142.05	–
MILP	–	15	446.9	–	–	–	–	–	–	–	0 (28.70)
GA-GP	36	14	445.4	**429.3**	73.33	10.00	624.23	**863.67**	78.23	**108.21**	–
GA-GQ	–	16	445.5	424.6	63.33	**20.00**	124.68	**233.67**	17.93	**34.19**	–
GA-UP	–	13	445.3	420.4	46.67	6.67	721.85	1303.00	59.90	223.23	–
GA-MP	–	12	445.5	**437.6**	96.67	**36.67**	191.27	920.45	137.18	231.17	–
MILP	–	19	646.9	–	–	–	–	–	–	–	0 (212.38)
GA-GP	48	23	645.4	**627.5**	86.67	10.00	526.73	**506.67**	233.30	**205.10**	–
GA-GQ	–	20	645.5	606.7	30.00	**16.67**	211.00	**336.60**	66.66	**99.33**	–
GA-UP	–	20	642.8	589.7	13.33	3.33	2929.75	1431.00	1008.85	705.23	–
GA-MP	–	18	645.4	**628.9**	90.00	**16.67**	385.81	808.80	110.55	240.91	–
MILP	–	23	846.9	–	–	–	–	–	–	–	0 (6226.95)
GA-GP	60	27	845.1	**823.2**	90.00	10.00	500.67	**620.67**	631.92	**789.02**	–
GA-GQ	–	26	845.5	799.4	43.33	10.00	278.31	**353.00**	112.97	**155.99**	–
GA-UP	–	–	–.–	–.–	0.00	0.00	∞	∞	∞	∞	–
GA-MP	–	22	845.5	**822.5**	80.00	**13.33**	782.33	1349.75	558.81	958.91	–

the unit and mixed-based approaches from the snapping phenomenon's exploitation, the subsequent sections (Subsects. 3 and 3) present more complex challenges.

Example 2. Beyond how the snapping phenomenon might reduce dimensionality, the study also explores how a unit-based approach might improve the SOTA. Table 5 presents the findings from Example 2. The results reveal difficulty converging for GA-GP, as indicated by a zero SR@99, which could be attributed to oscillations around the event points. However, GA-GQ, GA-MP and GA-UP overcome this micro-misalignment of batches by leveraging the snapping phenomenon. GA-GQ outperforms GA-GP across most metrics, demonstrating superior population convergence, thus requiring fewer iterations, indicating faster performance and improved exploration and exploitation.

In Table 5, observe that the MBF for the unit-based approaches GA-UP and GA-MP do not always lead to better solutions over GA-GQ. Nonetheless,

Table 5. Results for Example 2 (best has grey backdrop; top two in bold; red missed SR@99).

Model	Hours	EP	BOV	MBF	SR		AES		Efficiency		RG
					95	99	95	99	95	99	
MILP	–	5	1840.2	–	–	–	–	–	–	–	0 (0.86)
GA-GP	8	5	1838.7	1838.4	100.00	100.00	657.40	657.40	16.10	16.10	–
GA-GQ	–	4	1840.2	**1840.2**	100.00	100.00	22.60	**33.86**	0.22	**0.29**	–
GA-UP	–	5	1840.2	1817.1	96.66	70.00	232.89	626.95	13.21	36.37	–
GA-MP	–	4	1840.2	**1840.2**	100.00	100.00	45.70	**148.56**	0.42	**1.35**	–
MILP	–	7	2628.1	–	–	–	–	–	–	–	0 (1.11)
GA-GP	10	7	**2592.6**	**2592.2**	100.00	0.00	296.33	∞	17.80	∞	–
GA-GQ	–	7	2628.2	**2606.9**	100.00	**26.67**	47.76	**84.87**	5.45	**9.92**	–
GA-UP	–	7	2628.1	2527.8	80.00	16.67	515.08	708.40	51.03	73.83	–
GA-MP	–	6	2628.2	2581.2	96.67	**53.33**	110.48	**283.93**	18.68	**43.62**	–
MILP	–	9	3463.3	–	–	–	–	–	–	–	0 (5.77)
GA-GP	12	9	**3349.6**	3264.4	36.66	0.00	364.36	∞	55.84	∞	–
GA-GQ	–	8	3463.6	3297.3	66.66	**13.33**	99.55	**1049.50**	18.02	**92.09**	–
GA-UP	–	7	3461.6	**3305.0**	56.66	**13.33**	319.41	1943.50	74.95	360.56	–
GA-MP	–	8	3463.6	**3394.5**	90.00	**46.67**	151.59	**466.90**	48.32	**161.30**	–

GA-MP appears to hold a significant advantage in terms of SR@99. Due to the increased dimensionality, we also note that GA-UP and GA-MP have less Efficiency@99 and have higher AES@99.

The results for Example 2 support an argument that the increased complexity in the batch process exploiting the snapping phenomenon leads to better solutions over the GA-GP. GA-GQ, GA-UP and GA-MP all implement the snapping phenomenon. As such, it is impossible to determine if the positive results result from the unit-based approach or just the snapping phenomenon. In the following Subsection, the more complex Example 3 will seek to unpack these outstanding questions further.

Example 3. Example 3 presents a more complex framework with numerous local minima around the global minima. The objective is to assess the necessity for a unit-based approach beyond exploiting the snapping phenomenon, a concept that GA-MP already indicated in Example 2. Results from this example are shown in Table 6. The notable observation here is the struggle of both GA-GP and GA-GQ to converge, as illustrated by an SR@99 value of zero.

Looking at Table 6, it becomes clear that GA-UP and GA-MP outperform GA-GQ and GA-GP regarding SR@99 and MBF. GA-UP and GA-MP take the longest to converge (AES@99 and Efficiency@99). Although GA-UP and GA-MP use the least event points due to using more iterations, they still show convergence, not getting trapped, whereas the others terminate in the local minima.

The outcome from Example 3 is that unit-based approaches can be advantageous for more complex problems. Specifically, the performance of global-based

Table 6. Results for Example 3 (best has grey backdrop; top two in bold; red missed SR@99).

Model	Hours	EP	BOV	MBF	SR (%) 95	99	AES 95	99	Efficiency (s) 95	99	RG
MILP	–	5	1498.6	–	–	–	–	–	–	–	0 (0.47)
GA-GP	8	5	1498.5	1497.8	100.00	**100.00**	322.53	322.53	4.56	4.56	–
GA-GQ	–	5	1498.6	1449.6	93.33	83.33	74.82	95.24	2.59	**3.48**	–
GA-UP	–	4	1498.2	**1498.1**	100.00	**100.00**	46.76	**87.33**	1.66	**2.99**	–
GA-MP	–	5	1498.6	**1498.6**	100.00	**100.00**	44.60	**87.70**	2.09	4.15	–
MILP	–	7	1962.7	–	–	–	–	–	–	–	0 (10.74)
GA-GP	10	8	**1899.5**	1877.3	66.67	0.00	245.75	∞	10.59	∞	–
GA-GQ	–	7	1957.2	1793.4	26.67	3.33	75.75	**137.00**	6.07	**10.30**	–
GA-UP	–	6	1960.5	**1896.1**	70.00	**30.00**	126.57	810.77	9.26	47.07	–
GA-MP	–	6	1962.2	1889.9	76.66	**23.33**	84.87	**377.43**	2.84	**26.85**	–
MILP	–	8	2658.5	–	–	–	–	–	–	–	0 (16.78)
GA-GP	12	8	**2562.1**	2479.7	20.00	0.00	208.00	∞	14.80	∞	–
GA-GQ	–	7	**2610.9**	2395.4	20.00	0.00	171.16	∞	16.89	∞	–
GA-UP	–	7	2658.4	**2570.3**	76.67	**26.67**	279.91	924.13	58.83	**131.81**	–
GA-MP	–	7	2658.5	2541.4	66.67	20.00	213.55	**397.67**	42.14	**72.19**	–
MILP	–	9	3231.4	–	–	–	–	–	–	–	0 (32.83)
GA-GP	14	10	3230.9	**3123.5**	50.00	**43.33**	442.20	**465.54**	54.97	**57.79**	–
GA-GQ	–	10	3231.3	2952.8	26.67	16.67	261.50	748.40	89.97	140.99	–
GA-UP	–	8	3231.3	**3087.7**	56.67	20.00	323.82	694.50	111.86	182.46	–
GA-MP	–	7	3231.3	3084.1	56.67	20.00	254.82	**236.67**	121.27	**118.55**	–
MILP	–	10	3738.4	–	–	–	–	–	–	–	0 (309.06)
GA-GP	16	12	**3670.7**	3637.6	100.00	0.00	62.20	∞	43.62	∞	–
GA-GQ	–	11	3730.4	3535.8	63.33	10.00	221.84	**380.33**	185.00	**264.85**	–
GA-UP	–	8	3735.5	**3664.2**	96.67	**33.33**	212.58	634.00	187.65	**461.04**	–
GA-MP	–	10	3738.4	**3653.1**	96.66	30.00	332.06	**505.33**	452.65	708.94	–

approaches (GA-GP and GA-GQ) begins to deteriorate in these complex scenarios. One aspect worth noting is the exceptions to the rules across all examples. For instance, GA-UP seems to outperform the GA-MP in Example 3 compared to Example 2.

4 Conclusion

This study aimed to investigate the potential benefits of a unit-based approach for improving convergence in global-based methods and to evaluate the possibility of exploiting the snapping phenomenon to reduce dimensionality in batch process optimisation. The results and discussions show that exploiting the snapping phenomenon can improve convergence in batch process optimisation. The findings also suggest that unit-based approaches outperform global-based approaches when dealing with complex problems. Nevertheless, unit-based approaches may suffer in high dimensional spaces, i.e., when handling many event points. Therefore, a mixed approach that combines strategies proves advantageous.

The results of this study underline the importance of two key contributions. Firstly, the usefulness of unit-based approaches in batch process optimisation. Secondly, the potential to exploit specific characteristics of batch process optimisation, such as the snapping phenomenon, through GAs.

The presented results are based on empirical evidence. While the value of exploiting the snapping phenomenon and using unit-based approaches has been demonstrated, further mathematical analysis would be needed to understand the intrinsic dimensionalities of the various representations. Given these limitations, these results may not be generalizable beyond the examples presented.

References

1. Bowditch, Z., Woolway, M., van Zyl, T.: Comparative metaheuristic performance for the scheduling of multipurpose batch plants. In: 2019 6th International Conference on Soft Computing & Machine Intelligence (ISCMI), pp. 121–125. IEEE (2019)
2. Eiben, A., Smith, J.: Introduction to Evolutionary Computing. Springer, Berlin (2015)
3. Eshelman, L.J., Schaffer, J.D.: Real-coded genetic algorithms and interval-schemata. In: Foundations of Genetic Algorithms, vol. 2, pp. 187–202. Elsevier (1993)
4. Georgiadis, G.P., Elekidis, A.P., Georgiadis, M.C.: Optimization-based scheduling for the process industries: from theory to real-life industrial applications. Processes **7**(7), 438 (2019)
5. Harjunkoski, I., et al.: Scope for industrial applications of production scheduling models and solution methods. Comput. Chem. Eng. **62**, 161–193 (2014)
6. He, Y., Hui, C.W.: Rule-evolutionary approach for single-stage multiproduct scheduling with parallel units. Ind. Eng. Chem. Res. **45**(13), 4679–4692 (2006)
7. He, Y., Hui, C.W.: Genetic algorithm based on heuristic rules for high-constrained large-size single-stage multi-product scheduling with parallel units. Chem. Eng. Process. **46**(11), 1175–1191 (2007)
8. He, Y., Hui, C.W.: A binary coding genetic algorithm for multi-purpose process scheduling: a case study. Chem. Eng. Sci. **65**(16), 4816–4828 (2010)
9. Ierapetritou, M., Floudas, C.: Effective continuous-time formulation for short-term scheduling. 1. Multipurpose batch processes. Ind. Eng. Chem. Res. **37**(11), 4341–4359 (1998)
10. Kondili, E., Pantelides, C., Sargent, R.: A general algorithm for short-term scheduling of batch operations-i. MILP formulation. Comput. Chem. Eng. **17**(2), 211–227 (1993)
11. Li, Z., Ierapetritou, M.: Process scheduling under uncertainty: review and challenges. Comput. Chem. Eng. **32**(4–5), 715–727 (2008)
12. Mahfoud, S.W.: Niching methods for genetic algorithms. University of Illinois at Urbana-Champaign (1995)
13. Maravelias, C.T.: General framework and modeling approach classification for chemical production scheduling. AIChE J. **58**(6), 1812–1828 (2012)
14. Maravelias, C.T., Sung, C.: Integration of production planning and scheduling: overview, challenges and opportunities. Comput. Chem. Eng. **33**(12), 1919–1930 (2009)

15. Méndez, C.A., Cerdá, J., Grossmann, I.E., Harjunkoski, I., Fahl, M.: State-of-the-art review of optimization methods for short-term scheduling of batch processes. Comput. Chem. Eng. **30**(6), 913–946 (2006)
16. Seid, R., Majozi, T.: A robust mathematical formulation for multipurpose batch plants. Chem. Eng. Sci. **68**(1), 36–53 (2012)
17. Shaik, M.A., Janak, S.L., Floudas, C.A.: Continuous-time models for short-term scheduling of multipurpose batch plants: a comparative study. Ind. Eng. Chem. Res. **45**(18), 6190–6209 (2006)
18. Sundaramoorthy, A., Karimi, I.: A simpler better slot-based continuous-time formulation for short-term scheduling in multipurpose batch plants. Chem. Eng. Sci. **60**(10), 2679–2702 (2005)
19. Sundaramoorthy, A., Maravelias, C.T.: Computational study of network-based mixed-integer programming approaches for chemical production scheduling. Ind. Eng. Chem. Res. **50**(9), 5023–5040 (2011)
20. Woolway, M., Majozi, T.: A novel metaheuristic framework for the scheduling of multipurpose batch plants. Chem. Eng. Sci. **192**, 678–687 (2018)
21. Woolway, M., Majozi, T.: On the application of a metaheuristic suite with parallel implementations for the scheduling of multipurpose batch plants. Comput. Chem. Eng. **126**, 371–390 (2019)
22. van Zyl, T., Woolway, M.: Makespan minimisation for multipurpose batch plants using metaheuristic approaches. In: 2020 7th International Conference on Soft Computing & Machine Intelligence (ISCMI), pp. 56–60. IEEE (2020)

Investigating the Extent and Usability of Webtext Available in South Africa's Official Languages

Febe de Wet[1]([✉])(iD), Roald Eiselen[2](iD), Erwin Schillack[1](iD),
and Martin Puttkammer[2](iD)

[1] Department of Electrical and Electronic Engineering, Stellenbosch University,
Stellenbosch, South Africa
{fdw,22145540}@sun.ac.za
[2] Centre for Text Technology, North-West University, Potchefstroom, South Africa
{Roald.Eiselen,Martin.Puttkammer}@nwu.ac.za

Abstract. Large collections of text data are a central part of many aspects of natural language processing (NLP) development, and the availability of such data is a prerequisite for advances in the field. With the proliferation of very large, multi-lingual, web-based data sets over the last decade, it is often overlooked that there are still many languages that are not well represented in these data sets, and for which collecting even moderately sized data sets remains a challenge. Furthermore, a lack of data for a language in one or more of the core data sources, such as Wikipedia, often leads to such languages being further excluded in other collections of web-based data or having a detrimental effect on the quality of the data collected. The systematic review and investigation of the quality of these data sets are relatively limited, and the quality of the data is primarily extrinsically evaluated by measuring the improvements on downstream tasks, rather than implicitly evaluating the data. This paper reports on some of the text data that is currently available for South Africa's official languages (except for English and South African Sign Language) in various widely available web-derived corpora. The aim of the study was to harvest text from the web that could be used as prompts on the Mozilla Common Voice Platform. Towards this aim the extent of the resources available in each language as well as the degree of overlap between different sources were quantified. Results show that there remain several South African languages for which web-based corpora are still severely limited and that, for languages with at least some web presence, the majority of the text is of disputable quality.

Keywords: Under-resourced languages · text data · web text · South African languages · Mozilla Common Voice Platform

---

Supported by the German Federal Ministry of Economic Cooperation and Development (BMZ), represented by the GIZ project FAIR Forward - Artificial Intelligence for All.

1   Introduction

Over the last decade natural language processing has undergone a resurgence with substantial quality improvements in nearly all areas[1]. These improvements have primarily been driven by the introduction of new deep learning algorithms, the availability of huge amounts of training data, and the processing capacity to leverage these algorithms and data. Unfortunately most of the consequential improvements, such as those powering many of the modern generative models, are limited to a small set of languages where the availability of data and compute power allow for performance that some equate to so-called artificial intelligence. Most of these models are trained on some (sub)set of web-derived data, such as Wikipedia, reddit, Twitter, online news sites, and the like. Although it is now mostly taken for granted in languages with extensive language resources, there remains a large number of languages that are not as well represented and for which collecting even moderately sized data sets remains a challenge. While it is often assumed that most languages have at least some presence on the web, there are also still many who are not represented on the web at all.

Moreover, just being present on the web does not guarantee that the data is of good quality and suitable for NLP research and development. Many web texts are extremely domain specific and, in many languages, online text is often of either a religious or governmental nature. Both the quality and the specificity of these texts influence the usability of the data, be it for prompt generation or NLP research and development.

This paper provides an overview of the text data that was collected from various online data sources and repositories for South Africa's official languages[2]. The corpora were used to select sentences that could be used as prompts on the Mozilla Common Voice (MCV) platform[3]. In addition to listing the sources of text data, the availability of text resources in each language is quantified and the quality of the text is gauged in terms of the language reported and overlap between the corpora.

The next section addresses some of the risks, challenges, and pitfalls associated with using "data harvested from the internet" in NLP research, in general but also more specifically for under-resourced languages like those spoken in South Africa. The text sources that could be found for South Africa's official languages are presented in Sect. 3. Section 4 introduces the techniques that were used to clean and analyse the data to provide some descriptive statistics of the online resources. The section also reports on the usability of the data as a source of text prompts, while Sect. 5 outlines the final process followed to create the MCV text prompts. Section 6 gives an outlook on the NLP future of South Africa's languages, given the text data that is currently available on the web.

---

[1] https://nlpprogress.com/.
[2] With the exception of English (which is one of the most well-resourced languages in the world) and the newly recognised South African sign language.
[3] https://commonvoice.mozilla.org/en.

2   Background

There have been several efforts to collect corpora of text data in the context of NLP for the South African indigenous languages [6,8,15,16], yet there remain significant barriers for both collecting and using these resources. The most significant of these is the copyright restrictions in South Africa where fair use is substantially more restrictive than it is, for example, in the United States. As two examples, both the Viva Afrikaans corpora[4] (365 million words) and isiZulu National Corpus[5] (20 million words), the largest available corpora for these languages, are not freely available due to copyright restrictions on subsets of the data. Most of the corpora that are freely available rely on the availability of data from the South African government, which makes available various types of information in all of the South African languages, mostly through their respective web sites.

This reliance on government data links closely with the second problem, the relatively limited scope of the data. Since most of the data is based on governmental communications, this is not necessarily an accurate reflection of typical language use, but rather reflects very formal communication and distribution of information. Even though government sources are generally of adequate quality as they are edited and publication ready, to collect more general domain and larger corpora, other data sets need to be used.

The availability of very large web-derived data sets has proliferated over the last decade, and has made it significantly easier for anyone to access large collections of textual data [18]. It is well known that the quality of these data sets is generally substantially worse than hand-curated data sets [13], but the advantages of building NLP technologies based on these corpora for under-resourced languages generally outweigh the quality concerns [14]. The systematic review and investigation of the quality of these data sets is relatively limited, and the quality of the data is primarily extrinsically evaluated by measuring the improvements in downstream tasks, rather than implicitly evaluating the data itself. In their recent investigation of the quality of several of the more widely used data collections, [14] showed that there are substantial problems with the quality of the data, especially for lower-resourced languages, which may well be detrimental to development activities in such languages.

One of the areas in which the use of large data sets is most common is in the creation of language models and word embeddings. In these applications the size and representivity of the data are directly correlated to how well the models represent the characteristics of the language. This relationship has become even more important with the increased use of deep learning techniques, where these pre-trained language models and embeddings are at the core of most improvements in state-of-the-art NLP. It is not uncommon to build these models with several billion words [5], or even hundreds of billions [4], often with limited consideration of the quality of the data.

---

[4] https://viva-afrikaans.org/.
[5] https://iznc.ukzn.ac.za/.

As the extent of the data decreases, it becomes more important to ensure that the quality of the data is high, since the effect of incorrect and low quality data becomes more pronounced, because it is no longer just noise, but can become part of the signal. Several authors have investigated various filtering techniques for cleaning web-based data, including language identification, but it is not always clear what the quality of these processes are, especially where lower resourced languages are concerned. South Africa's official languages (with the exception of English) are all severely under-resourced and are therefore susceptible to the risks associated with low-quality data.

The Mozilla Foundation has established an open-source project called Mozilla Common Voice with the objective of gathering and disseminating a sizable data set of human voices in numerous languages. The primary purpose of this data set is to train and enhance automatic speech recognition (ASR) systems, which are employed in voice assistants and speech-to-text software. People from all across the world are invited to contribute their voices by reading sentences and phrases on the Common Voice platform. These contributions are documented and used to compile a varied and sizable collection of spoken language that is recorded in several languages.

For a language to be *launched* on the MCV platform, specific sections of the Common Voice website need to be translated and at least 5 000[6] sentences in the target language have to be collected and be available in the open domain under CC0 licensing[7]. Once *launched* status is achieved, the sentences are used as prompts for speech data collection. The aim of the project reported on here was to collect a minimum of 5 000 usable sentences for 10 South African languages[8].

3   Data

Given that government domain data was the primary source of data for previous speech related projects [2,3], our initial strategy was to use data outside of the government domain as source for the MCV speech prompts. For this reason the NCHLT corpora, which only includes data from governmental websites, was initially excluded as a potential data source. Text was collected from other web sources including Wikipedia, the Leipzig Corpora Collection [10], the Open Parallel Corpus [20] and FLORES-200 [11] data sets, as well as Common Crawl[9]. However, data collected from the South African governmental web domains (*.gov.za) remain an important source of text data in South Africa's official languages and the NCHLT corpora are therefore included in the analysis in Sect. 3 as a point of reference.

---

[6] This requirement has since been adjusted to accommodate under-resourced languages.

[7] https://creativecommons.org/share-your-work/public-domain/cc0/.

[8] Afrikaans (Afr), isiNdebele (Nbl), isiXhosa (Xho), isiZulu (Zul), Sepedi (Nso), Sesotho (Sot), Setswana (Tsn), Siswati (Ssw), Tshivenda (Ven), and Xitsonga (Tso).

[9] https://commoncrawl.org/.

3.1 Wikipedia

Generally speaking, the text data scraped from *Wikipedia* serves as an excellent resource for language modeling. The majority of Wikipedia articles are well-written and go through a collaborative editing process, producing accurate and trustworthy information. The content offers thorough coverage of many different disciplines and a broad range of subjects. Wikipedia articles are organized and structured using sections and headings, which helps give the text data context and structure.

Although Wikipedia data can be a useful tool for language modeling, there are some drawbacks that must be taken into account. The fact that Wikipedia articles are created and maintained by volunteers means that they may still contain errors or inaccurate information. The likelihood of informational biases and inaccuracies is a major cause for concern. Wikipedia articles may contain errors, out-of-date information, or personal opinions because they are collectively edited. This may contaminate the language model with noise and false information.

Furthermore, Wikipedia tends to utilize a formal and comprehensive writing style that may not accurately reflect the different types of informal language used in daily interactions or other types of content. As a result, the capacity of models derived from Wikipedia content to produce organic and contextually relevant responses may be constrained. Another limitation of Wikipedia content in most resource-scarce languages is that, although the articles tend to cover a wide range of topics, they usually do not provide the same in-depth information as is typically covered in English. For example, the article on "English language"[10] in English[11] is 33 pages long while the same article in Tshivenda[12] only consists of 21 words. More complex topics are often also missing completely, for example the article on "Gravity"[13] is only present for Afrikaans and isiZulu. Wikipedia sources therefore often lack the in-depth, domain-specific or specialized knowledge that is available in English. In the South African context this is exacerbated by the fact that for many of the official languages there is still relatively limited terminology available for various technical concepts. This may, in turn, discourage contributors from creating new pages and terminology for fear of providing incorrect or inaccurate information and terminology. To overcome these drawbacks thorough preparation, validation, and supplementation with a variety of trustworthy data sets are required.

3.2 Leipzig Corpora Collection

The *Leipzig Corpora Collection (LCC)* project has compiled 969 monolingual corpora for 292 different languages, including 11 of South Africa's 12 official languages. Each corpus has a website with comprehensive information, including

---

[10] Visited 13 October 2023.
[11] https://en.wikipedia.org/wiki/English_language.
[12] https://ve.wikipedia.org/wiki/English.
[13] https://en.wikipedia.org/wiki/Gravity.

the language(s), genre(s), and corpus size. Users can select the particular corpus in which they are interested and then download it. Choosing the preferred format (such as plain text or XML) and defining any special criteria or filters are standard steps in the download process.

3.3 Open Parallel Corpus

The *Open Parallel Corpus (OPUS)* is a widely utilized resource for machine learning and natural language processing (NLP). It offers a large assortment of parallel corpora in more than 200 languages with roughly 3.2 billion sentences [19], making it a useful resource for a variety of language-related studies and applications. Languages that are both widely spoken and under-resourced are included in the OPUS corpus. The platform collects text information from webpages, books, movie subtitles, and other text-based resources. Parallel corpora are collections of translations of the same content into other languages that are created by carefully selecting and aligning these sources. This allows for better text identification when the target language is unknown.

OPUS consists of multiple sub-corpora which include: CCAligned, GNOME, XLEnt, tico19 v2020, Tatoeba, CCMatrix, memat, KDE4, XhosaNavy, QED, SPC, TED2020, ELRC2922, and lastly OpenSUbtitles. One of the core strengths of the corpus is its linguistic diversity. Despite the range of languages included in the OPUS corpus, it is vital to keep in mind that the availability and coverage of data may differ between various language pairs. Depending on the resources available and the contributors' areas of interest, some language pairs may include data that is more comprehensive and of higher quality than others. The corpus has a very limited scope when only the particular domain or language is taken into account.

A preliminary inspection of the OPUS data revealed that a substantial amount of machine-translated content was present in the corpus. Using machine-translated material to build a language corpus carries some risk. Despite advances in machine translation over the last decade, mistakes and inconsistencies still occur frequently, especially when dealing with morphologically complex and under-resourced languages. The reliability and quality of the data may be impacted if this results in the inclusion of false or misleading information in the corpus. Additionally, machine translation may ignore key linguistic and cultural aspects due to contextual and cultural differences with the source language. The integrity and utility of the corpus may therefore be compromised by relying mainly on machine-translated text, which will reduce its value for use in research and linguistic analysis.

3.4 FLORES-200

FLORES-200 is a multilingual corpus that contains 9 of South Africa's 12 official languages. It provides a small, but extremely useful set of data for NLP due to its parallel nature. Approximately 2 000 additional sentences per language were obtained from the FLORES-200 data set. A major advantage of this data source was the quality. No extra filtering or processing of the data was required because the translated text is of an extremely high quality.

3.5   Common Crawl

The *Common Crawl (CC)* data set is likely the largest text data set currently available consisting of (bi-)monthly crawls of the web, and is available as downloads in multiple formats, including text from each site. The text component of the data amounts to approximately 10Tb (zipped). For the purposes of this study, we only considered a single instance of the CC data (12/2022). Part of the common crawl collection process is language identification to categorise the predominant languages in each web page. The result of this process is that all pages have one or more language categorisations assigned in the order of predominance of a language on a particular page. A single page can therefore have a classification as xho, zul, eng, which would indicate that most of the text in the page is isiXhosa, while there may also be isiZulu and English data on the page.

3.6   Gov.za

The South African government aims to promote parity of esteem and equitable treatment of all official languages and to facilitate equitable access to services and information of national government (Use of Official Languages Act (Act 12 of 2012)). Making information available in the mother-tongue of its citizens supports effective public service delivery and to enhance this effort, the government established language units that translate large quantities of information. The information is mainly made available on the circa 560 websites of the South African government web domain (*.gov.za).

The Gov.za data was used to compile some of the earliest and largest collections of text available for South Africa's languages. One of these collections is the NCHLT text corpora released in 2013 [8]. Although limited in scope and extent, these corpora typically consist of approximately 1 million tokens per language and have since been used in a wide variety of development projects such as text to speech systems [1], ASR systems [2,3], machine translation systems [16], and data annotation [7,9].

4   Data Collection and Pre-processing

Initially, each corpus was downloaded "as is" from the web. As a first step the language of each data set was verified. Subsequently, unique sentences were extracted and the extent to which the different data sets overlap with the Common Crawl data was determined. The unique sentence sets were pre-processed before attempting to harvest prompts for the MCV platform from the text.

4.1   Language Identification

Although a specific language is already specified as metadata in each corpus, the first step after collecting the text was to verify the language using automatic Language Identification[14] (LID) [12,17]. This step is necessary as the language identifier used in the Common Crawl corpus is trained using Wikipedia data and, for example, isiNdebele is not included in Wikipedia. Furthermore, the language families are very similar and the amount of data available on Wikipedia for Siswati, for example, is not large enough for accurate language identification.

The LID tool provides a threshold metric as a percentage of the number of n-grams in the corpus, or sentence, present in the language model. A threshold of 60% was used to verify the reported language. The verification revealed that many files contain sentences in languages other than the target language and these sentences were discarded. Table 1 shows the original sentence count, the sentence count after LID was applied[15] as well as the unique sentence count after LID was applied to the different corpora[16].

From the results in the Table 1 it is clear that a substantial amount of data in the CC corpus identified as a particular language is not actually from that language, especially for isiXhosa which only retains 4% of the data. Further manual investigation supports this assumption, as there are large amounts of data that are either just location names (Port Alfred, Malibu, Berlin) in a line, or are merely English phrases where one of the words has a morpheme attached that is part of one of the agglutinative languages, e.g. *yeWindowsPress Theme* or *Nceda unique cookies*. This seems especially true of various industrial producer websites where product names are "inflected", rather than translated in full. It was also found that there seems to be a lot of machine translated data where part of the sentence is translated, but many of the technical terms are still in English. The syntactic structure of these sentences indicate that it was unlikely to have been translated by a professional translator. Given the low quality of the sentences identified as not in the target language, it was decided not to include them in the other identified language corpora either.

4.2   Overlap with Common Crawl

Since CC is by far the largest web data source available, it is expected that most other web-based corpora should have substantial overlap with CC. To verify this assumption, the degree of overlap between each corpus and CC was evaluated. The expectation is that most sentences found in a corpus as well as in CC, were likely to be duplicates in the other web-based corpora as well. Combining corpora with a large overlap, especially if the corpora contain data from the same source, could be detrimental as the combined size would be a misrepresentation. Each sentence of each corpus was compared to CC and direct matches were

---

[14] https://hdl.handle.net/20.500.12185/350.

[15] Not reported for NCHLT as the LID was trained on NCHLT data.

[16] For Nbl only LCC and NCHLT are reported and FLORES-200 is omitted for Ven as these data sets are not available for these languages.

Table 1. Total sentences per corpus, original (first row), after LID applied (second row) and unique (third row).

Language	Wikipedia	LCC	OPUS	FLORES-200	NCHLT	Common Crawl
Afr	1 498 160	2 206 947	11 241 208	2 237	114 850	54 819 242
	1 040 234	2 039 424	9 605 094	2 134	n/a	23 383 669
	994 034	1 989 434	4 206 189	2 134	73 108	5 056 915
Nbl		334			62 571	
	n/a	282	n/a	n/a	n/a	n/a
		282			43 053	
Nso	29 047	29 853	8 213	2 246	100 130	453 098
	10 837	28 469	5 818	2 156	n/a	68 697
	10 724	26 584	3 931	2 155	87 453	19 889
Sot	7 423	20 057	1 182	2 246	81 119	2 988 789
	2 772	19 410	1 039	2 159	n/a	493 528
	2 743	17 809	932	2 157	68 363	208 043
Ssw	5 051	10 504	26 212	2 251	68 751	75 702
	2 308	9 967	3 575	2 121	n/a	1 5068
	2 274	9 921	2 454	2 119	50 950	13 837
Tsn	11 391	38 262	77 548	2 281	54 018	1 120 675
	7 420	34 958	4 631	2 157	n/a	95 895
	7 124	34 313	3 407	2 157	49 847	37 276
Tso	6 670	20 556	3 462	2 251	63 981	50 8347
	4 584	19 226	2 215	2 123	n/a	57 915
	4 547	18 215	1 864	2 123	53 869	22 233
Ven	3 900	19 123	1 993		47 653	89 752
	1 700	18 101	1 584	n/a	n/a	21 490
	1 614	16 893	1 345		39 300	17 844
Xho	18 847	85 099	1 320 330	2 251	78 425	94 882 812
	10 313	79 135	235 948	1 822	n/a	2 945 024
	9 981	69 829	172 714	1 820	60 547	413 870
Zul	55 548	111 767	215 117	2 253	108 743	9 826 789
	23 219	107 044	96 490	2 136	n/a	4 191 908
	22 973	100 475	75 777	2 136	102 196	471 203

removed. Table 2 provides a summary of the sentences unique to each corpus when compared to CC as well as the total tokens of the corpora[17].

---

[17] FLORES-200 is omitted for Ven.

Table 2. Number of unique (compared to Common Crawl) sentences (first row) and tokens per corpus (second row).

Language	Wikipedia	LCC	OPUS	FLORES-200	NCHLT
Afr	720 160	1 874 767	3 938 216	2 134	72 722
	15 975 617	36 327 584	62 622 134	48 603	1 723 721
Nso	8 416	4 006	3 922	2 155	87 385
	184 799	504 455	38 509	59 489	2 139 442
Sot	2 451	17 535	862	2 157	68 334
	63 370	420 749	16 825	60 113	1 730 818
Ssw	2 050	8 947	2 224	2 068	50 897
	28 792	146 818	16 772	34 177	691 428
Tsn	6 368	33 093	3 085	2 123	49 840
	211 046	824 543	37 764	55 266	1 286 934
Tso	3 434	16 103	1 683	2 123	53 863
	80 092	355 602	25 374	55 266	1 222 602
Ven	1 571	14 981	1 246	n/a	39 290
	27 455	327 554	21 426		900 740
Xho	8 563	66 413	170 056	1 820	60 475
	136 486	1 096 202	2 121 342	29 497	100 595
Zul	20 161	100 453	72 683	2 136	102 149
	399 070	1 576 731	909 566	34 769	1 620 176

Comparing Table 2 to the other tables in this paper immediately reveals the absence of isiNdebele. This observation is a result of the fact that there is no isiNdebele in the CC data. This situation likely has its origin from the fact that isiNdebele does not have any Wikipedia data and the Common Crawl processing utilises a fastText-based language identifier that has been trained on Wikipedia data. Since the language is not represented in Wikipedia, the language identifier therefore cannot be trained for or classify data as isiNdebele. This is a further indication of the fact that, if a language does not have one particular resource available, there are downstream consequences for its availability in other resources, such as the Common Crawl.

The values in Table 2 show some rather anomalous results for the different languages. Both the FLORES-200 and NCHLT corpora have very little overlap with the CC data. For other corpora however, the languages with the fewest sentences have the largest overlap with the CC data, where languages such as Nso, Sot, and Tso, have more than 50%, and in some instances over 80% overlap with CC. For Afr, Xho, and Zul, there is substantially less overlap in the Wikipedia, LCC and OPUS corpora, mostly less than 20%. Given that the various web corpora are collected at different points in time, these differences may be a reflection of how users create data on the web for the different languages. For those languages that already have the least amount of data, the data is also more static and changes less over time, while for Afrikaans or isiZulu Wikipedia there is substantially more changes that are introduced, and therefor more unique sentences. This is a further indication that those languages that have the smallest footprint on the web, will likely remain that way for the foreseeable future.

4.3   Pre-processing

The Common Voice community provides a set of text processing tools to harvest data from Wikipedia. The tools had to be adapted before they could be used in this project, because the default English rule set only allows sentences with ASCII characters. However, most of the South African languages include diacritic markers encoded by UTF-8 characters. The rule set therefore had to be adapted to accept within-language UTF-8 characters and reject irrelevant ones.

One of the Mozilla selection rules specifies that only three sentences may be copied from a Wikipedia page, but only if the article contains 10 or more sentences. Many articles in South African languages did not meet the 10-sentence limit and, as a result, no sentences could be harvested from them. Another limitation that became evident is the lack of lists of "disallowed words" in most of the languages, which are used as profanity filters on submitted data.

Using the MCV guideline to select potential prompts from the unique sentences (of which the target language had also been verified) in the Wikipedia data yielded less than a thousand sentences per language. Moreover, no isiNdebele data was found because, at the time of writing, the language did not have a presence on Wikipedia. Given the disappointing results for Wikipedia, the LCC, OPUS and FLORES-200 corpora were considered as alternatives.

The data for each language was merged into a raw corpus and basic clean-up, for example replacing smart quotes with straight quotes, removing bullets and empty spaces at the beginning of lines, etc. was performed on all the data. The raw corpora were subsequently separated into sentences and frequency lists were generated using CTexTools 2 [17]. Sentences or segments that did not include useful data (e.g. lines containing only telephone numbers or punctuation) as well as lines that did not constitute a well formed sentence (starting with optional punctuation or numbering, then a capital letter and ending with sentence ending punctuation) were removed from the sentence separated corpora. The sentences were also filtered based on the minimum (3) and maximum (14) allowed number of words and the maximum number of characters allowed (100). Sentences with numbers were also removed, as stipulated in the Mozilla guidelines[18,19].

Words that are spelled incorrectly can easily be mispronounced and as the sentences would be read and recorded, all sentences were ranked according to the percentage of correctly spelled words they contain. The frequency lists were

---

[18] https://commonvoice.mozilla.org/en/guidelines.

[19] Although these guidelines are restrictive, it demonstrates requirements of a simple real-world application, e.g. ASR.

checked for correct spelling using commercially available spelling checkers developed by the Centre for Text Technology[20]. Using the spell-checked lists, all remaining sentences were ranked according to the percentage correctly spelled words they contain. To ensure better coverage of the languages, the highest ranked sentences were selected and compared using a Levenshtein edit distance module[21]. Only sentences with more than 90% correctly spelled words and less than 70% overlap were included in the final, clean corpus. Following this procedure resulted in corpora ranging from only 657 usable sentences for isiNdebele to 54 129 sentences for Afrikaans. Only the data sets for Setswana and Afrikaans satisfied the 5 000-sentence requirement. An inspection of selected text showed that the sentences contained offensive language as well as religious scripts.

5   Gov.za Revisited

Due to these findings, the data obtained from the LCC, OPUS and FLORES-200 corpora was deemed unsuitable as a source of MCV prompts and a decision was made to revisit the gov.za data from which the NCHLT corpora were derived. To expand the existing NCHLT corpora, data was again collected from the *.gov.za domain. Data was crawled using HTTrack[22] from websites which permit such activities and whose data is in the public domain. The crawled data consisted of several file types that were converted into a uniform, text format encoded in UTF-8. The conversion of PDFs was done using pdf2xml[23], DOCs were extracted using Win32-OLE[24] and HTMLs were converted using a HTML parser module[25].

The language of each text file was identified using a language identification tool for South African languages [12,17] and all files for a specific language were merged into a raw corpus. These corpora were again processed, cleaned and filtered using the same procedure as described in Sect. 4.3. This resulted in corpora ranging from 5 849 usable sentences for isiNdebele to 81 665 sentences for Afrikaans. The initial corpus sizes and those after LID, unique sentence extraction and pre-processing are provided in Table 3.

The MCV platform allows for two types of sentence submissions. Sentences can either be uploaded manually or in bulk. However, bulk submissions are required to meet quality control (QC) measures that should be obtained according to a specific sampling strategy. Unfortunately the project budget could not accommodate the prescribed QC procedure. To get at least some indication of the quality of the text we randomly selected 200 sentences per language and had these verified by professional language editors. The editors were asked to flag sentences that contained spelling and/or grammar errors, might be difficult to

---

[20] https://sadilar.org/index.php/en/applications/spelling-checkers-for-south-african-languages.
[21] https://metacpan.org/pod/Text::LevenshteinXS.
[22] https://www.httrack.com/.
[23] https://metacpan.org/pod/Text::PDF2XML.
[24] https://metacpan.org/pod/Win32::OLE.
[25] https://metacpan.org/pod/HTML::Parser.

Table 3. Number of sentences per language harvested from *.gov.za.

Language	Raw data	Clean data	% Sentences flagged
Afr	608 216	81 665	24
Nbl	77 264	5 849	23
Nso	127 948	16 094	24
Ssw	74 541	7 650	25
Sot	120 542	15 528	10
Tsn	117 758	16 197	11
Tso	114 626	16 366	34
Ven	98 501	8 661	14
Xho	233 252	15 133	12
Zul	245 942	19 355	6

read, or could potentially be offensive. The values in the last column of Table 3 are the percentage of sentences that were flagged per language. The majority of the identified errors were related to spelling and grammar mistakes. No sentences were labelled as potentially offensive.

The results in Table 3 show that there are substantial differences between the different languages. Given the variation in judgement that is often observed in language editing tasks, it would have been ideal to perform more than one evaluation to confirm this trend. Unfortunately the project budget only allowed for a one round of evaluations.

The sentences harvested from the government websites were subsequently uploaded to the MCV platform as potential prompts. However, before a submitted sentence is accepted as a prompt, it has to be approved by at least two members of the relevant language community. Language practitioners were hired to verify as many sentences as the remainder of the project budget permitted. The 5 000-sentence requirement has since been adjusted by Mozilla and we managed to have an adequate number of sentences verified by two people in all 10 target languages to achieve *launched* status.

6   Discussion and Conclusions

The initial intent of data selection for the MCV project was to move away from the textual resources traditionally used in the South African context for both text and speech technology development. The hope was that using other open, web-based data sets, a different collection of sentences could be used to create prompts for the MCV platform. With this in mind, we describe the different available web-based corpora, as well as the outcomes of our investigations into the quality of the data of the respective sources. Our investigation of six different web-based corpora indicated that there are several shortcomings and risks associated with using these sources.

Wikipedia may be one of the most widely used web-based corpora in NLP, but for the majority of South African languages, the currently available articles are not useful in the NLP context. This is primarily due to a large number of very short and near duplicate articles in most of the languages with a small amount of data. Despite the linguistic richness of the OPUS corpus, not all language pairs have the same data availability and quality. Furthermore, there also seems to be a large amount of machine translated text, which introduces the risks of non-grammatical or just purely nonsense data being present in the corpus. Utilising such data jeopardizes the corpus' integrity and usefulness.

When considering the largest of these web-based data sets, Common Crawl, some more general problems with web data are encountered. For most of the languages, the CC language identification does not function accurately enough. This is especially true for isiXhosa, where only 4% of the sentences are retained when using a language identifier specifically developed only for the South African languages, and based on representative data than only Wikipedia. There are also a large number of sentences that appear to be machine translated, which in turn creates a feedback loop of machine created data being used as input to other NLP processes. Moreover, substantial parts of the data originate from either religious material or government data, which have very specific characteristics that are not ideal for development in the NLP context. Last but not least, isiNdebele's absence from Wikipedia makes it difficult to identify the language in the CC corpus because it lacks data. This observation emphasises the fact that the lack of an essential resource, such as Wikipedia, has a detrimental impact on the availability of other, more general resources.

Two aspects that have not been addressed in this work still remain for future work. Firstly, an investigation needs to determine whether additional sentences from the Common Crawl data can be included in a project such as MCV, where the restrictive nature of the requirements may be able to filter out sentences that are of lower quality. Secondly, the impact of high and low quality data sources on other downstream tasks, such as language modelling, and automatic linguistic analysis, should be evaluated to determine whether it is more important to have smaller high quality data sets, or larger lower quality data sets.

The shortcomings and risks mentioned in this paper clearly indicate that relying entirely on web data is not an option for languages with limited resources. While web data can offer insightful information, its quality, coverage, and trustworthiness for low-resource languages are still questionable. To effectively support under-resourced languages, language preservation and resource development activities must take a comprehensive approach that integrates web data with other sources. Ensuring the survival and expansion of low-resource languages in the digital age will require a dedicated effort that goes beyond the constraints of web-based text resources.

References

1. Badenhorst, J., Heerden, C.V., Davel, M., Barnard, E.: Collecting and evaluating speech recognition corpora for 11 South African languages. Lang. Resour. Eval. **3**(45), 289–309 (2011). https://doi.org/10.1007/s10579-011-9152-1
2. Barnard, E., Davel, M.H., van Heerden, C., de Wet, F., Badenhorst, J.: The NCHLT speech corpus of the South African languages. In: Proceedings of the 4th Workshop on Spoken Language Technologies for Under-resourced Languages, St Petersburg, Russia, pp. 194–200 (2014)
3. Barnard, E., Davel, M.H., Van Huyssteen, G.B.: Speech technology for information access: a South African case study. In: AAAI Spring Symposium Series, pp. 8–13 (2010)
4. Brown, T., et al.: Language models are few-shot learners. Adv. Neural. Inf. Process. Syst. **33**, 1877–1901 (2020)
5. Devlin, J., Chang, M.W., Lee, K., Toutanova, K.: BERT: pre-training of deep bidirectional transformers for language understanding. arXiv:1810.04805 [cs] (2019)
6. Dlamini, S., Jembere, E., Pillay, A., van Niekerk, B.: isiZulu word embeddings. In: 2021 Conference on Information Communications Technology and Society (ICTAS), pp. 121–126. IEEE (2021)
7. Eiselen, R.: South African language resources: phrase chunking. In: Proceedings of the Tenth International Conference on Language Resources and Evaluation (LREC 2016), pp. 689–693 (2016)
8. Eiselen, R., Puttkammer, M.J.: Developing text resources for ten South African languages. In: Proceedings of Language Resource and Evaluation (LREC), Reykjavik, Iceland, pp. 3698–3703 (2014)
9. Gaustad, T., Puttkammer, M.J.: Linguistically annotated dataset for four official South African languages with a conjunctive orthography: isiNdebele, isiXhosa, isiZulu, and Siswati. Data Brief **41**, 107994 (2022). https://doi.org/10.1016/j.dib.2022.107994, https://www.sciencedirect.com/science/article/pii/S2352340922002050
10. Goldhahn, D., Eckart, T., Quasthoff, U.: Building large monolingual dictionaries at the Leipzig corpora collection: from 100 to 200 languages. In: Proceedings of the Eighth International Conference on Language Resources and Evaluation (LREC) (2012)
11. Goyal, N., et al.: The FLORES-101 evaluation benchmark for low-resource and multilingual machine translation. Trans. Assoc. Comput. Linguist. **10**, 522–538 (2022)
12. Hocking, J.: Language identification for South African languages. In: Proceedings of the Annual Pattern Recognition Association of South Africa and Robotics and Mechatronics International Conference (PRASA-RobMech). PRASA (2014)
13. Koehn, P., Chaudhary, V., El-Kishky, A., Goyal, N., Chen, P.J., Guzmán, F.: Findings of the WMT 2020 shared task on parallel corpus filtering and alignment. In: Proceedings of the Fifth Conference on Machine Translation, pp. 726–742 (2020)
14. Kreutzer, J., et al.: Quality at a glance: an audit of web-crawled multilingual datasets. Trans. Assoc. Comput. Linguist. **10**, 50–72 (2022)
15. Marivate, V., et al.: Investigating an approach for low resource language dataset creation, curation and classification: Setswana and Sepedi. In: First workshop on Resources for African Indigenous Languages (RAIL), pp. 15–20. ELRA (2020)

16. McKellar, C.A., Puttkammer, M.J.: Dataset for comparable evaluation of machine translation between 11 South African languages. Data Brief **29**, 105146 (2020). https://doi.org/10.1016/j.dib.2020.105146, https://www.sciencedirect.com/science/article/pii/S2352340920300408
17. Puttkammer, M., Eiselen, R., Hocking, J., Koen, F.: NLP web services for resource-scarce languages. In: Proceedings of the 56th Annual Meeting of the Association for Computational Linguistics (ACL), pp. 43–49. Association for Computational Linguistics (2018)
18. Raffel, C., et al.: Exploring the limits of transfer learning with a unified text-to-text transformer. J. Mach. Learn. Res. **21**(140), 1–67 (2020). https://jmlr.org/papers/v21/20-074.html
19. Tiedemann, J.: Opus-parallel corpora for everyone. In: EAMT (Projects/Products) (2016)
20. Tiedemann, J., Nygaard, L.: The OPUS corpus - parallel and free: https://logos.uio.no/opus. In: Proceedings of the Fourth International Conference on Language Resources and Evaluation (LREC'04). European Language Resources Association (ELRA), Lisbon (2004). https://www.lrec-conf.org/proceedings/lrec2004/pdf/320.pdf

Voice Conversion for Stuttered Speech, Instruments, Unseen Languages and Textually Described Voices

Matthew Baas[(✉)] and Herman Kamper

MediaLab, Electrical and Electronic Engineering, Stellenbosch University,
Stellenbosch, South Africa
{20786379,kamperh}@sun.ac.za

Abstract. Voice conversion aims to convert source speech into a target voice using recordings of the target speaker as a reference. Newer models are producing increasingly realistic output. But what happens when models are fed with non-standard data, such as speech from a user with a speech impairment? We investigate how a recent voice conversion model performs on non-standard downstream voice conversion tasks. We use a simple but robust approach called k-nearest neighbors voice conversion (kNN-VC). We look at four non-standard applications: stuttered voice conversion, cross-lingual voice conversion, musical instrument conversion, and text-to-voice conversion. The latter involves converting to a target voice specified through a text description, e.g. "a young man with a high-pitched voice". Compared to an established baseline, we find that kNN-VC retains high performance in stuttered and cross-lingual voice conversion. Results are more mixed for the musical instrument and text-to-voice conversion tasks. E.g., kNN-VC works well on some instruments like drums but not on others. Nevertheless, this shows that voice conversion models – and kNN-VC in particular – are increasingly applicable in a range of non-standard downstream tasks. But there are still limitations when samples are very far from the training distribution. Code, samples, trained models: https://rf5.github.io/sacair2023-knnvc-demo/.

Keywords: Voice conversion · Speech processing · Speech synthesis · Instrument conversion · Stuttered speech

1 Introduction

The conventional goal of voice conversion is to change an input utterance to seem as though it is spoken by a different speaker while retaining the same words [20]. Older models could only convert to and from speakers seen during training, while more recent models can also handle unseen speakers [18]. There are early indications that recent models can even deal with non-speech inputs, e.g., using animal sounds as the target audio [29]. To assess how far voice conversion models have come, this work explores how a recent model can be applied to several non-standard downstream tasks.

Concretely, we apply the recent k-nearest neighbors voice conversion (kNN-VC) approach [3] to four non-standard tasks. kNN-VC is a simple method where each feature in the source speech is replaced with its nearest neighbors in the target data. The approach is underpinned by a large self-supervised speech model that is used as the feature extractor [6]. Because no part of the model's design explicitly assumes that the source or target data is regular human speech, we believe that kNN-VC would be particularly well-suited to non-standard downstream voice conversion tasks.

We look at four tasks. The first is stuttered voice conversion, where the target speech comes from a speaker with a stuttering speech impairment [9]. Secondly, we use kNN-VC for cross-lingual voice conversion where the source and target speakers use different languages [34]. Both the input and output languages are also unseen during training. Thirdly – going even further out-of-domain – we consider completely non-speech input: we apply kNN-VC to musical instrument conversion, where the source and target "utterances" come from different musical instruments. Fourthly, we use kNN-VC to perform text-to-voice conversion where, instead of using a reference recording, the target voice is specified through a user-provided textual description, e.g., "an elderly woman with a velvety and resonant low voice". This last task is performed by training a small multi-layer perceptron to map a vector representation of the target speaker description to a distribution over known speakers.

On these four non-standard tasks, we compare kNN-VC to an existing established voice conversion system called FreeVC [16]. We find that kNN-VC retains high performance in stuttered and cross-lingual conversion, outperforming the baseline in terms of similarity to the target speaker. But quality is poorer in instrument and text-to-voice conversion, where kNN-VC doesn't consistently outperform the baselines; despite the degraded performance, the conversions are still compelling considering how far out-of-domain music and text are for a model that has only seen speech during training. Taken together, our experiments indicate that new voice conversion models – and kNN-VC in particular – are increasingly applicable to non-standard downstream tasks, with a single model able to perform a range of tasks that it wasn't trained for. However, there are still limitations in generalization when the model is presented with data very far from the training distribution. We invite the reader to listen to samples and use our code: https://rf5.github.io/sacair2023-knnvc-demo/.

The paper is organized as follows. We first present the background of kNN-VC and the tasks we explore in Sect. 2. The four sections that follow then each investigates one of the tasks in detail. Finally, we conclude in Sect. 7.

2 Background

Voice conversion has a long history, but research has been accelerating in the last decade due to advances in generative machine learning [27]. The setup for most voice conversion methods is the same: the model takes in two inputs and produces one output. The first input is the source speech which specifies the linguistic content to appear in the output, and the second is the reference speech which specifies the target speaker [20]. Most models attempt to learn to disentangle content from speaker identity, so that output can be produced with the same linguistic content as the source but sounding as though the words are spoken by the target speaker specified by the reference.

There are various methods to perform this disentanglement, ranging from annotated text for linguistic content alignment [5] to various forms of information bottlenecks [8, 16, 32]. We are interested in applying a well-performing voice conversion model to downstream applications, not in developing a new model ourselves. We therefore use a particularly simple but robust method: kNN-VC.

2.1   K-Nearest Neighbors Voice Conversion (kNN-VC)

kNN-VC was inspired by ideas from concatenative speech synthesis, where an output waveform is produced by stitching together segments from different utterances [10, 12]. But instead of using raw speech waveforms directly, kNN-VC relies on high-level speech features obtained from a large neural network. It specifically uses the WavLM encoder for feature extraction. WavLM [6] is a pretrained, self-supervised speech representation model that transforms input speech into a vector sequence, producing a single vector for every 20 ms of input audio. Intuitively, the model is trained by masking parts of the input speech and then trying to predict what the vectors should be in the masked regions. The result is an encoder that maps raw speech to features that are linearly predictive of various aspects of speech [6]. kNN-VC uses the activations from layer six of WavLM, which are linearly predictive of speaker identity [3]. It is these features that underpin the simple kNN-VC approach.

Voice conversion is performed in kNN-VC by mapping each WavLM vector from the source utterance to the mean of its k nearest vectors from the reference utterance. The resulting vector sequence is then converted back into a waveform using a vocoder [3], specifically a HiFi-GAN [14] vocoder adapted to map WavLM features to audio. If there are multiple reference utterances for the target speaker, the features from each are simply collated together into a pool of features known as the matching set. While the WavLM encoder and HiFi-GAN vocoder have only seen speech during training, kNN-VC's design does not make any specific design assumptions restricting it to speech. This, combined with the generality of the non-parametric kNN algorithm, makes kNN-VC a good choice for testing it on non-standard conversion tasks.

2.2   Non-standard Voice Conversion Tasks

For each of the four tasks we explore, the conversion process can be formulated as a specific setting of a conversion model's source and reference data. Concretely, the specification for the source, reference and output of a model for each task is given in Table 1. Here we give more background on each of the four tasks.

Stuttered Voice Conversion. In our first task, we want to convert input speech so that it inherits the voice of a speaker having a stuttering impairment (as shown in Table 1). Imagine that a student needs to create a presentation for their thesis, but the student suffers from a speech impairment. If we could solve the stuttered speech conversion task, then someone else could read a prepared script which would then be converted to the student's voice. This would result in an output resembling the student's voice but without stutters (since the source was read by a fluent speaker). Prior efforts to correct stuttered speech have used energy correlations of speech segments to attempt to remove repetitions from stuttered speech [1]. However, this approach attempts to directly edit stuttered speech into a non-stuttered form, while we focus on stuttered voice conversion which would enable a stuttering speaker to map their voice onto an existing fluent utterance. Related efforts in [9] attempt to train a generative adversarial network to generate fluent speech for a given set of speakers with a range of speech impairments. But the approach struggles with stuttered speech because the model fails to appropriately change the prosody. A voice conversion approach can address this since prosody is inherited from the source utterance (and not the stuttering reference).

Cross-lingual Voice Conversion. Here the source and reference utterances come from different, potentially unseen, languages (Table 1). This is the most well-understood of the four voice conversion tasks, with the Voice Conversion Competition in 2020 [34] having a dedicated track for cross-lingual conversion. Converting speech between languages unseen during training would improve access to speech technologies by speakers of low-resource languages. But many existing methods are limited to languages seen during training [34]. Newer zero-shot conversion models like FreeVC [16] do not have this limitation but still struggle with completely unseen languages. We are particularly interested in seeing whether voice conversion models can generalize to unseen languages. In our experiments, we therefore use kNN-VC and FreeVC (both trained exclusively on English) to convert utterances between several non-English language pairs.

Musical Instrument Conversion. In this difficult task, we apply voice conversion to convert music played by one instrument to sound as though it is played by another. In this case, the source and reference are both non-speech pieces of music from a single instrument (Table 1). Most prior work in this area – unsurprisingly – uses models trained on music. The goal is typically to give a

Table 1. Voice conversion setup for the four non-standard downstream tasks.

Task	Source	Target reference	Output
Stuttered conversion	Non-stuttered speech	Stuttered speech from desired speaker	Non-stuttered speech in desired voice
Cross-lingual conversion	Language A utterance	Language B utterance	Language A utterance in voice B
Instrument conversion	Music with instrument A	Music with instrument B	"Content" from A using instrument B
Text-to-voice conversion	Source utterance A	Text description of a speaker B	Utterance A in voice B

user control over the style of the synthesized audio. E.g., [33] uses a variational autoencoder to model music style, while more recent studies like [4] use differentiable digital signal processing methods to allow a user to modify the output audio. Our goal is not to achieve state-of-the-art music synthesis, but rather to assess how well a voice conversion approach (trained on speech audio and not music) is able to generalize to out-of-domain tasks. So, for instrument conversion, we still use kNN-VC and again compare it to FreeVC – both models having only seen speech data.

Text-to-voice Conversion. The last task is one where, instead of using a reference utterance to specify the target speaker, a textual description is provided, e.g., "an older man with a British accent and a pleasing, deep voice". This is known as text-to-voice conversion or prompt-to-voice conversion [11]. This task is much less explored than the others. It was first introduced by Gölge and Davis [11] in 2022, who trained a specialized prompt-to-voice model. Their work is proprietary and they also specifically considered a text-to-speech task where the target voice is provided as a text description rather than a reference utterance. This motivates our introduction of a simple extension to kNN-VC to allow for textually described target voices in a voice conversion setting.

3   Stuttering Experiments

For stuttered voice conversion, we perform two experiments: a comprehensive large-scale evaluation and a smaller-scale practical conversion of monologue recordings. The former allows us to quantify kNN-VC's performance at stuttered conversion using the standard objective metrics of voice conversion, while the latter shows that it can be used in a real-world practical setting.

3.1   Performance Metrics

When evaluating voice conversion performance, two categories of metrics are used: ones which measure how *intelligible* the output is (i.e. how much of the linguistic content it retains form the source), and ones which measure the *speaker similarity* to the target speaker [34].

To measure intelligibility, we follow the same approach as in [3,22]: we assume we know the transcript of the source utterance, and then compare this to the transcript of the converted output. The transcript of the output is obtained using an existing high-quality automatic speech recognition system. We then compute a word/character error rate (W/CER) between the transcript of the converted output and the source: the lower the error rate, the better the conversion (i.e., more intelligible). The automatic speech recognition system we use is the pretrained Whisper `base` model using a decoding beam width of 5 [25].

For speaker similarity, we compute an equal error rate (EER) as in [22]. This metric is computed by calculating speaker similarity scores between pairs of real/real and real/generated utterances. The real/real pairs are assigned a label of 1, and real/generated pairs a label of 0. An equal error rate is then computed using the similarity scores and the labels. The speaker similarity score between two utterances is defined the cosine distance between speaker embeddings computed using a pretrained speaker embedding model [28]. The higher the EER, the less able we are to distinguish between the target speaker and converted output (i.e., better speaker similarity), up to a theoretical maximum of 50%.

3.2   Stuttering Events in Podcasts

We evaluate kNN-VC and FreeVC on a combination the LibriSpeech's `dev-clean` dataset [23] and the Stuttering Events in Podcasts dataset [15]. LibriSpeech consists of hundreds of hours of read books in English by speakers without any speech impairment, while the Stuttering Events dataset consists of data from podcasts where stuttering events have been annotated. Both datasets contain speech from many speakers, none of which are seen by kNN-VC or FreeVC. Using the annotations in the Stuttering Events dataset, we trim each utterance to only those segments where there is a sound or word repetition, i.e., we filter out all non-stuttered speech and only use stuttered instances as reference audio.

As explained in Sect. 2.2, we want the generated speech to match the identity of a reference speaker that stutters, but the content should match a fluent source utterance. So, for our evaluation, we convert each of the 200 utterances from LibriSpeech's `dev-clean` subset from the kNN-VC paper [3] to ten random speakers from the Stuttering Events dataset. Since each reference speaker has several short clips with stuttered speech, we sample up to 30 stuttered reference clips for each target speaker. These clips then serve as the matching set for conditioning kNN-VC when it is presented with a source utterance from LibriSpeech (see Sect. 2.1). For the FreeVC [16] baseline, we use the same input utterances but compute the mean speaker embedding of the stuttering clips to serve as the target. This is done because FreeVC's target conditioning mechanism doesn't use a matching set; instead, it uses a trained speaker encoder which maps a reference utterance to a single speaker embedding [16].

Table 2. Stuttered voice conversion performance, where the source is fluent speech and the target reference is from a speaker with a stuttering impairment.

Method	WER ↓	CER ↓	EER (%) ↑
dev-clean topline	4.63	1.60	—
FreeVC	**6.59**	**2.57**	25.6
kNN-VC	14.61	7.16	**46.7**

The result of this evaluation are presented in Table 2. The topline is an approach where we perform no conversion and simply evaluate the baseline transcription error rates of the Whisper model on the ground truth source utterances. We make two key observations from the results. First, voice conversion can be used effectively for this task, with high EER (good speaker similarity) and low

W/CER scores (good intelligibility). Second, there is a tradeoff between kNN-VC and FreeVC: FreeVC retains more of the desired linguistic content, with W/CERs that are less than half of the rates achieved by kNN-VC. Conversely, kNN-VC is substantially better in matching the target speaker, with a much higher EER. We suspect that the better speaker similarity of kNN-VC is due to the matching set mechanism that is more robust to non-standard speech compared to the speaker embedding approach of FreeVC. However, the matching approach of kNN-VC might also be what causes poorer intelligibility because the matching set contains several non-fluent phones and biphones.

3.3   Practical Evaluation

Having objectively evaluated kNN-VC's performance, we now apply it in two practical scenarios in a qualitative evaluation. First, we apply it to the example already mentioned in Sect. 2.2, where we want to generate the audio of a thesis presentation for a student that suffers from a stuttering impairment. The source to kNN-VC is provided by a fluent speaker reading a script provided by the student. In this case, the reference data was provided by the student, but stuttering was removed by hand. Second, we apply kNN-VC using stuttered speech from an actor as reference, specifically the actor Colin Firth acting as King George VI in the movie "The King's Speech". In both cases, to further improve output quality, we also apply the `voicefixer` upsampling model to remove some conversion artifacts [17]. We encourage the reader to listen to conversion samples at https://rf5.github.io/sacair2023-knnvc-demo/.

4   Cross-Lingual Voice Conversion

In cross-lingual voice conversion, we want to convert to a target speaker talking a different language from the source. The languages and identities of the source and target speakers are unseen during training. We again apply kNN-VC as our main model and use FreeVC as a baseline. Because the inputs and outputs in this task are still human speech, we use the same intelligibility and speaker similarity metrics as in Sect. 3.1.

4.1   Experimental Setup

We conduct a large-scale evaluation on the Multilingual LibriSpeech dataset [24]. The dataset consists of thousands of hours of audiobook recordings across eight European languages. For evaluation, we sample 16 utterances from three random speakers from the test subset of each language, yielding 384 evaluation utterances. Evaluation involves converting each of these utterances to every other speaker in the evaluation set, giving a total of just under 9000 unique source-target evaluation pairs covering all possible language combinations.

For measuring intelligibility as described in Sect. 3.1, we can still use Whisper `base` to transcribe the original and converted utterances because Whisper is a multilingual speech recognition model that can be presented with input in any of the European languages considered here [25]. Similarly, speaker similarity is evaluated using the same speaker embedding model, since these models are trained to ignore linguistic content and only extract speaker information. We can therefore calculate an EER exactly as in Sect. 3.1.

Table 3. Cross-lingual voice conversion performance on the Multilingual Librispeech test dataset.

Method	WER ↓	CER ↓	EER (%) ↑
test set topline	21.5	7.1	—
FreeVC	34.1	13.5	7.7
kNN-VC	**33.9**	**13.1**	**25.0**

4.2 Results

The cross-lingual voice conversion results on the Multilingual LibriSpeech dataset are presented in Table 3. Unlike the stuttered conversion experiments, we see that kNN-VC is consistently more intelligible and more similar to the target speaker than FreeVC.

How do cross-lingual conversion compare to converting between speakers in English (the language on which both FreeVC and kNN-VC are trained)? If we compare the results in Table 3 to the monolingual scores reported in [3,16], we see that both the intelligibility and speaker similarity is lower when performing cross-lingual voice conversion compared to monolingual conversion. This degradation is to be expected, given that these voice conversion models might not have seen all the phones necessary for converting between unseen language pairs. E.g., when converting Polish to Spanish with kNN-VC, the Spanish matching set might not have all the phones necessary to represent the source Polish utterance.

Nevertheless, for kNN-VC in particular, the cross-lingual performance is still compelling, with the CER and EER scores in Table 3 coming close to FreeVC's performance on English [16]. This shows that voice conversion models trained only on English can be applied succesfully to unseen languages. We encourage the reader to listen to a selection of conversion samples on our demo website.

5 Musical Instrument Conversion

Moving on to harder tasks, we now consider a conversion problem involving non-speech audio. Specifically, we want to see whether we can use kNN-VC to convert a piece of music played with one instrument to sound as though it is played with another – without fine-tuning or adapting the model. To this end, we use a dataset consisting of several music pieces, each played on a single instrument.

We then attempt to convert a song from one instrument to another by using all the audio from the target instrument as our matching set in kNN-VC. This effectively treats the recordings from the target instrument as "utterances" from a target "speaker".

5.1 Experimental Setup

We use the Kaggle Musical Instrument's Sound Dataset [21]. The dataset consists of short music pieces played with one of four instruments: drums, violin, piano, or guitar. For our music conversion evaluation set, we only consider recordings that are between 10 and 90 s long. Evaluation involves converting each recording in the test set to every other instrument, resulting in 153 synthesized output audio recordings that are evaluated. For kNN-VC, all the recordings from the target instrument are used as the matching set. For FreeVC, a mean "speaker" embedding is obtained from the target instrument audio.

The intelligibility and speaker similarity metrics from Sect. 3.1 are not applicable to instrument conversion. We therefore use a metric from the music synthesis domain: Fréchet audio distance (FAD) [13]. This metric uses a music classification model (trained to classify the tags of scraped YouTube videos) to compare two recordings, giving a single number which roughly indicates how similar the music in the recordings is [13]. This is achieved by computing an "Inception score" between the intermediary features produced by the classifier for two sets of audio inputs. In our case we measure the FAD between the converted outputs and the ground truth recordings from the target instrument. E.g., to measure FAD for drums, we compare all the outputs where drums are the target instrument to the real recordings of drums. We use an open-source implementation of FAD[1].

5.2 Results

Table 4. Musical instrument conversion performance on the Musical Instrument's Sound Dataset, measured in FAD [13].

Method	Target instrument FAD ↓			
	Drums	Violin	Piano	Guitar
FreeVC	20.11	**21.07**	**20.03**	23.22
kNN-VC	**9.81**	22.44	23.95	**18.01**

The musical conversion performance for each instrument is presented in Table 4. FAD values greater than 10 are considered to be very poor [13]. So we

---

[1] https://github.com/gudgud96/frechet-audio-distance.

can see that FreeVC performs very poorly, regardless of the instrument. kNN-VC also performs poorly for most target instruments, except when converting to drums where the best overall FAD is achieved. While a FAD of 9.8 is still poor compared to typical scores in music enhancement [13], qualitatively kNN-VC's conversion to drums sounds much closer to the target than any of the other conversions.

We are still unsure why kNN-VC performs so much better when converting to drums. We speculate that this might be because drum audio has lower frequency content that might be in a similar range to that of human speech – at least compared to the high-frequency content from guitars, pianos, and violins. The encoder and vocoder of kNN-VC have only seen speech, and it is therefore likely that the WavLM features would be better at capturing instrument information with a similar frequency range to speech.

Overall, the results indicate that the kNN-VC and FreeVC cannot perform instrument conversion, except in specific instances. This therefore remains an open question for future work; one idea would be to incorporate some of the techniques used in music enhancement and synthesis [4,33].

6   Text-to-voice Conversion

The last task we look at is text-to-voice conversion, where the target speaker is specified using a textual description instead of a reference utterance. Unlike the previous tasks, kNN-VC cannot be directly used with text inputs. So we propose a small extension.

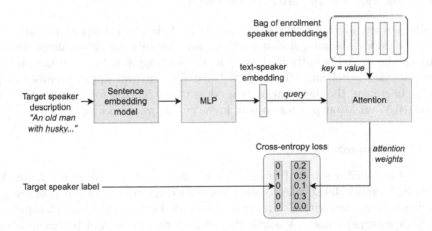

Fig. 1. For text-to-voice conversion, we use an approach where a text description of a voice is converted to a distribution over enrollment speakers. Only the multi-layer perceptron (MLP) and attention layer are trained; the sentence embedding model and speaker embeddings are fixed. The parameters are optimized to predict the target speaker label from the provided textual description.

6.1   Extending KNN-VC for Textual Voice Descriptions

We extend kNN-VC to work with text descriptions using an insight from Cheng [7]: during synthesis with kNN-VC, we can construct combinations of voices by linearly interpolating the converted WavLM features between those from different target speakers, before vocoding. We therefore rephrase the text-to-voice conversion task as finding a probability distribution over a set of enrollment speakers that best matches the provided text description. If we knew this distribution, we could use the probabilities as weighting coefficients for interpolating the converted output to best match the target speaker description.

Figure 1 illustrates our approach for predicting the distribution over speakers from a target speaker description. We construct a small text-to-voice matching network that maps a text description to a distribution over a set of enrollment speakers. The network consists of three components:

1. A sentence embedding model [26], which converts a text description into a single vector capturing the target speaker identity. We use the pretrained model from [26] directly.
2. A bag of speaker embeddings computed using a pretrained speaker embedding model [31]; a single speaker embedding is obtained for each enrollment speaker by averaging the embeddings from that speaker's utterances. We use an open-source pretrained speaker embedding model[2].
3. A multi-layer perceptron (MLP) with three layers and an attention [30] head. The MLP maps the sentence embeddings [26] to a "text-speaker embedding", which is used as the query over the bag of enrollment embeddings (acting as keys and values in the attention operation).

The output of this small matching network is the attention weights over the keys/values, defining a distribution over the enrollment speakers. During training, the model is optimized using a cross-entropy loss between the one-hot target speaker label and the predicted distribution over speaker embeddings. During inference, the distribution is used as the weighting for interpolating the converted WavLM output features of kNN-VC before vocoding.

6.2   Experimental Setup

We use the LibriSpeech [23] `train-clean-100` subset for this experiment. We manually labelled 100 speakers with a textual description of the speaker's voice, e.g., "a young woman with a squeaky and animated voice, speaking in a rapid and highly expressive manner". We split this set into 90 training and 10 test speakers. The MLP and attention layer from Fig. 1 is then trained for 300 steps using AdamW [19] with a learning rate of $5 \cdot 10^{-5}$ and weight decay of $3 \cdot 10^{-3}$. The MLP consists of three layers with LeakyReLU activations and layer normalization [2]

---

[2] https://github.com/RF5/simple-speaker-embedding.

between each layer. Since the sentence and speaker embeddings respectively have 384 and 256 dimensions, the output dimensions from the three MLP layers are 1024, 768 and 256. The attention is a standard scaled dot-product multi-head attention with 16 heads [30].

The evaluation task involves converting the same set of test-clean utterances as in [3] to each of the ten test speaker descriptions. First the network in Fig. 1 maps the description of each test target speaker to a distribution over the 90 enrollment speakers. Then we perform kNN-VC conversion to each enrollment speaker as target. Finally, we interpolate the output WavLM features in the ratio provided by the attention weights of the model in Fig. 1 before vocoding to achieve the final converted output. Using the set of converted outputs, speech inputs, and original reference utterances, we assess performance with the same metrics as in Sect. 3.1.

6.3 Results

The text-to-voice conversion results are presented in Table 5. We see that intelligibility is not harmed when using textually described target voices – it actually improves slightly compared to the 1-minute topline. This makes sense since the target is based on a combination of the training speakers, for which we have much more data to use as matching data compared to the topline. However, this comes at a cost: the similarity to the target speaker is much worse compared to the topline (lower EER). While not certain, we hypothesize this is largely due to the highly subjective nature of mapping utterances from a speaker to a textual description of that speaker's voice. E.g., what one person might label as a "squeaky and animated voice" might differ greatly from how another person describes the same voice. This is compounded by the sensitivity of the speaker embedding model used to evaluate EER, since it is trained to be sensitive enough to discern even similar speakers of the same gender and age. Overall, the results indicate that we can perform text-to-voice conversion to some degree, but there is still more work required to accurately map to the desired target speaker.

Table 5. Text-to-voice conversion performance on LibriSpeech, where unseen source speakers are converted to target speakers specified with a textual description. Topline performance when 1 min of speech audio is used to specify the target speaker (instead of text) is provided as comparison.

Method	WER ↓	CER ↓	EER (%) ↑
1 min speech kNN-VC topline	7.73	3.09	**35.05**
kNN-VC three-layer MLP	**6.32**	**2.51**	3.15

7   Conclusion

We explored the generalization capabilities of voice conversion models by applying a specific model (kNN-VC) to several non-standard conversion tasks. In stuttered voice conversion and cross-lingual voice conversion, kNN-VC retained high performance, achieving convincing results compared to a baseline. For music conversion, however, the results were more mixed, with kNN-VC being largely unable to produce high-quality conversions except when converting to drums. The results for textually described voice conversion were also mixed: the output was still intelligible but was a poor match to the target speaker compared to a standard system using audio rather than text to specify the target speaker. Taken together, we showed that an existing voice conversion system can be applied to a range of non-standard downstream voice conversion tasks without major changes to the model. Future work will consider how voice conversion approaches can be improved to explicitly deal with inputs that are very far from the data they have been exposed to during training.

Acknowledgements. We would like to thank Mikkel du Plessis for feedback on the stuttered voice conversion experiments in Sect. 3.3, and Chris-Mari Schreuder for help with the textual descriptions of voices for Sect. 6.

References

1. Arjun, K., Karthik, S., Kamalnath, D., Chanda, P., Tripathi, S.: Automatic correction of stutter in disfluent speech. Procedia Comput. Sci. **171**, 1363–1370 (2020)
2. Ba, J.L., Kiros, J.R., Hinton, G.E.: Layer normalization. arXiv preprint arXiv:1607.06450 (2016)
3. Baas, M., van Niekerk, B., Kamper, H.: Voice Conversion With Just Nearest Neighbors. In: Interspeech (2023)
4. Barahona-Ríos, A., Collins, T.: NoiseBandNet: controllable time-varying neural synthesis of sound effects using filterbanks. arXiv preprint arXiv:2307.08007 (2023)
5. Casanova, E., Weber, J., Shulby, C.D., Junior, A.C., Gölge, E., Ponti, M.A.: YourTTS: towards zero-shot multi-speaker TTS and zero-shot voice conversion for everyone. In: PMLR (2022)
6. Chen, S., Wang, C., Wu, Y., Liu, S., Chen, Z., Chen, Z., et al.: WavLM: large-scale self-supervised pre-training for full stack speech processing. IEEE J. Sel. Top. Signal Process. **16**(6), 1505–1518 (2022)
7. Cheng, E.: Morphing between voices using kNN-VC (2023). https://eccheng.github.io/ml/audio/vc/2023/07/04/knn-vc-morph.html. Accessed 07 Aug 2023
8. Chou, J.c., Lee, H.Y.: One-shot voice conversion by separating speaker and content representations with instance normalization. In: Interspeech (2019)
9. Chu, M., Yang, M., Xu, C., Ma, Y., Wang, J., Fan, Z., et al.: E-DGAN: an encoder-decoder generative adversarial network based method for pathological to normal voice conversion. IEEE J. Biomed. Health Inform. **27**(5), 2489–2500 (2023)

10. Fujii, K., Okawa, J., Suigetsu, K.: High-individuality voice conversion based on concatenative speech synthesis. Int. J. Electr. Comput. Eng. **1**(11), 1625–1630 (2007)
11. Gölge, E., Davis, K.: Introducing prompt-to-voice - describe it to hear it (2022), https://archive.is/MotSP, accessed: 2023-08-05
12. Jin, Z., Finkelstein, A., DiVerdi, S., Lu, J., Mysore, G.J.: Cute: a concatenative method for voice conversion using exemplar-based unit selection. In: ICASSP (2016)
13. Kilgour, K., Zuluaga, M., Roblek, D., Sharifi, M.: Fréchet audio distance: a reference-free metric for evaluating music enhancement algorithms. In: Interspeech (2019)
14. Kong, J., Kim, J., Bae, J.: HiFi-GAN: generative adversarial networks for efficient and high fidelity speech synthesis. In: NeurIPS (2020)
15. Lea, C., Mitra, V., Joshi, A., Kajarekar, S., Bigham, J.P.: SEP-28k: a dataset for stuttering event detection from podcasts with people who stutter. In: ICASSP (2021)
16. Li, J., Tu, W., Xiao, L.: FreeVC: towards high-quality text-free one-shot voice conversion. arXiv preprint arXiv:2210.15418 (2022)
17. Liu, H., Liu, X., Kong, Q., Tian, Q., Zhao, Y., Wang, D., et al.: VoiceFixer: a unified framework for high-fidelity speech restoration. In: Interspeech (2022)
18. Liu, S., Cao, Y., Wang, D., Wu, X., Liu, X., Meng, H.: Any-to-many voice conversion with location-relative sequence-to-sequence modeling. IEEE/ACM Trans. Audio Speech Lang. Process. **29**, 1717–1728 (2021)
19. Loshchilov, I., Hutter, F.: Decoupled weight decay regularization. In: ICLR (2019)
20. Mohammadi, S.H., Kain, A.: An overview of voice conversion systems. Speech Commun. **88**, 65–82 (2017)
21. Mohanty, S.: Musical instrument's sound dataset (2022). https://archive.is/BtIGp. Accessed 06 Aug 2023
22. van Niekerk, B., Carbonneau, M.A., Zaïdi, J., Baas, M., Seuté, H., Kamper, H.: A comparison of discrete and soft speech units for improved voice conversion. In: ICASSP (2022)
23. Panayotov, V., Chen, G., Povey, D., Khudanpur, S.: LibriSpeech: an ASR corpus based on public domain audio books. In: ICASSP (2015)
24. Pratap, V., Xu, Q., Sriram, A., Synnaeve, G., Collobert, R.: MLS: a large-scale multilingual dataset for speech research. ArXiv abs/2012.03411 (2020)
25. Radford, A., Kim, J.W., Xu, T., Brockman, G., McLeavey, C., Sutskever, I.: Robust speech recognition via large-scale weak supervision. arXiv preprint arXiv:2212.04356 (2022)
26. Reimers, N., Gurevych, I.: Sentence-BERT: sentence embeddings using siamese bert-networks. In: EMNLP (2019)
27. Sisman, B., Yamagishi, J., King, S., Li, H.: An overview of voice conversion and its challenges: from statistical modeling to deep learning. IEEE/ACM Trans. Audio Speech Lang. Process. **29**, 132–157 (2021)
28. Snyder, D., Garcia-Romero, D., Sell, G., Povey, D., Khudanpur, S.: X-Vectors: robust DNN embeddings for speaker recognition. In: ICASSP (2018)
29. Suzuki, K., Sakamoto, S., Taniguchi, T., Kameoka, H.: Speak like a dog: human to non-human creature voice conversion. In: APSIPA ASC (2022)
30. Vaswani, A., Shazeer, N., Parmar, N., Uszkoreit, J., Jones, L., Gomez, A.N., et al.: Attention is all you need. In: NeurIPS (2017)
31. Wan, L., Wang, Q., Papir, A., Moreno, I.L.: Generalized end-to-end loss for speaker verification. In: ICASSP (2018)

32. Wang, D., Deng, L., Yeung, Y.T., Chen, X., Liu, X., Meng, H.: VQMIVC: vector quantization and mutual information-based unsupervised speech representation disentanglement for one-shot voice conversion. In: Interspeech (2021)
33. Wu, S.L., Yang, Y.H.: MuseMorphose: full-song and fine-grained piano music style transfer with one transformer VAE. IEEE/ACM Trans. Audio Speech Lang. Process. **31**, 1953–1967 (2023)
34. Yi, Z., Huang, W.C., Tian, X., Yamagishi, J., Das, R.K., Kinnunen, T., et al.: Voice conversion challenge 2020: intra-lingual semi-parallel and cross-lingual voice conversion. In: Joint Workshop for the Blizzard Challenge and Voice Conversion Challenge (2020)

Extending Defeasible Reasoning Beyond Rational Closure

Luke Slater[1]([⊠])(iD) and Thomas Meyer[2](iD)

[1] University of Cape Town, Cape Town, South Africa
sltluk001@myuct.ac.za
[2] CAIR, University of Cape Town, Cape Town, South Africa
tmeyer@cs.uct.ac.za

Abstract. The KLM framework is a well known extension of classical logic for incorporating defeasible reasoning. Central to the KLM framework is rational closure, recognized as the most conservative approach to defeasible reasoning. Rational closure operates through two complementary paradigms: model-theoretic and formula-theoretic. This paper concentrates on the model-theoretic dimension, known as minimal ranked entailment. Unfortunately, the practical implementation of minimal ranked entailment remains largely impractical due to high computational and storage demands. To address this, we present reduced minimal ranked entailment, an optimization that employs reduced ordered binary decision diagrams to eliminate redundant information, thereby enhancing computational efficiency and reducing memory requirements. We further demonstrate that this optimized approach significantly facilitates the practical application and development of model-theoretic extensions of rational closure. This is illustrated through our own Bayesian refinement of minimal ranked entailment, which conceptualizes defeasible entailment as a form of conditional probability.

Keywords: knowledge representation and reasoning · defeasible reasoning · rational closure · bayesian reasoning

1 Introduction

The Kraus, Lehmann, and Magidor (KLM) framework [12,15,16] serves as a well known extension of classical logic for incorporating defeasible reasoning. Within this framework, defeasible entailment is facilitated through a ranking system grounded in typicality, operating via two distinct methodologies: model-theoretic and formula-theoretic. The model-theoretic method ranks potential scenarios or "worlds", while the formula-theoretic method ranks individual statements. Presently, two primary forms of defeasible entailment are recognized: rational closure [11,16,17] and lexicographic closure [15], each with well-defined definitions in both model-theoretic and formula-theoretic settings.

This paper focuses primarily on the model-theoretic facet of rational closure, known as minimal ranked entailment [7,10]. Regrettably, the practical implementation of any model-theoretic approach to defeasible entailment poses significant obstacles, mainly due to elevated computation and memory requirements, rendering it largely impractical. To mitigate this, we present reduced minimal ranked entailment, an optimized version of minimal ranked entailment. This optimization utilizes reduced ordered binary decision diagrams [1,5,14] to purge redundant information, thereby achieving marked improvements in computational efficiency and storage capacity.

Rational closure is acknowledged as the most conservative variant of defeasible entailment, making it a compelling foundation for the exploration of more advanced forms, of which lexicographic closure is a primary example [7]. We further demonstrate that reduced minimal ranked entailment can aid substantially in the practical application and development of model-theoretic extensions of rational closure. We illustrate this by presenting naive bayesian entailment, a Bayesian refinement of minimal ranked entailment, which treats defeasible entailment as a form of conditional probability, thereby providing a nuanced, statistical perspective on defeasible entailment.

2   Background

2.1   Propositional Logic

Syntax. Propositional logic focuses on fundamental units of knowledge known as propositions or *atoms*, represented here in typewriter text like c or cat. The set of all atoms is denoted as $\mathcal{P}$. Atoms combine with logical connectives to form *formulas*, symbolized by Greek letters such as α and β. Formulas are recursively defined as $\alpha ::= \top \mid \bot \mid p \mid \neg\alpha \mid \alpha \wedge \beta \mid \alpha \vee \beta \mid \alpha \to \beta \mid \alpha \leftrightarrow \beta$. The set of all such formulas constitutes the language of propositional logic, $\mathcal{L}$. A knowledge base $\mathcal{K}$ is a collection of statements in $\mathcal{L}$.

Semantics. Truth arises from *valuations*, which assign truth values to atoms in $\mathcal{P}$. Valuations are represented by lowercase letters like v, and visually depicted as sequences of atoms (e.g., p) and barred atoms (e.g., $\bar{p}$), where the former indicates true and the latter false. When some formula α is true under some valuation v, it is said to be *satisfied* by v, denoted $v \Vdash \alpha$. Such a valuation is termed a *model* of α. The set of models for a formula α is denoted as $[\![\alpha]\!]$.

Classical Entailment. Logical consequence, or entailment, refers to the situation where the truth of one or more statements necessary leads to the truth of one or more other statements.

Definition 1. *A knowledge base $\mathcal{K}$ classically entails α, denoted $\mathcal{K} \models \alpha$, iff* $[\![\mathcal{K}]\!] \subseteq [\![\alpha]\!]$ [3].

Verifying entailment using models is highly inefficient, the standard way to check entailment is known as reduction to unsatisfiability, and allows the use of a regular SAT solver. This works by conjoining all formulas in $\mathcal{K}$ with $\neg\alpha$, if the result is unsatisfiable (has no models), it indicates that $\mathcal{K} \models \alpha$.

2.2 KLM-Style Defeasible Reasoning

Consider the sea squirt. In its youth, this creature uses its basic brain to find a suitable spot-like a rock or coral-to attach to and filter-feed on whatever gunk drifts by. Upon attachment, it digests its brain, keeping only a simple nervous system.[1] It would be incorrect to categorically assert that sea squirts possess or lack brains. More appropriately, they *typically* do not.

Classical propositional logic, when applied to the above scenario, represented by the knowledge base $\mathcal{K} = \{y \rightarrow s, s \rightarrow \neg b, y \rightarrow b\}$, where y, s, and b correspond to young sea squirt, sea squirt, and brain, respectively, results in y being false in every model of $\mathcal{K}$. Kraus et al. [12], address this by introducing defeasible implication, expressed as $\alpha \mathrel{|\!\sim} \beta$, meaning 'if α, then typically β'. The semantics of $\mathrel{|\!\sim}$ comes from structures called ranked interpretations [16].

Definition 2. *A ranked interpretation is a function $\mathcal{R} : \mathcal{U} \mapsto \mathbb{N} \cup \{\infty\}$ with the property that for any $u \in \mathcal{U}$ where $\mathcal{R}(u) = i$, there must exist a $v \in \mathcal{U}$ such that $\mathcal{R}(v) = j$ for every $j < i$ [7].*

Classical formulas dictate which valuations are matched to either an integer in $\mathbb{N}$, or to ∞ (not possible), while defeasible implications delineate the precise integer ranking that should be assigned to a given, feasible valuation. The intuitive understanding is that valuations with a lower ranking are considered more typical. Figure 1 depicts a ranked interpretation for the defeasible knowledge base $\mathcal{K} = \{y \rightarrow s, s \mathrel{|\!\sim} \neg b, y \mathrel{|\!\sim} b\}$, and $\mathcal{P} = \{y, s, b\}$.

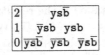

Fig. 1. Minimal ranked interpretation $\mathcal{R}_{RC}^{\mathcal{K}}$.

Definition 3. *A defeasible implication $\alpha \mathrel{|\!\sim} \beta$ is satisfied by a ranked interpretation $\mathcal{R}$, denoted as $\mathcal{R} \Vdash \alpha \mathrel{|\!\sim} \beta$, if the lowest ranked models of α are also models of β [7].*

---

[1] Often likened to the transition in an academic career, before and after securing a permanent university position.

Defeasible Entailment. In the KLM framework, the concept of defeasible entailment, symbolized by $\approx$, is not strictly defined. In this paper, we are concerned with defeasible entailment as defined using ranked interpretations.

Definition 4. *A defeasible implication $\alpha \mathrel|\!\sim \beta$ is defeasibly entailed by a defeasible knowledge base $\mathcal{K}$, for a particular ranked interpretation $\mathcal{R}$, denoted as $\mathcal{K} \approx_{\mathcal{R}} \alpha \mathrel|\!\sim \beta$, iff $\mathcal{R} \Vdash \alpha \mathrel|\!\sim \beta$* [7].

2.3   Rational Closure

Rational closure, introduced by Lehmann and Magidor [16], serves as the inaugural nonmonotonic framework for defeasible entailment. It captures the principle of *presumption of typicality* [15], which advocates for assuming information to be as typical as possible. Rational closure can be characterized in two complementary ways: syntactically (formula-theoretic) and semantically (model-theoretic).

Base Rank. A DI $\alpha \mathrel|\!\sim \beta$ implies that the validity of α generally leads to the validity of β. However, β may conflict with other defeasible information relating to α, rendering α classically untenable. In such cases, $\alpha \mathrel|\!\sim \beta$ is categorized as atypical. Being classified as atypical means that $\alpha \mathrel|\!\sim \beta$ acts as an exception (defeats) to other DIs in $\mathcal{K}$, necessitating a separate evaluation.[2] This separation is performed by the base rank procedure, which initiates at rank 0, identifies every DI in $\mathcal{K}$ that is defeated by other DIs, and places them at the current rank. It then advances the remaining atypical DIs to the next rank and repeats the process. The base rank of $\alpha \mathrel|\!\sim \beta$ is thus the lowest rank where α is not recognized as exceptional [10].

Freund [9] was the first to use the base rank procedure to propose and algorithm for determining if a DI $\alpha \mathrel|\!\sim \beta$ is in the rational closure of a defeasible knowledge base $\mathcal{K}$, denoted $\mathcal{K} \approx_{RC} \alpha \mathrel|\!\sim \beta$. Beginning at rank 0, DIs are systematically removed from $\mathcal{K}$ until it is classically consistent with α. The query is then considered defeasibly entailed if the resulting knowledge base classically entails the materialisation of $\alpha \mathrel|\!\sim \beta$ ($\alpha \rightarrow \beta$) [7]. We have included our own algorithms for implementing this approach in Appendix A (`BaseRank` and `Rational Closure`), which are slightly modified versions of those presented in [7].

Minimal Ranked Entailment. The semantic characterization of rational closure involves preferentially ranking every valid ranked interpretation for $\mathcal{K}$: $\mathcal{R}_1 \preceq_{\mathcal{K}} \mathcal{R}_2$ if for every $u \in \mathcal{U}$, $\mathcal{R}_1(u) \leq \mathcal{R}_2(u)$ [7]. Giordano et al. [10] demonstrated the existence of a minimal $\mathcal{R}$ in this ranking, denoted as $\mathcal{R}_{RC}^{\mathcal{K}}$.

Definition 5. *A DI $\alpha \mathrel|\!\sim \beta$ is in the rational closure of $\mathcal{K}$, denoted $\mathcal{K} \approx_{RC} \alpha \mathrel|\!\sim \beta$, if $\mathcal{R}_{RC}^{\mathcal{K}} \Vdash \alpha \mathrel|\!\sim \beta$.*

---

[2] A DI can also defeat a classical formula, though this would make it redundant.

There are two distinct levels of typicality at play here: one among the valuations within each individual ranked interpretation, and another between the individual ranked interpretations themselves. This notion of maximizing typicality resonates with the inherent focus on the presumption of typicality found in rational closure; it consistently uses the lowest-ranked $\mathcal{R}$. This feature also underpins its non-monotonic nature; the introduction of new information into $\mathcal{K}$ may alter the perception of what is considered most typical. Consequently, the minimal ranked interpretation is subject to change, thereby affecting the conclusions that can be subsequently drawn from it. We have included our own algorithms for implementing minimal ranked entailment in Appendix A (`MinimalRank` and `MinimalEntailment`).

3 Reduced Minimal Ranked Entailment

3.1 #P-complete

Implementing model-theoretic defeasible entailment requires significant computational resources and memory capacity. The task of identifying all satisfying solutions for a Boolean expression, known as #SAT, has the distinction of being the first problem shown to be #P-complete [19]. Often, the memory requirements of storing every valuation quickly emerge as a bottleneck, posing an even more substantial obstacle than the computational demands. Furthermore, entailment verification involves the individual assessment of each model, a process that can also escalate to exponential complexity in the worst-case scenario. For this reason, the BaseRank algorithm is frequently employed to execute rational closure, given its compatibility with standard SAT solvers designed for classical logic. To address these challenges, a compact model representation for each rank is needed, one that balances efficient computation with ease of entailment verification.

3.2 Reduced Ordered Binary Decision Diagrams

Binary Decision Diagrams (BDDs) are a Directed Acyclic Graph (DAG) structure pioneered by Lee [14], and later Akers [1], to provide compact representations of Boolean expressions. This paper centers on Reduced Ordered Binary Decision Diagrams (ROBDDs), a variant by Bryant [5]. Constructing a ROBDD involves selecting a variable ordering and branching iteratively from the root (first variable) into two paths (0 and 1) for each variable in the ordering. At each decision node, the variable assignment is evaluated against the expression, and accordingly, it branches to the appropriate terminal node (0 or 1). Redundant nodes and isomorphic substructures are eliminated through reduction rules, culminating in a compact, non-redundant representation of the initial expression. Figure 2 illustrates a ROBDD example. Examining the truth table, we observe that when x and y are true, the value of z becomes redundant, leading to the removal of the corresponding node. The same applies to the scenario where x is true, and y is false. In the finalized structure, every path from the root to 1 symbolizes a partial model of the original expression. When expanded, they collectively yield the set of complete models corresponding to the original expression. ROBDDs are vital in solving #SAT, efficiently minimizing computation and memory requirements.[3]

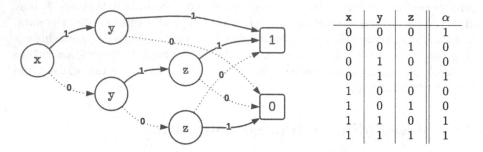

x	y	z	α
0	0	0	1
0	0	1	0
0	1	0	0
0	1	1	1
1	0	0	0
1	0	1	0
1	1	0	1
1	1	1	1

Fig. 2. ROBDD and truth table for $\alpha = (\neg x \wedge \neg y \wedge \neg z) \vee (x \wedge y) \vee (y \wedge z)$

ROBDD Procedures. ROBDDs provide a *canonical* form; under a predetermined variable ordering, equivalent expressions condense to identical ROBDDs [5]. This distinctive property is not only appealing but also highly relevant to this paper, as it greatly improves the efficiency of operations on ROBDDs. Bryant demonstrates that performing a Boolean operation on two ROBDDs has a complexity of $O(N_1 \cdot N_2)$, where N_1 and N_2 represent the respective node counts [5]. In cases where ROBDDs are compact, this offers a powerful method for entailment verification through reduction to unsatisfiability. There has been extensive research into the use of ROBDDs for symbolic model checking [6].

Bryant also illustrates various other procedures that can be applied to ROBDDs [5]. The `Satisfy-all` procedure returns the set $\mathcal{S}_f$ of partial models for a given ROBDD f, and has a complexity of $O(n \cdot |\mathcal{S}_f|)$, where n is the number of variables. The `Satisfy-one` procedure retrieves a single element from $\mathcal{S}_f$, with a complexity of $O(n)$. Lastly, the `Satisfy-count` procedure returns the total number of complete models represented by f, with a complexity of $O(N_f)$.

3.3   No Free Lunch

The 'No Free Lunch' (NFL) theorem asserts that the average performance of any two optimization algorithms will be identical across a wide set of problems [20], a concept relevant to model-theoretic defeasible entailment (#SAT). The compactness of a ROBDD is directly related to the chosen variable ordering. In some Boolean expressions, a specific ordering may produce a linear number of nodes, while a different order might lead to an exponential number in a

---

[3] Many BDD libraries also include optimizations like garbage collection, a process that frees up memory by removing nodes that are no longer in use.

worst-case. Figure 3 illustrates an example of a good and bad variable ordering, provided by Bryant [5]. Determining the best variable ordering is a complex task, shown to be NP-hard [4]. A trade-off thus exists between the construction of ranked interpretations and the verification of entailment. Investing more computational resources in finding the optimal variable ordering may decrease the effort required for entailment checking, and vice versa.

Determining optimal or near-optimal variable ordering for Boolean expressions is a subject of extensive research. Key approaches include the Sifting Algorithm, which starts with an initial variable ordering and iteratively refines it using heuristics [18]. Genetic Algorithms apply principles of natural selection to evolve a set of variable orderings over multiple generations, assessing each one's fitness based on the resulting diagram size [8]. Simulated Annealing employs a probabilistic approach, exploring the space of possible orderings by making random changes and accepting or rejecting them based on a temperature-dependent probability function [4]. Each of these techniques offers a unique balance between computational efficiency and the quality of the variable ordering.

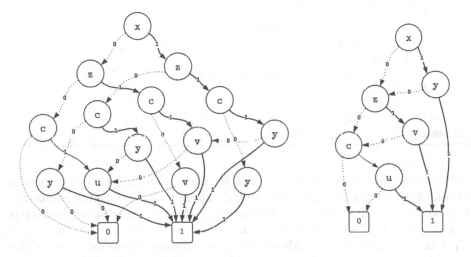

Fig. 3. Good (right) and bad (left) variable ordering for $\alpha = x \wedge y \vee z \wedge v \vee c \wedge u$

3.4 Algorithm

ReducedRank constructs a distinct ROBDD to represent the models for each rank within the associated minimal ranked interpretation. This ROBDD is then leveraged to identify all DIs whose antecedents are negated by that rank, achieved through reduction to unsatisfiability (where unsatisfiable expressions reduce to 0). In standard practice, a ROBDD is represented as a function, and denoted with the symbol f. Operations on ROBDDs are expressed using the conventional connectives of propositional logic. The Reduce function symbolizes the construction of a ROBDD given a specific formula and variable ordering, for which various efficient algorithms are well-established [1,5,14].

Utilizing a uniform variable ordering across all ranks yields a canonical form for defeasible knowledge bases (in the context of rational closure). This facilitates an efficient means of assessing formal equivalence. Conversely, if the objective is to attain a compact representation, employing distinct variable orderings for each rank is recommended. The Order function serves to specify the algorithm for determining variable ordering for a given expression. The output of this function is at the discretion of the implementer; it may produce a static ordering, generate a random sequence, or implement an established algorithm [4,8,18].

Algorithm 1: ReducedRank

Input: A knowledge base $\mathcal{K}$ and a set of atoms $\mathcal{P}$
Output: An ordered tuple $((f_0, \mathcal{P}_0), ..., (f_{n-1}, \mathcal{P}_{n-1}), n)$
1 $i := 0$;
2 $\gamma := \bigwedge(\mathcal{K} \setminus \{\alpha \hspace{1pt}\vdash\hspace{-3pt}\sim \beta \in \mathcal{K}\})$;
3 $\mathcal{B}_i := \{\alpha \rightarrow \beta \mid \alpha \hspace{1pt}\vdash\hspace{-3pt}\sim \beta \in \mathcal{K}\}$;
4 **repeat**
5      $\delta_i := \gamma \wedge (\bigwedge \mathcal{B}_i) \wedge (\bigwedge_{j<i}(\bigvee \neg \mathcal{B}_j))$;
6      $\mathcal{P}_i := \text{Order}(\delta_i, \mathcal{P})$;
7      $f_i := \text{Reduce}(\delta_i, \mathcal{P}_i)$;
8      $\mathcal{B}_{i+1} := \{\alpha \rightarrow \beta \in \mathcal{B}_i \mid f_i \wedge \text{Reduce}(\alpha, \mathcal{P}_i) = 0\}$;
9      $\mathcal{B}_i := \mathcal{B}_i \setminus \mathcal{B}_{i+1}$;
10     $i := i + 1$;
11 **until** $\mathcal{B}_i = \emptyset$ *or* $\mathcal{B}_{i-1} = \emptyset$;
12 **return** $((f_0, \mathcal{P}_0), ..., (f_{i-1}, \mathcal{P}_{i-1}))$;

Constructing a ROBDD for each rank involves formulating an exact representation of the models for that rank. In this context, γ symbolizes the constraints from the classical formulas in $\mathcal{K}$. Separating classical formulas from DIs diverges from the norm, where a classical formula α is typically represented as $\neg \alpha \hspace{1pt}\vdash\hspace{-3pt}\sim \perp$ [7]. This key optimization avoids redundant evaluations of classical formulas that would otherwise recur with each rank. The expression $\bigwedge \mathcal{B}_i$ refers to the constraints imposed by the DIs in the current rank, which all models must satisfy. Meanwhile, $\bigwedge_{j<i}(\bigvee \neg \mathcal{B}_j)$ conveys the constraints from the DIs in previous ranks, each model of the current rank must fail to satisfy at least one DI from each prior rank. The expression is subsequently passed to the Order function to ascertain the optimal variable ordering for the current rank, after which the ROBDD is constructed based on the selected ordering.

ReducedEntailment creates a ROBDD for α, represented by f_α and $f_{\alpha \rightarrow \beta}$, for the purpose of verifying entailment. f_α is consecutively conjoined with the ROBDD of each rank until the outcome does not reduce to 0 (meaning there exists a model in that rank that satisfies α). This ROBDD is then combined with $f_{\neg(\alpha \rightarrow \beta)}$. If the resulting ROBDD reduces to 0 (is unsatisfiable), it indicates that every model of that rank satisfying α also satisfies β.

Algorithm 2: ReducedEntailment

Input: A knowledge base $\mathcal{K}$, a set of atoms $\mathcal{P}$, and a DI $\alpha \mathrel{|\!\sim} \beta$

Output: true, if $\mathcal{K} \mathrel{\approx\!\!\!\mid} \alpha \mathrel{|\!\sim} \beta$, and **false**, otherwise

1 $i := 0$;

2 $((f_0, \mathcal{P}_0), ..., (f_{n-1}, \mathcal{P}_{n-1}), n) := \texttt{ReducedRank}(\mathcal{K}, \mathcal{P})$;

3 $f_\alpha := \texttt{Reduce}(\alpha, \mathcal{P}_0)$;

4 **while** $f_i \wedge f_\alpha = 0$ *and* $i < n - 1$ **do**

5     $i := i + 1$;

6     $f_\alpha := \texttt{Reduce}(\alpha, \mathcal{P}_i)$;

7 $f_{\neg(\alpha \to \beta)} := \texttt{Reduce}(\neg(\alpha \to \beta), \mathcal{P}_i)$;

8 **return** $f_i \wedge f_{\neg(\alpha \to \beta)} = 0$;

3.5 Analysis

Let $\mathcal{K} = \{p \to b, b \mathrel{|\!\sim} f, p \mathrel{|\!\sim} \neg f, b \mathrel{|\!\sim} w\}$. Figure 4 illustrates the minimal ranked interpretation for $\mathcal{K}$, a corresponding reduced version, as well as the base ranking. Examining rank 0, p is atypical, mandating $\bar{p}$ in every partial model. When b is false, the truth of f and w become redundant due to the semantics of implication, leaving only $\bar{b}\bar{p}$. However, if b is true, then both f and w must be true, leaving $fwb\bar{p}$. Even with a small knowledge base, a significant amount of redundant information can be eliminated.

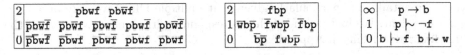

Fig. 4. Complete vs reduced minimal ranked interpretation $\mathcal{R}^{\mathcal{K}}_{RC}$

ROBDDs are especially effective with expressions containing notable structural regularity, a characteristic often found in ontology-based knowledge bases. As the number of formulas in a knowledge base grows, redundant information also tends to increase, especially in a structurally regular knowledge base. Let $\mathcal{K} = \{p \to b, b \mathrel{|\!\sim} f, p \mathrel{|\!\sim} \neg f, p \mathrel{|\!\sim} t, p \mathrel{|\!\sim} y, p \mathrel{|\!\sim} u, b \mathrel{|\!\sim} q, b \mathrel{|\!\sim} w, b \mathrel{|\!\sim} e, r \to p, r \mathrel{|\!\sim} f, r \mathrel{|\!\sim} z, r \mathrel{|\!\sim} x, r \mathrel{|\!\sim} c, n \to r, n \mathrel{|\!\sim} \neg f, n \mathrel{|\!\sim} j, n \mathrel{|\!\sim} k, n \mathrel{|\!\sim} l\}$. Figure 5 illustrates a corresponding reduced minimal ranked interpretation.

The corresponding complete version contains 40,960 complete models with 17 atoms each, while the reduced version contains only 25 partial models with at most 8 atoms each. It is clear that this significantly reduces memory requirements, without any loss of information. Importantly, despite the substantial increase in the number of formulas and atoms in $\mathcal{K}$, the count of partial models remains relatively stable. This trend can be expected to continue as more formulas and atoms are added to $\mathcal{K}$, as long as the degree of randomness in the knowledge remains low (with the proviso that substantial computational effort

4	jnr̄fbp	k̄jnr̄fbp	l̄kjnr̄fbp	nz̄rfbp	nx̄zrfbp	nc̄xzrfbp	ncxzrfbp
3		n̄rfbp	l̄kjnr̄fbp	nz̄rfbp	nx̄zrfbp	nc̄xzrfbp	
2		nt̄rfbp	nȳtr̄fbp	nuytr̄fbp	n̄rfbp	nc̄xzrfbp	
1		n̄rfbp	nrqfbp	nrwqfbp	nrewqfbp	nuytr̄fbp	
0			n̄rfbp	n̄rfbp	nrewqfbp		

Fig. 5. Reduced minimal ranked interpretation $\mathcal{R}_{RC}^{\mathcal{K}}$.

may be necessary to identify the optimal ordering). Furthermore, it is evident that this approach offers a more efficient means of verifying entailment in comparison to its complete counterpart.

An idiosyncrasy arises between the base rank of a DI $\alpha \mathrel|\!\sim \beta$ and its influence on the overall reducibility of a ranked interpretation. Specifically, the truth value of β becomes redundant in all preceding ranks. Consequently, any information exclusively related to the truth of $\alpha \mathrel|\!\sim \beta$ can be replaced with simply $\neg\alpha$. This generalizes to other DIs that share the same α. For instance, the base rank of $n \mathrel|\!\sim j, n \mathrel|\!\sim k$, and $n \mathrel|\!\sim l$ is 3. The atoms j, k, and l are solely connected to the rest of the knowledge through n, rendering them redundant in ranks 2, 1, and 0; hence, they are entirely absent.

This observation reveals that the `BaseRank` algorithm performs redundant computation by including the material counterpart $\alpha \rightarrow \beta$ for a DI in all ranks preceding its base rank. A more efficient approach would involve including only $\neg\alpha$. Such a modification would significantly enhance the efficiency of verifying entailment, particularly given that multiple DIs may share an identical antecedent. We have included the corresponding irredundant versions of the `BaseRank` algorithms in Appendix B.

4   Naive Bayes Entailment

4.1   Motivation

The application of reduced minimal ranked entailment is not limited to rational closure alone. Representing each rank as a ROBDD introduces a symbolic system for processing and manipulating the models of each rank using ROBDD procedures and Boolean algebra. This has broad utility, such as creating model-theoretic extensions of rational closure, which is an active area of research [7]. In this paper, we illustrate this through our own Bayesian refinement of minimal ranked entailment, which conceptualises defeasible entailment as a form of conditional probability.

Bayesian reasoning is a well known methodology in Statistics and Artificial Intelligence, first introduced by Thomas Bayes in 1763 [2], and later refined by his colleague, Pierre-Simone Laplace, in 1812 [13]. At the core of Bayesian reasoning is Bayes' Rule:

$$P(H|E) = \frac{P(E|H)P(H)}{P(E|H)P(H) + P(E|\neg H)P(\neg H)} \tag{1}$$

$P(H|E)$ represents the *posterior* probability of some hypothesis H given some evidence E. $P(E|H)P(H)$ is the *likelihood* of the evidence given the hypothesis, multiplied by the *prior* probability of the hypothesis. $P(E|\neg H)P(\neg H)$ is the likelihood of the evidence given the negation of the hypothesis, multiplied by the prior probability of the negation of the hypothesis.

4.2   Algorithm

The most straightforward approach would be to apply Bayes' Rule directly to $\alpha \mathrel{|\!\sim} \beta$:

$$P(\beta \mid \alpha) = \frac{\frac{f \wedge f_\alpha \wedge f_\beta}{f \wedge f_\alpha} \cdot \frac{f \wedge f_\beta}{f}}{\frac{f \wedge f_\alpha \wedge f_\beta}{f \wedge f_\alpha} \cdot \frac{f \wedge f_\beta}{f} + \frac{f \wedge f_\alpha \wedge f_{\neg\beta}}{f \wedge f_\alpha} \cdot \frac{f \wedge f_{\neg\beta}}{f}} \tag{2}$$

In this context, f denotes the ROBDD corresponding to the lowest rank that satisfies α, whereas f_α and f_β represent the ROBDDs for α and β, respectively. However, utilizing this approach would diverge too far from the core tenets of defeasible entailment.

In the exploration of defeasible entailment from a statistical standpoint, the aim is to identify the most statistically probable minimal worlds that satisfy α, and evaluate their compliance with β. α can encapsulate any formula in $\mathcal{L}$, thereby representing a variety of scenarios or evidential fragments. These fragments are equivalent to the models of α.[4] In each minimal world compatible with a given fragment, the remaining atoms constitute potential hypotheses. For instance, if $\alpha = \mathsf{a}$, and a minimal world satisfying a is $\mathsf{a}\overline{\mathsf{b}}\mathsf{c}\overline{\mathsf{d}}$, the corresponding probability is $P(\overline{\mathsf{b}} \wedge \mathsf{c} \wedge \overline{\mathsf{d}} \mid \mathsf{a})$. Accurate computation of such conjunctions necessitates understanding the conditional dependencies among atoms. Although it is theoretically possible to derive these dependencies from formulas in $\mathcal{K}$, this task is complex and falls outside the scope of this paper. However, an assumption of conditional independence, as exemplified by the naive bayes classifier, can often produce useful results. This simplifies the calculation to the product of individual conditional probabilities:

$$P(\overline{\mathsf{b}} \wedge \mathsf{c} \wedge \overline{\mathsf{d}} \mid \mathsf{a}) = P(\overline{\mathsf{b}} \mid \mathsf{a}) * P(\mathsf{c} \mid \mathsf{a}) * P(\overline{\mathsf{d}} \mid \mathsf{a}) \tag{3}$$

For each fragment, the analysis unfolds in three sequential steps:

1. Identify the minimal partial models that are consistent with the fragment.
2. Quantify the true, and by implication, false occurrences of each remaining atom within their corresponding complete models.
3. Employ naive bayesian calculations to ascertain the most probable complete models.

ROBDD procedures are highly effective for executing steps 1 and 2. The identification of minimal partial models follows the same procedure employed in

---

[4] It is crucial to emphasize that this refers to the models of α devoid of extraneous information, meaning they contain exclusively the atoms found in α.

Algorithm 3: NaiveBayesEntailment

Input: An ordered tuple $((f_0, \mathcal{P}_0), ..., (f_{n-1}, \mathcal{P}_{n-1}), n)$, a set of propositions $\mathcal{P}$,
and a DI $\alpha \mathrel{|\!\sim} \beta$

Output: true, if $\mathcal{K} \approx_{NB} \alpha \mathrel{|\!\sim} \beta$, and **false**, otherwise

1 $i := 0$;
2 $P := 0$;
3 $\mathcal{U}_P := \emptyset$;
4 $f_\alpha := \texttt{Reduce}(\alpha, \mathcal{P}_i)$;
5 **while** $f_i \wedge f_\alpha = 0$ **do**
6      $i := i + 1$;
7      $f_\alpha := \texttt{Reduce}(\alpha, \mathcal{P}_i)$;
8 $f_h := f_i \wedge f_\alpha$;
9 $\mathcal{H} := \texttt{Satisfy-all}(f_h)$;
10 **for** $\tilde{e} \in [\![\alpha]\!]$ **do**
11      $f_{\tilde{e}} := \texttt{Reduce}(\bigwedge \tilde{e}, \mathcal{P}_i)$;
12      $total, count_\mathcal{P} := \texttt{Count}(f_{\tilde{e}} \wedge f_h, \mathcal{P}_i)$;
13   **for** $\tilde{w} \in \mathcal{H}$ **do**
14   **if** $\tilde{w} \cap \neg\tilde{e} = \emptyset$ **then**
15              $P_{\tilde{w}} := 0$;
16   **for** $p \in \tilde{w} \setminus \tilde{e}$ **do**
17   **if** $\tilde{w} \setminus \tilde{e} \Vdash p$ **then**
18                      $P_{\tilde{w}} := P_{\tilde{w}} + \log(\frac{count_p}{total})$;
19   **else**
20                      $P_{\tilde{w}} := P_{\tilde{w}} + \log(\frac{total - count_p}{total})$;
21              $P_{\tilde{w}}, \mathcal{U}_{\tilde{w}} := \texttt{Expand}(P_{\tilde{w}}, \tilde{w} \cup \tilde{e}, \mathcal{P} \setminus (\tilde{w} \cup \tilde{e}), total, count_\mathcal{P})$;
22   **if** $P_{\tilde{w}} < P$ **then**
23                  $P := P_{\tilde{w}}$;
24                  $\mathcal{U}_P := \mathcal{U}_{\tilde{w}}$;
25   **else if** $P_{\tilde{w}} = P$ **then**
26                  $\mathcal{U}_P := \mathcal{U}_P \cup \mathcal{U}_{\tilde{w}}$;
27 **return** $\forall u \in \mathcal{U}_P, u \Vdash \beta$;

ReducedEntailment. The ROBDD corresponding to the minimal satisfying rank is then conjoined with f_α, yielding a resultant f_h that delineates the hypothesis space for α. The Satisfy-all procedure is then invoked to obtain the set of partial models within this hypothesis space. Partial models are designated with a tilde symbol to differentiate from complete models. In this context, a partial model serves to explicitly encapsulate a collection of literals (negated or non-negated atoms). To facilitate set-based comparisons with partial models derived from a ROBDD, each model of α is likewise represented as a partial model $\tilde{e}$.

The algorithm iterates over each model of α to identify the most probable complete models that they align with. During this procedure, the highest calculated probability is retained in P, and the complete models sharing this probability are stored in $\mathcal{U}_P$. Prior to conducting probabilistic calculations, requisite counting operations must be executed. The current evidence space $f_{\tilde{e}}$ is conjoined with the hypothesis space f_h to isolate the segment of the hypothesis space that aligns with this evidence space, which is then passed to Count.

Algorithm 4: Count

Input: f, $\mathcal{P}_i$
1 $total :=$ Satisfy-count(f);
2 **for** $p \in \mathcal{P}_i$ **do**
3      $f_p :=$ Reduce$(p, \mathcal{P}_i)$;
4      $count_p :=$ Satisfy-count$(f \wedge f_p)$;
5 **return** $total, count_\mathcal{P}$;

The Count method executes the required counting operations within a specific evidence-hypothesis space, denoted by f. The Satisfy-count procedure is directly applied to f to enumerate the total number of complete models. For each atom in $\mathcal{P}$, the method conjoins f with f_p and employs the Satisfy-count procedure to tally the total number of true occurrences for each atom. In practical implementation, these values would be stored in an array/dictionary ($count_\mathcal{P}$). This detail has been omitted here for brevity.

Algorithm 5: Expand

Input: P, $\tilde{u}$, $\mathcal{P}$, $total$, $count_\mathcal{P}$
1 **if** $\mathcal{P} = \emptyset$ **then**
2   **return** $P, \{\tilde{u}\}$;
3 $p :- p \in \mathcal{P}$;
4 $P_l, \mathcal{U}_l :=$ Expand$(P + \log(\frac{count_p}{total}), \tilde{u} \cup p, \mathcal{P} \setminus p, total, count_\mathcal{P})$;
5 $P_r, \mathcal{U}_r :=$ Expand$(P + \log(\frac{total - count_p}{total}), \tilde{u} \cup \neg p, \mathcal{P} \setminus p, total, count_\mathcal{P})$;
6 **if** $P_l < P_r$ **then**
7   **return** $P_l, \mathcal{U}_l$;
8 **else if** $P_l > P_r$ **then**
9   **return** $P_r, \mathcal{U}_r$;
10 **else**
11   **return** $P_l, \mathcal{U}_l \cup \mathcal{U}_r$;

The Expand method essentially re-expands a given partial model while concurrently executing the requisite probabilistic calculations for each corresponding complete model. To optimize computational efficiency and avert floating-point underflow, these calculations are conducted in logarithmic space. Subsequently, the method returns a set comprising the most probable complete models. This set is then evaluated against $\mathcal{U}_P$, the set containing the overall most probable complete models, and is either amalgamated with it or supersedes it. Upon completion of the last piece of evidence, each model within $\mathcal{U}_P$ undergoes evaluation to ascertain its compatibility with the consequent. Should they satisfy the consequent, the query is deemed to be probabilistically defeasibly entailed.

4.3   Analysis

In the study of different forms of defeasible entailment, Lehmann and Magidor propose a set of rationality properties known as the KLM postulates [7,12]. A defeasible entailment relation that adheres to these principles is termed *LM-rational*. Lehmann and Magidor proved that if a defeasible entailment relation can be derived from a ranked interpretation, then it is LM-rational [12]. Naive Bayes Entailment identifies the most statistically probable minimal models tailored to a specific query, implying that the model rankings are query-dependent. Consequently, it cannot be represented by a ranked interpretation and is thus not considered rational. Despite this, we will illustrate that naive bayes entailment can still provide meaningful results by providing a more quantitative perspective of defeasible entailment that neither rational nor lexicographic closure can furnish.

Example 1. Consider the defeasible knowledge base $\mathcal{K} = \{$penguin $\rightarrow$ bird, bird $\mid\!\sim$ flies, penguin $\mid\!\sim$ ¬flies, bird $\mid\!\sim$ wings, emperor $\rightarrow$ penguin, emperor $\mid\!\sim$ wings, king $\rightarrow$ penguin, king $\mid\!\sim$ wings, chinstrap $\rightarrow$ penguin, chinstrap $\mid\!\sim$ ¬wings$\}$. Figure 6 illustrates a corresponding reduced minimal ranked interpretation. Consider the query $\mathcal{K} \mid\!\approx$ penguin $\mid\!\sim$ wings, for which standard rational closure would return false. However, naive bayes entailment discerns $\overline{\text{ckewf}}$bp as the most probable model, resulting in the evaluation of $\mathcal{K} \mid\!\approx_{NB}$ penguin $\mid\!\sim$ wings as true.

2	$\text{ke}\overline{\text{wf}}\text{bp}$	$\text{e}\overline{\text{wf}}\text{bp}$	$\text{cw}\overline{\text{f}}\text{bp}$	fbp
1	$\overline{\text{ckewf}}\text{bp}$	$\overline{\text{ckewf}}\text{bp}$	$\overline{\text{ckewf}}\text{bp}$	$\overline{\text{kewf}}\text{bp}$	$\overline{\text{cwf}}\text{bp}$
0	$\overline{\text{ckewbp}}$	$\overline{\text{ckewbp}}$	$\overline{\text{ckewf}}\text{bp}$

Fig. 6. Reduced minimal ranked interpretation $\mathcal{R}^{\mathcal{K}}_{RC}$.

An intriguing question arises from this: How does employing Bayesian reasoning to logical truth assignments yield any valuable results? The key lies not in the possibilities but rather in the constraints. The minimal partial **penguin** models, namely $\overline{\text{kewf}}$bp and $\overline{\text{cwf}}$bp, reveal two distinct types of **penguin**: those with wings (**emperor** and **king**) and those without (**chinstrap**). If a penguin lacks wings, it cannot be **emperor** or **king**, thus reducing the scenarios in which a penguin does not have wings by half, twice over. Conversely, if the penguin has wings, it cannot be **chinstrap**, cutting the scenarios in which a penguin does have wings only once. Consequently, there are twice as many scenarios in which a penguin possesses wings compared to those in which it does not.

Example 2. Let $\mathcal{K} = \{$dove $\rightarrow$ bird, dove $\mid\!\sim$ wings, dove $\mid\!\sim$ flies, swan $\rightarrow$ bird, swan $\mid\!\sim$ wings, swan $\mid\!\sim$ flies, penguin $\rightarrow$ bird, penguin $\mid\!\sim$ ¬wings, penguin $\mid\!\sim$ ¬flies, chicken $\rightarrow$ bird, chicken $\mid\!\sim$ wings, chicken $\mid\!\sim$ ¬flies$\}$.

This knowledge contains information about four specific types of birds: two species that both have wings and can fly (dove and swan), and two species that cannot fly. Among the non-flying birds, chickens possess wings, while penguins do not. Notably, $\mathcal{K}$ does not contain any exceptional information; it does not explicitly state that bird $\mid\!\sim$ wings, thus $\mathcal{K} \not\approx$ bird $\mid\!\sim$ wings evaluates to false. NaiveBayesEntailment identifies $\overline{s}d\overline{c}fwb\overline{p}$ as the most probable model for bird, making $\mathcal{K} \approx_{NB}$ bird $\mid\!\sim$ wings evaluate to true. It therefore also follows that $\mathcal{K} \approx_{NB}$ bird $\mid\!\sim$ flies, which might appear counter-intuitive at first glance, but is statistically correct. Since three of the four birds have wings, it is statistically typical for a bird to have wings. However, the number of birds that can and cannot fly are equal, meaning that the ability to fly has no impact on defining what is statistically typical for a bird. Among the statistically typical birds, two can fly, and one cannot.

Consider the addition of bird $\mid\!\sim$ flies to $\mathcal{K}$. Both penguin and chicken are now explicitly recognized as atypical birds, and segregated from those considered typical. Naive bayes entailment identifies $sd\overline{c}w\overline{p}\overline{f}b$, $\overline{s}d\overline{c}w\overline{p}\overline{f}b$, $sd\overline{c}w\overline{p}\overline{f}b$, $\overline{s}d\overline{c}w\overline{p}\overline{f}b$, $sd\overline{c}w\overline{p}\overline{f}b$, $\overline{s}d\overline{c}w\overline{p}\overline{f}b$, $sd\overline{c}w\overline{p}\overline{f}b$, and $\overline{s}d\overline{c}w\overline{p}\overline{f}b$ as the most probable models for bird$\wedge\neg$flies. As a result, both $\mathcal{K} \approx_{NB}$ (bird$\wedge\neg$flies) $\mid\!\sim$ wings and $\mathcal{K} \approx_{NB}$ (bird$\wedge\neg$flies) $\mid\!\sim$ $\neg$wings will evaluate to false. Lastly, consider the addition of kiwi $\rightarrow$ bird, kiwi $\mid\!\sim$ wings, and kiwi $\mid\!\sim$ $\neg$flies to $\mathcal{K}$. NaiveBayesEntailment identifies valuations $s\overline{c}d\overline{p}\overline{f}wb\overline{k}$, $s\overline{c}d\overline{p}fwb\overline{k}$, $\overline{s}d\overline{c}w\overline{p}\overline{f}b\overline{k}$, and $\overline{s}d\overline{c}w\overline{p}\overline{f}b\overline{k}$ as most probable for b $\wedge$ $\neg$flies, making $\mathcal{K} \approx_{NB}$ (bird $\wedge$ $\neg$flies) $\mid\!\sim$ wings true.

5   Conclusion

Reduced entailment marks a pivotal reallocation of computational effort. Its efficacy lies in channeling computational resources into the preprocessing phase for the one-time construction of ranked interpretations. This often leads to a substantial reduction in the computational load needed for entailment verification, which is generally done multiple times. Additionally, we have introduced a method that markedly diminishes the storage requirements for these ranked interpretations. It is plausible that numerous scenarios exist wherein reduced entailment outperforms its base-ranked counterpart as well. However, empirical validation through comprehensive experimentation is necessary to substantiate this comparative advantage.

We have illustrated that reduced entailment serves purposes beyond mere optimization. This offers a rigorously studied and efficacious framework for developing model-theoretic extensions of rational closure. Specifically, it enables the direct mathematical manipulation of information within each rank through the application of Boolean algebra. Future research directions include the formulation of reduced lexicographic closure and the empirical evaluation of the comparative merits between reduced entailment and the BaseRank algorithm. Each methodology presents unique advantages, and a nuanced understanding of these could pave the way for a more adaptive algorithm capable of selecting the optimal strategy for specific problem instances.

A   Rational Closure Algorithms

Algorithm 6: BaseRank

Input: A knowledge base $\mathcal{K}$

Output: An ordered tuple $(\mathcal{R}_0, ..., \mathcal{R}_{n-1}, \mathcal{R}_\infty, n)$

1 $i := 0$;

2 $\mathcal{R}_\infty := \mathcal{K} \setminus \{\alpha \mathrel{\vert\!\sim} \beta \in \mathcal{K}\}$;

3 $\mathcal{R}_i := \{\alpha \to \beta \mid \alpha \mathrel{\vert\!\sim} \beta \in \mathcal{K}\}$;

4 **repeat**

5      $\mathcal{R}_{i+1} := \{\alpha \to \beta \in \mathcal{R}_i \mid \mathcal{R}_\infty \cup \mathcal{R}_i \models \neg\alpha\}$;

6      $\mathcal{R}_i := \mathcal{R}_i \setminus \mathcal{R}_{i+1}$;

7      $i := i + 1$;

8 **until** $\mathcal{R}_i = \emptyset$ **or** $\mathcal{R}_{i-1} = \emptyset$;

9 **if** $\mathcal{R}_{i-1} = \emptyset$ **then**

10      $\mathcal{R}_\infty := \mathcal{R}_\infty \cup \mathcal{R}_i$;

11      $i := i - 1$;

12 **return** $(\mathcal{R}_0, ..., \mathcal{R}_{i-1}, \mathcal{R}_\infty, i)$;

Algorithm 7: RationalClosure

Input: A knowledge base $\mathcal{K}$ and a DI $\alpha \mathrel{\vert\!\sim} \beta$

Output: **true**, if $\mathcal{K} \mathrel{\approx\!\!\!\mid\sim} \alpha \mathrel{\vert\!\sim} \beta$, and **false**, otherwise

1 $i := 0$;

2 $(\mathcal{R}_0, ..., \mathcal{R}_{n-1}, \mathcal{R}_\infty, n) := \mathtt{BaseRank}(\mathcal{K})$;

3 $\mathcal{R} := \bigcup_{i=0}^{j<n} \mathcal{R}_j$;

4 **while** $\mathcal{R}_\infty \cup \mathcal{R} \models \neg\alpha$ **and** $\mathcal{R} \neq \emptyset$ **do**

5      $\mathcal{R} := \mathcal{R} \setminus \mathcal{R}_i$;

6      $i := i + 1$;

7 **return** $\mathcal{R}_\infty \cup \mathcal{R} \models \alpha \to \beta$;

Algorithm 8: MinimalRank

Input: A knowledge base $\mathcal{K}$

Output: An ordered tuple $(\mathcal{R}_0, ..., \mathcal{R}_{n-1}, n)$

1 $i := 0$;

2 $\mathcal{B}_i := \{\alpha \to \beta \mid \alpha \mathrel{\vert\!\sim} \beta \in \mathcal{K}\}$;

3 $\mathcal{U} := [\![\mathcal{K} \setminus \{\alpha \mathrel{\vert\!\sim} \beta \in \mathcal{K}\}]\!]$;

4 **repeat**

5      $\mathcal{R}_i := [\![\mathcal{B}_i]\!] \cap \mathcal{U}$;

6      $\mathcal{B}_{i+1} := \{\alpha \to \beta \in \mathcal{B}_i \mid \mathcal{R}_i \cap [\![\alpha]\!] = \emptyset\}$;

7      $\mathcal{U} := \mathcal{U} \setminus \mathcal{R}_i$;

8      $i := i + 1$;

9 **until** $\mathcal{B}_i = \mathcal{B}_{i-1}$;

10 **return** $(\mathcal{R}_0, ..., \mathcal{R}_{i-1}, i)$;

Algorithm 9: MinimalEntailment

Input: A knowledge base $\mathcal{K}$, and a DI $\alpha \mathrel{|\!\sim} \beta$
Output: **true**, if $\mathcal{K} \mathrel{\approx\!\!\!\!\!\!\!\not\;} \alpha \mathrel{|\!\sim} \beta$, and **false**, otherwise

1   $i := 0$;
2   $(\mathcal{R}_0, ..., \mathcal{R}_{n-1}, n) := \texttt{MinimalRank}(\mathcal{K})$;
3   **while** $\mathcal{R}_0 \cap [\![\alpha]\!] = \emptyset$ **and** $i < n-2$ **do**
4     $i := i + 1$;
5   **return** $\mathcal{R}_i \cap [\![\alpha]\!] \subseteq [\![\beta]\!]$;

B   Irredundant Base Rank Algorithms

Algorithm 10: IrredundantBaseRank

Input: A knowledge base $\mathcal{K}$
Output: An ordered tuple $(\mathcal{R}_0, ..., \mathcal{R}_{n-1}, \mathcal{R}_\infty, n)$

1   $i := 0$;
2   $\mathcal{R}_\infty := \mathcal{K} \setminus \{\alpha \mathrel{|\!\sim} \beta \in \mathcal{K}\}$;
3   $\mathcal{R}_i := \{\alpha \to \beta \mid \alpha \mathrel{|\!\sim} \beta \in \mathcal{K}\}$;
4   **repeat**
5     $\mathcal{R}_{i+1} := \{\alpha \to \beta \in \mathcal{R}_i \mid \mathcal{R}_\infty \cup \mathcal{R}_i \models \neg\alpha\}$;
6     $\mathcal{R}_i := (\mathcal{R}_i \setminus \mathcal{R}_{i+1}) \cup \{\neg\alpha \mid \alpha \to \beta \in \mathcal{R}_{i+1}\}$;
7     $i := i + 1$;
8   **until** $\mathcal{R}_i = \emptyset$ **or** $\{\alpha \to \beta \in \mathcal{R}_{i-1}\} = \emptyset$;
9   **if** $\{\alpha \to \beta \in \mathcal{R}_{i-1}\} = \emptyset$ **then**
10    $\mathcal{R}_\infty := \mathcal{R}_\infty \cup \mathcal{R}_i$;
11    $i := i - 1$;
12   **return** $(\mathcal{R}_0, ..., \mathcal{R}_{i-1}, \mathcal{R}_\infty, i)$;

Algorithm 11: IrredundantRationalClosure

Input: A knowledge base $\mathcal{K}$ and a DI $\alpha \mathrel{|\!\sim} \beta$
Output: **true**, if $\mathcal{K} \mathrel{\approx\!\!\!\!\!\!\!\not\;} \alpha \mathrel{|\!\sim} \beta$, and **false**, otherwise

1   $i := 0$;
2   $(\mathcal{R}_0, ..., \mathcal{R}_{n-1}, \mathcal{R}_\infty, n) := \texttt{IrredundantBaseRank}(\mathcal{K})$;
3   **while** $\mathcal{R}_\infty \cup \mathcal{R}_i \models \neg\alpha$ **and** $i < n$ **do**
4     $i := i + 1$;
5   **if** $i < n$ **then**
6   **return** $\mathcal{R}_\infty \cup \mathcal{R}_i \models \alpha \to \beta$;
7   **else**
8   **return** $\mathcal{R}_\infty \models \alpha \to \beta$;

C   Proof of Properties for Naive Bayes Entailment

Theorem 1. *Naive bayes entailment is not LM-rational.*

Reflexivity. *Reflexivity* stipulates that all formulas should be defeasible consequences of themselves, a condition generally met by any defeasible entailment relation.

$$(Ref) \quad \mathcal{K} \mathrel{\approx\!\!\!\!\mid} \alpha \mathrel{\mid\!\sim} \alpha \tag{4}$$

Lemma 1. *Naive bayes entailment satisfies reflexivity.*

Proof. Naive bayes entailment identifies the most statistically probable minimal models consistent with any evidential fragment in $[\![\alpha]\!]$. It is therefore guaranteed to return a set of models that each satisfy α.

Left Logical Equivalence. *Left logical equivalence* posits that if two formulas are logically equivalent, then their respective defeasible consequences must also be identical.

$$(LLE) \quad \frac{\alpha \equiv \beta, \ \mathcal{K} \mathrel{\approx\!\!\!\!\mid} \alpha \mathrel{\mid\!\sim} \beta}{\mathcal{K} \mathrel{\approx\!\!\!\!\mid} \beta \mathrel{\mid\!\sim} \gamma} \tag{5}$$

Lemma 2. *Naive bayes entailment satisfies left logical equivalence.*

Proof. Should $\alpha \equiv \beta$, it follows by definition that $[\![\alpha]\!] = [\![\beta]\!]$, indicating identical evidential fragments. Naive bayes entailment will therefore return the same models for both α and β.

Right Weakening. *Right weakening* asserts that if β is a defeasible consequence of α, then any formula that is logically equivalent to β must likewise be a defeasible consequence of α.

$$(RW) \quad \frac{\mathcal{K} \mathrel{\approx\!\!\!\!\mid} \alpha \mathrel{\mid\!\sim} \beta, \ \beta \models \gamma}{\mathcal{K} \mathrel{\approx\!\!\!\!\mid} \alpha \mathrel{\mid\!\sim} \gamma} \tag{6}$$

Lemma 3. *Naive bayes entailment satisfies right weakening.*

Proof. If $\mathcal{K} \mathrel{\approx\!\!\!\!\mid} \alpha \mathrel{\mid\!\sim} \beta$, then all models returned by naive bayes entailment satisfy β. Given that $\beta \models \gamma$, it follows by definition that $[\![\beta]\!] \subseteq [\![\gamma]\!]$, thereby ensuring that each model also satisfies γ.

And. *And* dictates that the conjunction of any defeasible consequences of a formula should likewise be a defeasible consequence of that formula.

$$(And) \quad \frac{\mathcal{K} \mathrel{\approx\!\!\!\!\mid} \alpha \mathrel{\mid\!\sim} \beta, \ \mathcal{K} \mathrel{\approx\!\!\!\!\mid} \alpha \mathrel{\mid\!\sim} \gamma}{\mathcal{K} \mathrel{\approx\!\!\!\!\mid} \alpha \mathrel{\mid\!\sim} \beta \wedge \gamma} \tag{7}$$

Lemma 4. *Naive bayes entailment satisfies and.*

Proof. If every model returned by naive bayes entailment for α satisfies both β and γ, then they are guaranteed to satisfy $\beta \wedge \gamma$.

Or. *Or* states that a defeasible consequence of two individual formulas should also be a defeasible consequence of their disjunction.

$$(Or) \quad \frac{\mathcal{K} \approx \alpha \mathrel{|\!\sim} \gamma, \ \ \mathcal{K} \approx \beta \mathrel{|\!\sim} \gamma}{\mathcal{K} \approx \alpha \vee \beta \mathrel{|\!\sim} \gamma} \tag{8}$$

Lemma 5. *Naive bayes entailment does not satisfy or.*

Proof. Consider $\mathcal{K} = \{\text{a} \mathrel{|\!\sim} \text{b}, \text{c} \rightarrow \text{a}, \text{c} \mathrel{|\!\sim} \neg \text{b}, \text{d} \rightarrow \text{c}, \text{d} \mathrel{|\!\sim} \text{e}, \text{f} \rightarrow \text{a}, \text{f} \mathrel{|\!\sim} \neg \text{b}\}$, Fig. 7 shows a corresponding reduced $\mathcal{R}^{\mathcal{K}}_{RC}$. Naive bayes entailment identifies $\text{fed}\overline{\text{c}}\overline{\text{b}}\text{a}$ and $\overline{\text{f}}\text{ed}\overline{\text{c}}\overline{\text{b}}\text{a}$ as most probable for c, making $\mathcal{K} \approx_{NB} \text{c} \mathrel{|\!\sim} \text{e}$ true. Naive bayes entailment also identifies $\text{fed}\overline{\text{c}}\overline{\text{b}}\text{a}$ as most probable for f, making $\mathcal{K} \approx_{NB} \text{f} \mathrel{|\!\sim} \text{e}$ true. However, the most probable models for c $\vee$ f are $\text{fed}\overline{\text{c}}\overline{\text{b}}\text{a}$ and $\overline{\text{f}}\text{ed}\overline{\text{c}}\overline{\text{b}}\text{a}$, making $\mathcal{K} \approx_{NB} \text{c} \vee \text{f} \mathrel{|\!\sim} \text{e}$ false.

2	$\overline{\text{e}}\text{dcba}$	$\text{fd}\overline{\text{c}}\text{ba}$	cba
1	$\text{d}\overline{\text{c}}\text{ba}$	$\overline{\text{d}}\text{cba}$	edcba
0	$\text{fd}\overline{\text{c}}\overline{\text{b}}\text{a}$	$\text{fd}\overline{\text{c}}\text{ba}$	$\overline{\text{f}}\text{d}\overline{\text{c}}\text{ba}$

Fig. 7. Reduced minimal ranked interpretation $\mathcal{R}^{\mathcal{K}}_{RC}$.

Cautious Monotonicity. *Cautious monotonicity* posits that the integration of newly deduced information should not negate or undermine any previously inferable conclusions.

$$(CM) \quad \frac{\mathcal{K} \approx \alpha \mathrel{|\!\sim} \beta, \ \ \mathcal{K} \approx \alpha \mathrel{|\!\sim} \gamma}{\mathcal{K} \approx \alpha \wedge \beta \mathrel{|\!\sim} \gamma} \tag{9}$$

Lemma 6. *Naive bayes entailment does not satisfy cautious monotonicity.*

Proof. Consider again the knowledge from the previous proof. Both $\mathcal{K} \approx_{NB} \text{c} \mathrel{|\!\sim} \text{e}$ and $\mathcal{K} \approx_{NB} \text{c} \mathrel{|\!\sim} \neg \text{d}$ are true. However, naive bayes entailment identifies $\text{fed}\overline{\text{c}}\overline{\text{b}}\text{a}$, $\overline{\text{f}}\text{ed}\overline{\text{c}}\overline{\text{b}}\text{a}$, $\text{fed}\text{c}\overline{\text{b}}\text{a}$, and $\overline{\text{f}}\text{ed}\text{c}\overline{\text{b}}\text{a}$ as most probable for c $\wedge$ e, making $\mathcal{K} \mid \approx_{NB} \text{c} \wedge \text{e} \mathrel{|\!\sim} \neg \text{d}$ false.

Rational Monotonicity. *Rational monotonicity* asserts that incorporating information which was not previously negated by existing knowledge should not cause the retraction of any prior conclusions.

$$(RM) \quad \frac{\mathcal{K} \approx \alpha \mathrel{|\!\sim} \gamma, \ \ \mathcal{K} \not\approx \alpha \mathrel{|\!\sim} \neg \beta}{\mathcal{K} \approx \alpha \wedge \beta \mathrel{|\!\sim} \gamma} \tag{10}$$

Lemma 7. *Naive bayes entailment does not satisfy rational monotonicity.*

Proof. Proving that naive bayes entailment doesn't satisfy cautious monotonicity concurrently establishes its non-compliance with rational monotonicity.

References

1. Akers: Binary decision diagrams. IEEE Trans. Comput. **C-27**(6), 509–516 (1978). https://doi.org/10.1109/TC.1978.1675141
2. Bayes, T.: An essay towards solving a problem in the doctrine of chances. Philos. Trans. Roy. Soc. London **53**, 370–418 (1763)
3. Ben-Ari, M.: Propositional logic: formulas, models, tableaux. In: Ben-Ari, M. (ed.) Mathematical Logic for Computer Science, pp. 7–47. Springer, London (2012). https://doi.org/10.1007/978-1-4471-4129-7_2
4. Bollig, B., Wegener, I.: Improving the variable ordering of OBDDs is NP-complete. IEEE Trans. Comput. **45**(9), 993–1002 (1996). https://doi.org/10.1109/12.537122
5. Bryant, R.E.: Graph-based algorithms for Boolean function manipulation. IEEE Trans. Comput. **C-35**(8), 677–691 (1986). https://doi.org/10.1109/TC.1986.1676819
6. Bryant, R.E.: Binary decision diagrams: an algorithmic basis for symbolic model checking (2019). https://api.semanticscholar.org/CorpusID:202740601
7. Casini, G., Meyer, T., Varzinczak, I.: Taking defeasible entailment beyond rational closure. In: Calimeri, F., Leone, N., Manna, M. (eds.) JELIA 2019. LNCS (LNAI), vol. 11468, pp. 182–197. Springer, Cham (2019). https://doi.org/10.1007/978-3-030-19570-0_12
8. Drechsler, R., Becker, B., Göckel, N.: Genetic algorithm for variable ordering of OBDDs. In: IEE Proceedings of the Computers and Digital Techniques, vol. 143, pp. 364–368 (1996). https://doi.org/10.1049/ip-cdt:19960789
9. Freund, M.: Preferential reasoning in the perspective of Poole default logic. Artif. Intell. **98**(1–2), 209–235 (1998)
10. Giordano, L., Gliozzi, V., Olivetti, N., Pozzato, G.: Semantic characterization of rational closure: from propositional logic to description logics. Artif. Intell. **226**, 1–33 (2015). https://doi.org/10.1016/j.artint.2015.05.001, https://www.sciencedirect.com/science/article/pii/S0004370215000673
11. Goldszmidt, M., Pearl, J.: System-Z+: a formalism for reasoning with variable-strength defaults. In: Proceedings of the Ninth National Conference on Artificial Intelligence, AAAI 1991, vol. 1, p. 399–404. AAAI Press (1991)
12. Kraus, S., Lehmann, D., Magidor, M.: Nonmonotonic reasoning, preferential models and cumulative logics. Artif. Intell. **44**(1), 167–207 (1990). https://doi.org/10.1016/0004-3702(90)90101-5, https://www.sciencedirect.com/science/article/pii/0004370290901015
13. Laplace, P.S.: Théorie Analytique des Probabilités. Courcier (1812)
14. Lee, C.Y.: Representation of switching circuits by binary-decision programs. Bell Syst. Tech. J. **38**(4), 985–999 (1959). https://doi.org/10.1002/j.1538-7305.1959.tb01585.x
15. Lehmann, D.: Another perspective on default reasoning. Ann. Math. Artif. Intell. **15**, 61–82 (1995)
16. Lehmann, D., Magidor, M.: What does a conditional knowledge base entail? Artif. Intell. **55**(1), 1–60 (1992). https://doi.org/10.1016/0004-3702(92)90041-U, https://www.sciencedirect.com/science/article/pii/000437029290041U
17. Pearl, J.: System Z: a natural ordering of defaults with tractable applications to nonmonotonic reasoning. In: Proceedings of the 3rd Conference on Theoretical Aspects of Reasoning about Knowledge, TARK 1990, pp. 121–135. Morgan Kaufmann Publishers Inc., San Francisco (1990)

18. Rudell, R.: Dynamic variable ordering for ordered binary decision diagrams. In: Proceedings of 1993 International Conference on Computer Aided Design (ICCAD), pp. 42–47 (1993). https://doi.org/10.1109/ICCAD.1993.580029
19. Valiant, L.G.: The complexity of enumeration and reliability problems. SIAM J. Comput. **8**(3), 410–421 (1979). https://doi.org/10.1137/0208032
20. Wolpert, D., Macready, W.: No free lunch theorems for optimization. IEEE Trans. Evol. Comput. **1**(1), 67–82 (1997). https://doi.org/10.1109/4235.585893

Sequence Based Deep Neural Networks for Channel Estimation in Vehicular Communication Systems

Simbarashe A. Ngorima[1,2,3]([✉])(ID), Albert S. J. Helberg[1](ID),
and Marelie H. Davel[1,2,3](ID)

[1] Faculty of Engineering, North-West University, Potchefstroom, South Africa
aldringorima@gmail.com, albert.helberg@nwu.ac.za
[2] Centre for Artificial Intelligence Research, Pretoria, South Africa
[3] National Institute for Theoretical and Computational Sciences,
Stellenbosch, South Africa

Abstract. Channel estimation is a critical component of vehicular communications systems, especially in high-mobility scenarios. The IEEE 802.11p standard uses preamble-based channel estimation, which is not sufficient in these situations. Recent work has proposed using deep neural networks for channel estimation in IEEE 802.11p. While these methods improved on earlier baselines they still can perform poorly, especially in very high mobility scenarios. This study proposes a novel approach that uses two independent LSTM cells in parallel and averages their outputs to update cell states. The proposed approach improves normalised mean square error, surpassing existing deep learning approaches in very high mobility scenarios.

Keywords: Channel estimation · deep learning · dual-cell LSTM · IEEE 802.11p · vehicular channels

1 Introduction

The international wireless communication standard IEEE 802.11p was introduced for vehicle-to-vehicle communication. However, the original IEEE 802.11p framework did not take into account high mobility and rapid channel changes, which can affect system performance at high speeds [15]. Reliable wireless vehicular communication is challenging due to the need for accurate channel state information (CSI) in high-mobility scenarios, where estimates can quickly become outdated.

IEEE 802.11p employs a data pilot-aided (DPA) approach for channel estimation, with techniques such as spectral temporal averaging (STA) [3] and time-domain reliable test frequency domain interpolation (TRFI) [11] estimators to improve accuracy. However, in high signal-to-noise ratios (SNRs), STA may not effectively account for frequency and time variability while TRFI's assumption of high correlation between successive transmitted signals may not hold true.

A. Pillay et al. (Eds.): SACAIR 2023, CCIS 1976, pp. 172–186, 2023.
https://doi.org/10.1007/978-3-031-49002-6_12

The ability of a deep neural network (DNN) to generalize in a robust manner has made it an attractive option to improve channel estimation. DNN algorithms have been investigated as a means of improving traditional channel estimation techniques such as STA and TRFI. These DNN-based postprocessors have excelled at capturing time-varying behaviours in complex propagation environments. An auto-encoder (AE) DNN successfully handled the inherent error propagation challenges in DPA estimation [8]. Another approach combined STA for initial linear channel estimation with a deep neural network (DNN) as a nonlinear postprocessor [5]. In Gizzini et al. [6] a TRFI-DNN was introduced that combines TRFI and a DNN. TRFI-DNN has been shown to be more effective than other DNN methods compared with.

Performance-wise, AE-DNN [8] has a relatively low bit error rate (BER). However, it is computationally complex and does not effectively learn the time-frequency channel correlation. STA-DNN [5] outperforms both AE-DNN and TRFI-DNN [6] in low SNRs of up to 20 dB at a velocity of 120 km/h. However, its performance starts to deteriorate as the SNR increases. TRFI-DNN, on the other hand, outperforms AE-DNN by approximately 3 dB for BER = 10^{-3}. In scenarios where mobility is higher than 120 km/h, STA-DNN starts to experience an error floor at even lower SNRs of 17 dB. However, the challenge of dealing with a rapidly changing channel remains for both temporal filtering and DNN postprocessing.

To estimate wireless channels in high-mobility situations, we propose using a dual-cell long-short-term memory (LSTM) network. This network can capture short-term temporal dependencies and causal correlations. A DPA block and a temporal averaging (TA) block are used to further reduce the noise in the dual-cell LSTM estimates.

The remainder of the paper is structured as follows: Sect. 2 reviews IEEE 802.11p and channel estimation, Sect. 3 introduces the dual-cell LSTM approach, Sect. 4 describes the experiments, Sect. 5 presents results and analysis, and Sect. 6 concludes.

2   Background

This section provides a brief overview of the IEEE 802.11p standard specification. It also discusses channel estimation system models and techniques within the context of IEEE 802.11p, serving as the foundation for improving vehicular communication systems.

2.1   IEEE 802.11p Standard Specifications and Frame Structure

The IEEE 802.11p protocol employs a technique referred to as Orthogonal Frequency Division Multiplexing (OFDM) to transfer data via radio channels. OFDM divides the available bandwidth into several closely spaced subcarrier frequencies, allowing many signals to be sent simultaneously. During transmission, some subcarriers carry pilot signals that are known by both the transmitter and the receiver. The receivers use pilots to analyse changes in the communication channel. As vehicles travel at different speeds, factors such as distance, barriers, and multipath effects constantly change the way signals propagate. These pilot signals can be used to track time-varying channel conditions and obtain the CSI. This CSI can then be used to improve the decoding of transmitted user data, ensuring efficient communication.

OFDM signal consists of modulation symbols sent on an active subcarrier during a particular time interval. OFDM frames are made up of several payload OFDM signals that are preceded by preamble symbols that fulfil various key roles. The preamble enables the receiver to synchronise, establish an initial coarse channel estimate, and mark the beginning of incoming frames.

In IEEE 802.11p, only 52 of the 64 available subcarriers are used for data transmission and pilot symbols, as illustrated in Fig. 1. The rest of the subcarriers are kept for other purposes, such as guard bands and direct current (DC) offset. To ensure accurate channel tracking, pilot signals are inserted into specific subcarriers.

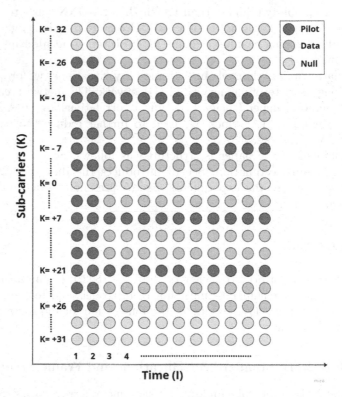

Fig. 1. Subcarrier arrangements in a IEEE802.11p transmission [5].

2.2   Channel Estimation System Model

This study assumes perfect synchronisation and only considers a frame structure comprising two extended preambles at the start, followed by I OFDM data symbols. Following the terminology used in [4], the expression of the transmitted and received OFDM symbol can be represented as follows:

$$\tilde{y}_i[k] = \begin{cases} \tilde{h}_{i,d}[k]\tilde{x}_{i,d}[k] + \tilde{v}_{i,d}[k], & k \in \mathcal{K}_d \\ \tilde{h}_{i,p}[k]\tilde{x}_{i,p}[k] + \tilde{v}_{i,p}[k], & k \in \mathcal{K}_p \end{cases}, \tag{1}$$

where $\tilde{x}_i[k]$ denotes the OFDM symbol transmitted at the ith instance, affected by its corresponding time-varying frequency-domain channel response $\tilde{h}_i[k]$. $\tilde{y}_i[k]$ are the OFDM data symbols received. Furthermore, $\tilde{v}_i[k]$ denotes the frequency-domain equivalent of additive white Gaussian noise (AWGN) characterised by a variance of σ^2. The indices K_d and K_p refer to the sets of subcarriers dedicated to data and pilot signals, respectively.

Thus, the OFDM symbol received ($\tilde{y}_i[k]$) consists of both the data and the pilot subcarrier symbols affected by the channel response, with noise added. Channel estimation is the task of determining $\tilde{h}_i[k]$ for each OFDM symbol.

2.3   IEEE 802.11p Channel Estimation Schemes

Accurate channel estimation in OFDM systems such as IEEE 802.11p relies on well-allocated pilot subcarriers. However, the standard's limited number of pilots may not suffice for tracking channel variations effectively. Two proposed channel estimation schemes tackle this issue: one uses data subcarriers alongside pilots, while the other inserts additional pilots at the cost of a reduced transmission rate. Achieving reliable channel estimation requires a careful balance between accuracy and transmission efficiency.

Recently, models based on recurrent neural networks (RNNs) have shown promise in leveraging sequential dependencies in channel data for channel estimation in vehicular communication [4, 10, 13]. Among RNN architectures, LSTM networks can learn long-term dependencies and have shown excellent results in sequential tasks [9].

In the context of channel estimation, an LSTM has been used to directly map pilot sequences to channel responses [13], while Gizzini et al. proposed a long-short-term memory data pilot-aided temporal averaging (LSTM-DPA-TA) integration to improve estimates in high mobility situations [4]. Hou et al. [10] also employed a single-layer Gated Recurrent Unit (GRU) as a post-processing for the estimation of DPA.

DPA Estimation. The DPA estimation technique uses both pilot subcarriers and demapped data subcarriers to estimate the channel for the present OFDM symbol [8]. Following the notation in [4] the demapped data symbol $d_i[k]$ for each subcarrier $\mathcal{K}$ is given in Eq. 2 as follows:

$$d_i[k] = \mathfrak{D}\left(\frac{y_i[k]}{\hat{\tilde{h}}_{DPA_{i-1}}[k]}\right), \hat{\tilde{h}}_{DPA_0}[k] = \hat{\tilde{h}}_{LS}[k], k \in \mathcal{K}_{on}, \tag{2}$$

Here, $\mathfrak{D}(.)$ signifies the process of demapping to the closest constellation point based on the chosen modulation order, $\mathcal{K}_{on}$ are the active subcarriers, $\tilde{h}_{LS}[k]$ is the channel estimated using the Least Squares (LS) method from the received preambles, namely $y_1^{(p)}[k]$ and $y_2^{(p)}[k]$ such that:

$$\tilde{h}_{LS}[k] = \frac{y_1^{(p)}[k] + y_2^{(p)}[k]}{2p[k]}, k \in \mathcal{K}_{on}, \tag{3}$$

In this context, $p[k]$ denotes the predetermined frequency-domain preamble sequence. Subsequently, the final updates for the DPA channel estimation are carried out in the following manner:

$$\tilde{h}_{DPA_i}[k] = \frac{y_i[k]}{d_i[k]}, k \in \mathcal{K}_{on}. \tag{4}$$

It should be noted that DPA fundamental estimation serves as an initial reference for a majority of IEEE 802.11p channel estimation techniques [10].

TA Processing. Temporal averaging (TA) is a noise reduction technique in which the noise reduction ratio can be calculated analytically [4]. The TA method assumes that the noise terms of consecutive OFDM symbols are not correlated. TA analyses variations by averaging the channel estimations across OFDM symbols. Following the notation in [4], consider the estimate of the channel $h_{k,n}$ in the subcarrier k and symbol n. TA computes a moving estimate $\hat{H}_k$ of the channel frequency response h_k as the weighted sum:

$$\hat{H}_{k,n} = (1 - 1/\alpha)\hat{h}_{k-1,n} + (1/\alpha)\hat{h}_{k,n} \tag{5}$$

In this case, α denotes the window size over which the averages are calculated. A larger α incorporates more symbols. TA reduces noise through averaging, making it a good candidate for postprocessing.

STA-DNN Estimator. Previous work [5] recognised the problems in only using traditional STA estimates [3] and proposed a hybrid STA-DNN technique to solve them. The typical STA estimator averages the initial estimate of the DPA channel over frequency and time.

To compensate for the time-varying channel conditions, STA uses fixed average window widths and weights [3]. To counteract this, Gizzini et al. [5] suggest creating an STA-DNN estimator by feeding the STA estimate to a deep neural network (DNN). Their evaluation showed that the STA-DNN technique corrects errors and greatly increases the accuracy of the estimation over the traditional STA estimation [5]. Despite these improvements, there is still a significant error in high-mobility vehicle situations at low SNR [4].

LS Channel Estimate. In IEEE 802.11p, the basic LS channel estimation technique is used as initial channel estimation. Two successive training symbols T_1 and T_2 are used to obtain the initial frequency response of the channel of the kth subcarrier as follows:

$$\tilde{H}_0(k) = \frac{Y_{T1}(k) + Y_{T2}(k)}{2X_T(k)}, k \in \mathcal{K}_{\text{on}} \tag{6}$$

where $X_T(k)$ is the predefined training symbol of the kth subcarrier, $Y_{T1}(k)$ and $Y_{T2}(k)$ are the symbols in the frequency domain of the receiver. The LS channel estimation approach overlooks time variations, resulting in reduced accuracy as the OFDM symbol index increases. To maintain reliable performance in mobile scenarios, more advanced channel tracking techniques are essential, accounting for dynamic channel changes and ensuring accurate estimation.

The channel estimation schemes discussed are used in this work for comparison purposes. LS estimation is used as an initial estimation approach before our dual-cell LSTM method. This is followed by applying DPA as a post-processing step to refine our dual-cell LSTM estimate. A step-by-step procedure on how these techniques are used in our work is discussed in detail in Sect. 3.

3 Proposed Dual-Cell LSTM Estimation Method

LSTM networks are designed for sequential data with temporal dependencies. These networks have an architecture that allows them to understand the temporal dependencies in the data, enabling them to predict future data based on past observations [9]. The gated cell structure of LSTMs enables them to learn long-term temporal relations by regulating the input of new information, eliminating unnecessary past information, and updating the hidden state using selected cell state values [14]. LSTMs have been successfully used in different domains and applications such as time series forecasting [12] and speech recognition [7].

This section introduces a novel dual-cell LSTM architecture that uses two independent LSTM cells in parallel to capture sequential information from different temporal perspectives. This enables the model to fuse representations learnt from different, non-overlapping temporal perspectives at each timestep. The method also integrates DPA and TA as post-processing methods to reduce noise.

The dual-cell structure is distinct from a bidirectional LSTM as it is composed of two parallel LSTMs operating in the forward direction, while a bidirectional LSTM comprises two LSTMs, one processing the sequence in a forward direction and the other in a backward direction. Our hypothesis is that this dual-cell architecture is capable of capturing a more complete image of channel dynamics compared to other LSTM structures. The proposed protocol aims to efficiently address channel estimation in highly dynamic environments, such as vehicular communication.

As illustrated in Fig. 2, the inputs (x_{t-1}, x_{t-2}, x_{t-3}, etc.) are supplied directly to the independent LSTM cells at the same time. Following the typical LSTM

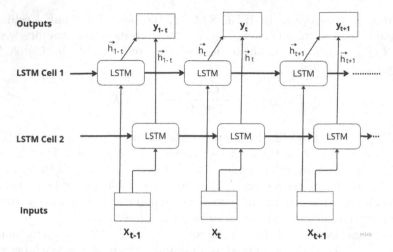

Fig. 2. Schematic representation of the dual-cell LSTM estimator. The architecture consists of two parallel LSTM cell chains, with each cell processing sequential inputs from x_{t-1} to x_t. The outputs from both chains are fused to produce the final output.

design [9], each LSTM cell consists of an input gate, a forget gate, an output gate, and a cell memory block. For both cells, the number of hidden units is set to H. At each time step t, the same input sequence x_t is fed into both cells to update their hidden states ($h_{1,t}$ and $h_{2,t}$) and cell states ($c_{1,t}$ and $c_{2,t}$). This processing takes place in parallel, without information exchange between cells. Finally, a combined output y_t is produced by averaging the hidden state outputs of both cells and passing them through a linear layer. This captures representations from the two processing streams, although independently and simultaneously. Averaging the outputs of the two cells has various benefits:

- Combining the representations learned by each cell helps to prevent overfitting. Cells may capture somewhat different features of the input sequence.
- Averaging provides a type of ensembling in which the combination of numerous models (cells) outperforms isolated ones.
- By averaging time-distributed outputs at each step, the model may use information processed in both cells at the same time.

3.1   Dual-Cell LSTM Estimation Algorithm

This section presents the algorithm for the proposed dual-cell LSTM model. (Also see Algorithm 1.)

The Dual-Cell LSTM model for the channel estimation method can be broken down into four phases:

- First, an LS channel estimation is performed. Initial channel estimation is obtained as illustrated in Sect. 2.3. This forms the input dataset ($\tilde{x}_i$) used to

Algorithm 1. Dual-Cell LSTM Model for Channel Estimation

Data: HLS_Structure, True_Channel_Structure,
   Step 1: LS estimation to obtain the initial channel estimation that is used as
the training dataset (HLS_Structure)
   Step 2: Load dataset
      -     Load True_Channel_Structure
      -     Load HLS_Structure
   Step 3: Prepare input and target matrices
      -     Construct Dataset_X by combining real and imaginary parts of
HLS_Structure.
      -     Construct Dataset_Y by combining real and imaginary parts of
True_Channel_Structure.
   Step 4: Initialize dual LSTM cell model
      -     Create two LSTM cells: cell1 and cell2
      -     Initialise hidden and cell states of both cells
   Step 5: Train model
while not converged **do**
      Step 5.1: Forward pass
      -     Feed Dataset_X to both cells
      -     Get outputs out1 and out2 from cell1 and cell2
      -     Calculate out_avg = (out1 + out2)/2
      -     Calculate loss between out_avg and Dataset_Y
      Step 5.2: Backpropagate loss
      -     Update cell parameters using backpropagation
end while
   Step 6: Channel estimation on new data
      -     Forward pass new samples through trained model
      -     Get channel estimates $e\hat{s}t = out\_avg$
   Step 7: Post-processing
      -     Perform DPA estimation on $e\hat{s}t$ to improve estimates
      -     Perform TA on DPA outputs
      -     Final estimated channel
   Stop 8: Calculate evaluation metrics
      -     Calculate NMSE, BER., on test dataset
   Step 9: Repeat step 4 for the required number of epochs
   Step 10: Save trained model

---

train the model. Following the notation presented in [4], the input $\bar{x}_i$ can be
obtained as follows:

$$\bar{x}_i = \begin{cases} \hat{\bar{h}}_{\text{LSTM}_{i-1,d}}[k], & k \in \mathcal{K}_d \\ \hat{\bar{h}}_{i-1,p}[k], & k \in \mathcal{K}_p \end{cases}. \tag{7}$$

The input $\tilde{\bar{x}}_i$ is computed by transforming LS estimate $\bar{x}_i$ from complex values
to real values. $\hat{\bar{h}}_{i-1,p}[k]$ is the LS estimated channel at the K_p subcarriers. $\tilde{\bar{x}}_i$
is input to the LSTM as:

$$\hat{\bar{h}}_{\text{LSTM}_{i,d}} = \Omega_{\text{LSTM}}(\tilde{\bar{x}}_i,\ \Theta), \tag{8}$$

where Ω_{LSTM} denotes the LSTM processing unit and Θ denotes the overall
weights.

- Second, the LSTM model is trained to estimate the channel characteristics from the received data.
- Third, the DPA post-processing technique is used to improve estimations, as follows:

$$d_{\text{LSTM}_i}[k] = \mathfrak{D}\left(\frac{y_i[k]}{\hat{\bar{h}}_{\text{LSTM}_{i-1}}[k]}\right), \hat{\bar{h}}_{\text{LSTM}_0}[k] = \hat{\bar{h}}_{\text{LS}}[k], \tag{9}$$

$$\hat{\bar{h}}_{\text{LSTM-DPA}_i}[k] = \frac{y_i[k]}{d_{\text{LSTM}_i}[k]}. \tag{10}$$

- Fourth, the TA technique is the final post-processing method: TA is applied to the estimated channel $\hat{\bar{h}}_{\text{LSTM-DPA}_i}[k]$ to further reduce the impact of AWGN noise as follows:

$$\hat{\bar{h}}_{\text{DNN-TA}_{i,d}} = \left(1 - \frac{1}{\alpha}\right)\hat{\bar{h}}_{\text{DNN-TA}_{i-1,d}} + \frac{1}{\alpha}\hat{\bar{h}}_{\text{LSTM-DPA}_{i,d}}. \tag{11}$$

A fixed α value of 2 is used this is based on the analysis done in [4].

Finally, estimates are evaluated using normalised mean squared error (NMSE) and bit error rate (BER) metrics.

4   Experimental Setup

This section discusses the channel model employed for data generation, the data preparation for LSTM and the hyperparameter optimisation process.

4.1   Channel Model

Vehicular channel models, widely examined in the literature [4–6], originate from channel measurements in metropolitan Atlanta, Georgia, USA [1]. We conducted an analysis of the vehicle-to-vehicle same direction with wall (VTV-SDWW) tapped delay line (TDL) vehicular channel model, which is used for communication between two vehicles travelling in the same direction with a central wall between them and maintaining a distance of 300–400 m between them.

Two mobility scenarios based on the VTV-SDWW TDL were adopted for comparison: (i) high mobility (V = 100 km/h, Doppler shift f_d = 550 Hz) and (ii) very high mobility (V = 200 km/h, f_d = 1,100 Hz). Simulation parameters included a frame size of 50 OFDM symbols, 16QAM modulation, and convolutional channel coding at a half-code rate. The dataset consisted of 12,000 training samples, 4,000 validation samples, and 2,000 testing samples. During training, a training SNR level of 40 dB is used to improve the generalization of the model. Each simulation generated a packet of 50 OFDM symbols sent across a simulated wireless channel, capturing varying channel conditions. In this study, a 'packet sample' represents one 50-OFDM symbol packet, serving as a single observation for estimation.

4.2   Data Preparation for LSTM

The dataset that was created according to the IEEE 802.11p standard (see Sect. 2). We excluded the signal field from the dataset, assuming optimal receiver synchronisation, for the sake of simplicity. As a result, the emphasis is on the presence of two extended training symbols within each transmitted frame, followed by a set of I Orthogonal Frequency Division Multiplexing (OFDM) data symbols. The received OFDM symbol is represented by Eq. 1.

We constructed $Dataset\_X$ and $Dataset\_Y$ matrices for the input and target data, respectively. For training data, we obtained a real-valued training dataset $Dataset\_X$ by concatenating the real and imaginary parts of the propagating channel ($HLS\_Structure$) into one vector. $Dataset\_Y$ is the target output matrix created similarly to $Dataset\_X$ by concatenating the real and imaginary parts of the actual or ground truth channel for specific positions, 1 to 48 for real and 49 to 96 for imaginary.

$Dataset\_X$ has dimensions $12,000 \times 50 \times 104$. Similarly, $Dataset\_Y$ has dimensions $12,000 \times 50 \times 96$. The first dimension is the size of the dataset. The second dimension denotes the number of OFDM symbols per frame, while the third is the sequence length, which is the size of the input data positions if it is $Dataset\_X$ or output if its $Dataset\_Y$. Our input size is 104 and our output size is 96 (as above).

4.3   Hyperparameter Optimisation

The hyperparameters of the dual-cell LSTM model were optimised using Optuna to find the values that minimise loss of validation during training. Optuna is a Bayesian optimisation framework that generates hyperparameter configurations using a tree-structured parzen estimator (TPE) to maintain a probabilistic model that links hyperparameters to measurable outcomes [2]. It terminates underperforming tests early to maximise search efficiency. The procedure is repeated until either the desired convergence is achieved or the maximum number of trials is reached.

We optimised a set of hyperparameters, namely the learning rate, the step size of the optimiser, the gamma of the $StepLR$ scheduler, the batch size, the weight decay and the dropout probability. The optimisation procedure was carried out as follows:

1. A dual LSTM model with fixed input size (104) was defined. This is the total number of input subcarriers used in this work ($K_{on} * 2$ where $K_{on} = 52$ active subcarriers). The LSTM size was 128.
2. The non-dominated Sorting Genetic Algorithm II (NSGA-II) sampler created an Optuna study for multi-objective optimisation.
3. An objective function trained the dual-cell LSTM for 100 epochs with specified hyperparameters (Table 1).
4. Optuna recommended hyperparameters within defined limits, and Adam optimiser initialised and trained the dual-cell LSTM.

5. Training and validation losses were tracked for each epoch, with the average validation loss used.
6. The study ran 150 optimisation trials to identify optimal hyperparameters based on the lowest validation loss.
7. The loss curve, weight, and bias plots were logged to the weights and bias[1] for visualisation.
8. Optimal trial hyperparameters (learning rate, step size, gamma, dropout, weight decay) were determined on the basis of the lowest validation loss.

The final model is trained for 250 epochs using a batch of 128. A summary of the parameters of the dual cell LSTM model used in this work is given in Table 1.

Table 1: Optuna Hyperparameter optimisation: 150 trials and 100 epochs

Hyperparameters	Search space	Final value
Learning rate	[1e−5, 1e−1]	0.01
Step size	[1, 10]	9
gamma	[0.1, 1]	0.6
Dropout probability	[0.0, 0.8]	0.002
Weight decay	[1e−6, 1e−2]	0.008
Batch size	[16, 32, 48, ..., 128]	128

5   Analysis and Simulation Results

We evaluate the following channel estimation methods, STA-DNN [5], LSTM-DNN-DPA [13], LSTM-DPA-TA [4], DPA-TA and the proposed dual-cell LSTM estimator against a perfect channel by calculating BER and NMSE in MATLAB. The hidden layer dimensions of the STA-DNN were 15-15-15 and 128-40 for the LSTM-DNN.

1. **LSTM-DNN-DPA:** This method involves cascading an LSTM and a multi-layer perceptron network (MLP) to estimate the channel and DPA as a post processor [13]. However, overfitting was not evaluated in this work, as cascaded models possess a high risk of overfitting due to an increased number of parameters.
2. **LSTM-DPA-TA (128):** This approach uses an LSTM model for channel tracking followed by DPA and TA for noise reduction. Overall performance is based on the precision of the initial LSTM estimate. The noise mitigation ratio of TA processing has been analytically derived in [4]. This method was not evaluated against other LSTM architectures.

---

[1] https://wandb.ai/.

3. **LSTM-DPA-TA (64):** This is the same LSTM-based method as above but using a smaller LSTM: size 64 instead of 128 [4]. A smaller LSTM size struggles to learn long-term dependencies in dynamic environments compared to a larger LSTM.
4. **STA-DNN:** This technique utilises DNNs to capture additional temporal and frequency correlations [5]. However, this method does not consider sequential data, thus limiting its capability.
5. **DPA-TA:** We established our lower bound by using only the DPA and TA techniques. The shortcoming of this approach is that it only uses conventional techniques which are limited to learning patterns in data, especially dynamic channels.
6. **Dual-cell LSTM:** We evaluate the performance of our dual-cell LSTM network approach against the other methods in the same scenarios discussed below.

5.1 High Mobility Scenario

In this section, we evaluate the performance of existing approaches and our proposed dual-cell LSTM method under high mobility conditions, characterised by a velocity (v) of 100 km/h and a Doppler frequency (f_d) of 550 Hz using two key metrics; BER and NMSE.

Clarification on Data Comparison: The results of existing methods have been extracted directly from Gizzen et al. [4], while the dataset used to develop the dual-cell LSTM method was generated in-house following the experimental setup detailed in that respective paper. The performance of the proposed method is compared against the stated published performance of the methods presented in [4]. We caution that in this comparison, the datasets generated according to Sect. 4.2 and the unpublished dataset in [4] may differ, although the described vehicular scenarios were precisely duplicated. To reflect this caution, we present the comparison in separate graphs in Figs. 3 and 4.

Figures 3a and 3b show the BER performance results of the DNN-based estimators and the classical DPA-TA estimator. Our experiments showed that the dual-cell LSTM was capable of dealing with a variety of SNR levels. At SNR levels between 0 and 12 dB, it performed similarly to LSTM-DPA-TA (128). When the SNR was between 27 and 40 dB, it outperformed all the models tested. Figures 3c and 3d compare the dual-cell LSTM NMSE performance with that of existing techniques. Lower NMSE values indicate a better estimate of the true channel response. At an NMSE of 10^{-2}, the dual-cell LSTM method shows better performance than LSTM-DA-TA (128), LSTM-DA-TA (64) and LSTM-DNN by approximately 7 dB, 8 dB, and 12 dB improvements, respectively. At 35 dB SNR, it reaches an NMSE of 10^{-3}.

5.2 Very High Mobility Scenario

This section evaluates channel estimation methods under very high mobility conditions of 200 km/h and a frequency Doppler of 1,100 Hz. BER and NMSE

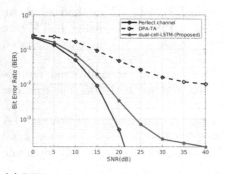
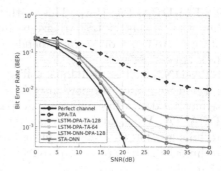

(a) BER performance of the proposed dual-cell LSTM method.

(b) BER performance results of existing methods.

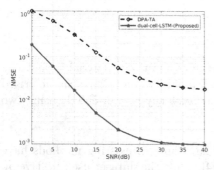
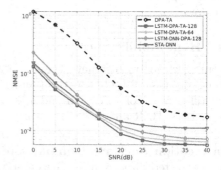

(c) NMSE performance of the proposed dual-cell LSTM method.

(d) NMSE performance of existing methods.

Fig. 3. BER and NMSE results under High Mobility conditions using the VTV-SDWW channel model at $v = 100$ km/h, $f_d = 550$ Hz.

metrics are used to evaluate performance. We again follow the experimental setup of Grizzini et al. [4] to simulate the very high mobility environment, and present results separately. Figures 4a and 4b plot the BER against SNR, where all estimators showed reduced performance due to fast fading that causes a decrease in the temporal correlations of the channel.

At SNR levels ranging from 0 to 15 dB, the LSTM-DPA-TA (128) approach surpasses the dual-cell LSTM technique by up to 0.5 dB. On the other hand, the BER of feedforward models such as STA-DNN deteriorates significantly more rapidly even at high SNR due to its inability to exploit temporal dependencies in highly dynamic channels. The corresponding NMSE vs SNR graphs are shown in Figs. 4c and 4d. Similarly, despite the challenges of extremely high mobility, dual-cell LSTM showed a capability to achieve a very low NMSE by a significant margin compared to the other existing DNN methods.

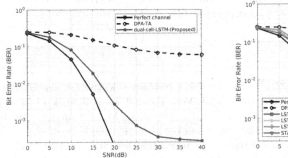 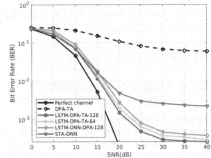

(a) BER performance of the proposed dual-cell LSTM method.

(b) BER performance results of existing methods.

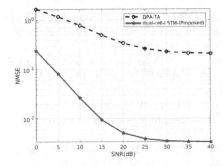 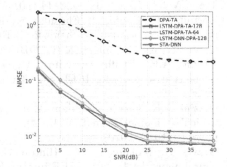

(c) NMSE performance of the proposed dual-cell LSTM method.

(d) NMSE performance of existing methods.

Fig. 4. BER and NMSE performance at Very High Mobility: VTV-SDWW channel model at V = 200km/h, f_d = 1,100 Hz.

6 Conclusion

This paper introduced a dual-cell LSTM network model for dynamic channel estimation in vehicular communication systems. A comparison was conducted with results available from the literature, with experimental setups duplicated. Our dual-cell LSTM model showed the potential to achieve very low NMSE between estimated and true channel responses across all tested SNR levels of magnitude close to 1 compared to other sequence models. In terms of BER, the dual LSTM method showed improved performance in high SNRs, starting at 30 dB, performing better than the other methods. This result supports our hypothesis that the combination of two LSTM methods in a parallel ensemble can outperform the existing single LSTM method and the feedforward methods. Future work will include evaluating the complexity and learning capacity of ensemble sequential learning methods.

Acknowledgements. The authors are grateful to NITheCS, the Telkom CoE at the NWU and Hensoldt South Africa, who supported this research.

References

1. Acosta-Marum, G., Ingram, M.A.: Six time-and frequency-selective empirical channel models for vehicular wireless LANs. IEEE Veh. Technol. Mag. **2**(4), 4–11 (2007)
2. Akiba, T., Sano, S., Yanase, T., Ohta, T., Koyama, M.: Optuna: a next-generation hyperparameter optimization framework. In: Proceedings of the 25th ACM SIGKDD International Conference on Knowledge Discovery & Data Mining, pp. 2623–2631 (2019)
3. Fernandez, J.A., Borries, K., Cheng, L., Kumar, B.V., Stancil, D.D., Bai, F.: Performance of the 802.11 p physical layer in vehicle-to-vehicle environments. IEEE Trans. Veh. Technol. **61**(1), 3–14 (2011)
4. Gizzini, A.K., Chafii, M., Ehsanfar, S., Shubair, R.M.: Temporal averaging LSTM-based channel estimation scheme for IEEE 802.11 p standard. In: 2021 IEEE Global Communications Conference (GLOBECOM), pp. 01–07. IEEE (2021)
5. Gizzini, A.K., Chafii, M., Nimr, A., Fettweis, G.: Deep learning based channel estimation schemes for IEEE 802.11 p standard. IEEE Access **8**, 113751–113765 (2020)
6. Gizzini, A.K., Chafii, M., Nimr, A., Fettweis, G.: Joint TRFI and deep learning for vehicular channel estimation. In: 2020 IEEE Globecom Workshops (GC Wkshps), pp. 1–6. IEEE (2020)
7. Graves, A., Mohamed, A.R., Hinton, G.: Speech recognition with deep recurrent neural networks. In: 2013 IEEE International Conference on Acoustics, Speech and Signal Processing, pp. 6645–6649. IEEE (2013)
8. Han, S., Oh, Y., Song, C.: A deep learning based channel estimation scheme for IEEE 802.11 p systems. In: ICC 2019–2019 IEEE International Conference on Communications (ICC), pp. 1–6. IEEE (2019)
9. Hochreiter, S., Schmidhuber, J.: Long short-term memory. Neural Comput. **9**(8), 1735–1780 (1997)
10. Hou, J., Liu, H., Zhang, Y., Wang, W., Wang, J.: GRU-based deep learning channel estimation scheme for the IEEE 802.11p standard. IEEE Wirel. Commun. Lett. **12**(5), 764–768 (2023). https://doi.org/10.1109/LWC.2022.3187110
11. Kim, Y.K., Oh, J.M., Shin, Y.H., Mun, C.: Time and frequency domain channel estimation scheme for IEEE 802.11 p. In: 17th International IEEE Conference on Intelligent Transportation Systems (ITSC), pp. 1085–1090. IEEE (2014)
12. Li, Y., Zhu, Z., Kong, D., Han, H., Zhao, Y.: EA-LSTM: Evolutionary attention-based LSTM for time series prediction. Knowl.-Based Syst. **181**, 104785 (2019)
13. Pan, J., Shan, H., Li, R., Wu, Y., Wu, W., Quek, T.Q.: Channel estimation based on deep learning in vehicle-to-everything environments. IEEE Commun. Lett. **25**(6), 1891–1895 (2021)
14. Staudemeyer, R.C., Morris, E.R.: Understanding LSTM - a tutorial into long short-term memory recurrent neural networks. ArXiv abs/1909.09586 (2019). https://api.semanticscholar.org/CorpusID:202712597
15. Zhao, Z., Cheng, X., Wen, M., Jiao, B., Wang, C.X.: Channel estimation schemes for IEEE 802.11 p standard. IEEE Intell. Transp. Syst. Mag. **5**(4), 38–49 (2013)

Comparative Study of Image Resolution Techniques in the Detection of Cancer Using Neural Networks

Oliver Nagaya[1,2]([✉]) [iD], Anban W. Pillay[1,2] [iD], and Edgar Jembere[1,2] [iD]

[1] School of Mathematics, Statistics and Computer Science, University of KwaZulu-Natal, Durban, South Africa
oliver@olivernagaya.com, {pillayw4,jemberee}@ukzn.ac.za
[2] Centre for Artificial Intelligence Research (CAIR), Durban, South Africa

Abstract. The quality and accuracy of deep learning models for image classification have historically relied on the resolution and quality of the images. Super-resolution has improved the resolution and detail of low-resolution lossy images. This study investigates the efficacy of image resolution techniques in enhancing the binary classification efficacy of cancers from histopathologic images using a deep-learning model. A comparative analysis was conducted across five cancer datasets, using ten image upscaling and downscaling resolutions, and utilising six performance metrics beyond accuracy to evaluate the model performance. The extensive dataset collection in cancer-based medical imaging, image upscaling and downscaling permutations, testing, and experimental pipeline results gave the study a cutting-edge understanding of image resolution in cancer-based histopathological image classification. Low-quality or super-resolved medical images do not improve the accuracy of the binary classification of good cancer-based histopathological images. The comprehensive image resolution permutations and their applicability across multiple cancer datasets have yielded state-of-the-art results that can help advance the thinking around super-resolution and image resolution considerations for binary classification of cancer-based histopathological medical images.

Keywords: Deep Learning · Image Classification · Convolutional Neural Networks (CNNs) · Generative Adversarial Network (GAN) · Data Augmentation

1 Introduction and Background

Recent work has shown that super-resolution techniques that improve the resolution of histopathological images [1] result in improved accuracy of the binary classification of cancer-based histopathological images [2]. This study tests this by performing a comprehensive comparative study across five cancer datasets. The work reports on classification performance across various image resolutions.

© The Author(s), under exclusive license to Springer Nature Switzerland AG 2023
A. Pillay et al. (Eds.): SACAIR 2023, CCIS 1976, pp. 187–202, 2023.
https://doi.org/10.1007/978-3-031-49002-6_13

Deep learning approaches have been used to enhance the processing of medical images, resulting in numerous solutions and advances [3]. Convolutional Neural Networks (CNNs) have gained significant attention and popularity, mostly attributed to their remarkable achievements in image classification [4]. Notably, there has been a substantial rise in research focused on CNNs, particularly in medical image analysis. Deep learning has emerged as a highly effective approach for constructing end-to-end models capable of classification across multiple modalities.

The use of deep learning techniques has significantly advanced progress in various computer vision challenges [5]. There has been a notable progression from a 26% error rate observed in machine learning models in 2011 to approximately 3% error rates achieved through deep learning techniques in computer vision [6]. This can be attributed to effectively utilising state-of-the-art graphics processing units (GPUs) for training [7]. Secondly, introducing the Rectified Linear Unit (ReLU) significantly accelerates convergence while maintaining high quality. Lastly, ample data resources, such as ImageNet, facilitate the training of larger models. However, these gains come at a cost of time and computational resources [8].

Image resolution typically pertains to pixel (picture element) resolution, which denotes the quantity of pixels in the image. The horizontal and vertical dimensions commonly denote the measurement of image resolution. An image resolution of 6000 × 4000 means that the image has a width of 6000 pixels and a height of 4000 pixels. The pixel count can be retrieved from the image metadata. Higher pixel counts correspond to higher levels of detail. The acronym PPI denotes "pixels per inch," which serves as a metric for quantifying the pixel density present in an image. Dots per inch (DPI) is a term used in printing, wherein the image is composed of dots rather than pixels, thereby sharing a similar underlying principle [9].

Lossless compression is a technique employed to reduce the size of a file. Importantly, this compression method allows for subsequent information decompression, resulting in an identical replication of the original file. Lossy compression is a viable alternative to lossless compression and is frequently employed in the compression of images. Lossy compression involves discarding certain details inside an image, leading to a significantly reduced file size [10].

There is potential for gaining a better understanding of the effects of the resolution of input images in deep learning applications for medical applications [11]. This is supported by [12] where image quality was a crucial factor in endoscopy image classification that could result in enhanced performance. In [13], the study aimed to examine, present, and analyse the impact of varying sizes of late gadolinium enhancement (LGE) magnetic resonance imaging (MRI) images on training datasets, and found a strong correlation between semi-automatic and completely automated quantification of myocardial infarction (MI) outcomes, particularly in the dataset comprising larger LGE-MRI pictures, compared to the dataset consisting of smaller LGE-MRI images.

Several papers have demonstrated that super-resolution algorithms have been effectively employed in medical imaging, demonstrating their potential to become an innovative and essential pre-processing and post-processing technique in clinical image interpretation [1]. In this work, we set to test the hypothesis that super-resolution will enhance the binary classification efficacy of cancers from histopathologic images using a deep-learning model. We developed a deep-learning model prototype, selected five cited cancer datasets, produced a set of binary classification results on the baselined dataset, super-resolved the images in each dataset and produced a set of binary classification results, downscaled the images in each dataset and produced a set of binary classification results and lastly super-resolved the downscaled images to produce a final set of binary classification results.

The subsequent sections of the paper are structured as follows. Section 2 presents a concise overview of prior research on image classification, super-resolution, image resolution and binary classification of cancer-based medical imaging. In Sect. 3, a comparative analysis of ten image upscaling and downscaling permutations is examined across five cancer datasets. Section 4 presents a comprehensive series of experimental comparisons of the image resolutions and existing state-of-the-art image classification approaches. Section 5 provides concluding thoughts and outlines potential avenues for future research.

2 Literature Review

Image classification is an effective solution that utilises machine learning techniques or deep learning for the purposes of image recognition and categorisation [4]. Deep learning methods undergo training using extensive datasets comprising millions of images.

2.1 Deep Learning for Cancer Diagnosis

Medical imaging serves as a significant and indispensable repository of essential information that is utilised by healthcare professionals [14]. In recent times, emerging technologies have brought forth numerous breakthroughs that enable the efficient utilisation of available information, hence facilitating the generation of enhanced analytical insights. Deep learning approaches have been increasingly utilised to analyse medical images within computer-assisted imaging contexts. These techniques have demonstrated numerous solutions and improvements in analysing such images by radiologists and other specialists [15].

One of the difficulties encountered in maintaining and using large images, with medical images cited as one of the key uses of large images, is the significant increase in the size of the deep learning model [6]. In [11], a significant reduction in image resolution resulted in the loss of valuable information that is necessary for accurate categorisation with respect to radiography. The classification performance of CNNs was considerably affected by poor image resolution for endoscopic deep learning applications, with even minor improvements in performance having the potential to impact patient treatment and results greatly [12]. Image resolution is important for medical images, as confirmed in [16] where the study conducted a systematic evaluation of the potential for improving clinical task performance through the utilisation of high-resolution images as opposed to downscaled low-resolution images, and highlighted the significance of image resolution and feature size in the development of neural networks to automate radiological image classification.

Image interpolation was discussed in [13] as a fundamental technique widely employed in various image-processing tasks to rescale, translate, shrink, or rotate an image. Three methods, in particular, were mentioned based on their acceptable performance and popularity – Nearest Neighbor Interpolation, Bicubic Interpolation and Lanczos Interpolation.

[17] comprehensively reviews image super-resolution methods, including degradation modelling-based algorithms, image pairs-based algorithms, domain translation-based algorithms, and self-learning-based algorithms.

The inherent trade-off in implementing CNNs arises from the fact that optimising CNNs using GPUs can be limited by memory constraints. Specifically, when employing a higher image resolution, the maximum batch size that can be effectively utilised may be reduced. Conversely, a larger batch size can facilitate more accurate gradient calculations with respect to the loss function.

The field of deep learning has demonstrated significant achievements in the domain of cancer diagnosis using medical images [15]. The utilisation of deep learning technology for the purpose of cancer diagnosis using medical images is now a prominent area of research within the domains of artificial intelligence (AI) and computer vision. Cancer detection has witnessed significant advancements in recent years, mostly driven by the quick progress in deep learning techniques. This progress is necessitated by the need for both high accuracy and timeliness in cancer diagnosis, considering the inherent intricacies and complexities associated with medical imaging [18].

2.2 Super-Resolution

Deep learning has concurrently also shown significant potential in super-resolution [19]. Super-resolution is the process by which a low-resolution image is enhanced to generate a high-resolution representation [7]. There is extensive discourse surrounding deep learning-based super-resolution methodologies in medical image processing. This is mostly due to their capacity to attain superior-quality, high-spatial-resolution images without necessitating supplementary scans. In [1], super-resolution algorithms were used for medical-based images with good success and found that the use of super-resolution algorithms could possess the capability to emerge as a novel fundamental pre-processing and post-processing procedure in clinical image analysis.

[2] used low-resolution, high-resolution, and super-resolution images to test the accuracy of the binary classification of histopathological images for a widely cited breast cancer dataset [20] and demonstrated that the approach had a beneficial effect on image resolution, with an increase in classification accuracy from 96.83% for low-quality images to 99.11% for super-resolution images.

Based on the findings around super-resolution and its potential to improve the binary classification of histopathological images, the study aimed to apply the experimentation across more cancer-based datasets and determine if the assumption around the positive benefit of super-resolution techniques could be applicable across cancer-based histopathological images.

3 Method

3.1 Introduction

The study investigated the effect of image resolution on the binary classification of cancer medical images and implemented an efficient deep-learning pipeline to report the accuracy of the binary classification of cancer-based histopathological images. Furthermore, the research aimed to examine the effects of image resolution changes on improving the overall precision of a multi-class binary classification. This section overviews the experimental procedures used to achieve the stated objectives.

3.2 Datasets

To rigorously evaluate the study's hypothesis, it was imperative to secure good-quality datasets. After careful evaluation, we selected a collated augmented dataset from Kaggle that focused on five widely cited cancer datasets [21]. Each dataset contained two classes of benign and malignant data images.

The first dataset was for Acute Lymphoblastic Leukemia (ALL) [22] and was discussed in [23], with two classes of 5,000 images in each class. The second and third dataset was for Lung and Colon Cancer [24]. The fourth dataset was for Breast Cancer [20]. The fifth and final dataset was for Kidney Cancer [25].

Each of the five cancer datasets contained 10,000 images with two classes of images with 5,000 images in each class, with the classes being benign and malignant. The supplementary information for each dataset is included below:

- **Breast Cancer** - The BreaKHis database contains microscopic biopsy images of benign and malignant breast tumors. Images were collected through a clinical study from January 2014 to December 2014. All patients referred to the Prevention and Diagnosis Laboratory, Brazil, during this period, with a clinical indication of breast cancer were invited to participate in the study. The institutional review board approved the study, and all patients gave written informed consent. All the data were anonymised [20].
- **Kidney Cancer** - This research deals with the three major renal disease categories: kidney stones, cysts and tumors. The study gathered and annotated to produce 12,446 CT whole abdomen and urogram images to construct an AI-based kidney diseases diagnostic system. This contributed to the AI community's research scope e.g., modelling digital-twin of renal functions. The collected images were exposed to exploratory data analysis, which revealed that the images from all the classes had the same type of mean colour distribution [25].
- **Lung Cancer** - Health Insurance Portability and Accountability Act of 1996, or the HIPAA Act (HIPAA) compliant and validated seven hundred and fifty total images of lung tissue (250 benign lung tissue, 250 lung adenocar-cinomas, and 250 lung squamous cell carcinomas). Augmentor is a Python library to enhance and create artificial image data. This library was used to expand the dataset to 25,000 images by the following augmentations: left and right rotations (up to 25 degrees, 1.0 probability) and horizontal and vertical flips (0.5 probability) [24].

- **Colon Cancer** - HIPAA compliant and validated 500 total images of colon tissue (250 benign colon tissue and 250 colon adenocarcinomas). Using Augmentor, the dataset was expanded to 25,000 images by the following augmentations: left and right rotations (up to 25 degrees, 1.0 probability) and by horizontal and vertical flips (0.5 probability) [24].
- **Acute Lymphoblastic Leukemia** - The images of this dataset were prepared in the bone marrow laboratory of Taleqani Hospital (Tehran, Iran). This dataset consisted of 3,256 peripheral blood smear (PBS) images from 89 suspected ALL patients, whose blood samples were prepared and stained by skillful laboratory staff. This dataset is divided into two classes benign and malignant. All the images were taken using a Zeiss camera in a micro-scope with 100 x magnification and saved as Joint Photographic Experts Group (JPEG) files. A specialist using the flow cytometry tool made the definitive determination of the types and subtypes of these cells [22].

3.3 Image Pre-processing

We used upscaling and downscaling functionality in our image pre-processing, to test our hypothesis of whether the quality and accuracy of deep learning models for image classification rely heavily on the resolution and quality of the images.

The Lanczos resampling method [26], used for downsizing the images, is a high-quality interpolation method known for preserving image quality [15]. This method was used because it minimises aliasing and produces smoother results than other methods such as bilinear or nearest-neighbor. Furthermore, Lanczos resampling is considered lossless or nearly lossless for practical purposes when downsizing images. The implementation ensured that the image downscaling preserved image quality as far as possible.

Other than upscaling and downscaling functionality, no other image pre-processing was explored and remains a possible addition to the experimentation in future work.

3.3.1 Image Upscaling and Super Resolution

We used the Enhanced Super-Resolution Generative Adversarial Networks (ESRGAN) super-resolution technique, which is an improvement over the original Super-Resolution Generative Adversarial Network (SRGAN), which in turn uses a more advanced architecture to achieve even better results. The SRGAN is a very effective super-resolution technique [27]. It is a deep learning-based approach that uses a generative adversarial network (GAN) to generate high-resolution images from low-resolution input images. SRGAN has shown significant improvements over previous super-resolution techniques regarding image quality and visual fidelity. It achieves this by training the GAN on a large dataset of low-resolution and high-resolution image pairs, which allows it to learn the mapping between the two domains.

In addition, SRGAN incorporates perceptual loss, which uses a pre-trained deep neural network to measure the similarity between the generated high-resolution image and the ground truth high-resolution image. This helps to ensure that the generated images not only have high resolution but also maintain the visual characteristics of the original high-resolution images. SRGAN is a state-of-the-art super-resolution technique with promising results in various applications, including image and video processing, medical imaging, and satellite imagery.

While ESRGAN can enhance the sharpness and overall quality of photos, it is not specifically made to eliminate JPEG compression artefacts. Lossy compression, which JPEG picture encoding uses, usually causes JPEG compression artefacts. Blocky patterns, ringing, and the loss of tiny details are a few examples of these artefacts. Although ESRGAN's main objective is to improve the visual quality of photos by boosting resolution and enhancing details, it is possible that some JPEG compression artefacts will remain after implementation.

3.4 Image Splitting

We split each dataset for the purpose of image classification into three distinct subsets: a training set, a validation set, and a test set.

Each dataset was converted to a Python Panda Dataframe from the list of images on disk, after which it was split using the scikit-learn Python library into different subsets for training, validation and testing purposes.

The larger subset for training and smaller subset for validation comprised 95% of the dataset, or 9,500 images; with 90% (or 8,550 images) being used for the training and 10% (or 950 images) used for the validation. A small subset of 5% (or 500 images) was used for the testing. This data split was consistent across all five cancer datasets.

The training set was employed for the purpose of model training, whereas the validation set was utilised to fine-tune the hyperparameters of the model and mitigate the risk of overfitting. The test set was employed to assess the ultimate performance of the model.

The hyperparameters used to fine-tune the model were:

- rescale – set to 1./255
- rotation_range – set to 40
- width_shift_range and height_shift_range – set to 0.2
- shear_range – set to 0.2
- zoom_range – set to 0.2
- horizontal_flip and vertical_flip – set to True.

The dataset was randomly selected from the shuffled dataset until the necessary proportions were attained for each set. The division of the data was performed once at the onset and maintained consistently throughout the course of the experimental procedure.

A 5-fold cross-validation was utilised to yield a more resilient evaluation of the model's performance since it assesses the model on various subsets of the data. This approach mitigates the potential for overfitting to a singular train-test split, and was needed to maintain a constant level of performance of the model across the dataset.

3.5 Image Classifiers

In the context of the CNNs employed, various image classifiers were tested before selecting ResNet50, a convolutional neural network architecture with significant depth categorised under the ResNet lineage. The ResNet50 architecture, initially introduced by [28], was used since it is the most versatile and dynamic model compared to other models. As per [29], the ResNet50 model is employed in the field of computer vision for the purpose of image classification and object recognition applications. As the level of input complexity escalates, the neural network model exhibits a corresponding increase in complexity. Nevertheless, it is worth noting that when the number of layers increases, there is a potential issue known as the vanishing gradient problem. This difficulty arises when the gradients get extremely small, making it difficult for the first layers to learn during the training phase effectively. To mitigate the issue at hand, the architectural design of ResNet incorporates skip connections, which effectively counteracts the problem of vanishing gradients.

ResNet50 does have the ability to capture more nuanced features from data, at the cost of higher computational and memory resources during both training and inference processes. As per [30], the ResNet structure is notably characterised by its simplicity, effectively addressing the issue of performance deterioration in deep convolutional neural networks when subjected to extremely deep conditions, and exhibits exceptional classification performance.

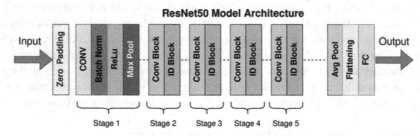

Fig. 1. ResNet50 Architecture [31]

3.6   The Proposed Model

The data is partitioned into three distinct subsets: a larger portion designated for training purposes (train), a smaller subset allocated for validation purposes (valid), and a tiny fraction specifically reserved for testing (test). The training subsets are employed for the purpose of training machine learning models, while the validation subset was utilised to refine the models and make informed decisions regarding hyperparameters. Lastly, the testing subset is employed to assess the ultimate performance of the trained models.

The prototype utilised the ImageDataGenerator class from the Keras library to generate data generators specifically designed for image data. Two data generators were established to process image data:

- The training data generator was responsible for implementing a range of augmentations and preprocessing techniques to improve the diversity and quality of the images used for training.
- The testing data generator was employed to assess the data.

The ResNet50 model was initialised with pre-trained weights from ImageNet, and designed to work with 224 × 224 pixel RGB images and includes all layers up to the point just before the classification layers. The architecture is represented as per [31] in Fig. 1. Additional layers are then appended onto the base model. The model was then subjected to additional training and refinement to optimise its performance for the study's classification objective.

The accuracy metric is not the best performance indicator, although this is acceptable for binary classification tasks. As per [18], the following metrics were also included for a more comprehensive evaluation - precision, recall, F1-score, the confusion matrix, Area Under the Curve (AUC) and Receiver Operating Characteristics (ROC) Curve. The ROC AUC curve measures performance for classification issues at various threshold levels. The AUC reflects the level or measure of separability, and the ROC is a probability curve.

3.7 Experiments Performed

It was important to test each dataset using the developed deep learning pipeline in a rigorous batch of tests. We decided to use image downscaling and upscaling to ensure adequate results could be deciphered. These included the following ten tests on each of the five cancer datasets.

- Original images with an image resolution of 512 × 512 pixels.
- Upscaled images with an image resolution of 2048 × 2048 pixels, using the original images with an image resolution of 512 × 512 pixels.
- Four downscaled image resolutions, using the original images with an image resolution of 512 × 512 pixels.
- Four upscaled image resolutions, using the downscaled images produced in step 3.

4 Results and Discussion

This section details the outcomes and analyses of each experiment conducted. A base-line set of results was compiled for comparison purposes.

4.1 Baseline Set of Results

Experimental results were completed on the original images of the five cancer datasets and are presented in Table 1, along with each of the measurement metrics used. This represented our first experimental result. The resolution for each image was 512 × 512 pixels.

The baseline results show that the accuracy produced by the experimental pipeline was good across all five cancer datasets. Based on these results, we sought to determine whether the quality of image resolution of these images would have a significant impact on the results being produced by the experimental prototype.

Table 1. Baseline results across five cancer datasets, with images of 512 × 512 pixels.

Cancer Dataset	Classes	Accuracy	Precision	Recall	F1-Score	ROC AUC
Breast Cancer	2	86.84%	0.85	0.88	0.86	0.94
Kidney Cancer	2	90.74%	0.92	0.88	0.89	0.97
Lung Cancer	2	99.26%	0.99	1.00	0.99	1.00
Colon Cancer	2	98.11%	0.97	0.99	0.98	1.00
ALL	2	95.05%	0.95	0.94	0.94	0.99

Table 2. Results using super-resolution across five cancer datasets with images of 2048 × 2048 pixels.

Cancer Dataset	Classes	Accuracy	Precision	Recall	F1-Score	ROC AUC
Breast Cancer	2	87.89%	0.87	0.89	0.87	0.94
Kidney Cancer	2	89.26%	0.90	0.89	0.89	0.96
Lung Cancer	2	99.16%	0.99	0.99	0.99	1.00
Colon Cancer	2	98.32%	0.97	0.99	0.98	1.00
ALL	2	94.63%	0.96	0.94	0.95	0.99

4.2 Using Super-Resolution and Image Upscaling on Baseline Medical Images

Our second experiment involved taking the baseline medical images and using the ESGRAN super-resolution technique to produce a dataset that could be tested in the experimental prototype. Table 2 shows these results, along with each performance metric used. The resolution for each image was 2048 × 2048 pixels.

From the experimental results, there is a clear indication that super-resolution can have a positive impact on the accuracy of binary classification of cancer-based images. This is collaborated by the supporting performance metrics. However, due to the already good results on the original images, the accuracy improvement from using the super-resolution technique is marginal. Considering the time and computational resources spent on producing the super-resolution images, it is certainly not a clear-cut decision that using super-resolution will result in a dramatic improvement in the accuracy of the binary classification of cancer-based medical images.

4.3 Impact of Image Downscaling

Our third experiment involved taking the baseline medical images and downscaling each image to produce a dataset that could be tested in the experimental prototype. Table 3 shows these results for colon cancer, along with the performance metrics used.

At an average accuracy of 91.10%, low-quality images do not significantly diminish the accuracy of the binary classification of cancer-based medical images. If this accuracy is compared against the 98.11% accuracy for the baselined original images, and 98.32%

for super-resolution images, we argue that the margin decrease in accuracy does not warrant the increase in time and computational resources. More encouraging regarding the potential widespread use of such models in the future is that low-quality images are not a deterrent to the potential results that can be gained from such models.

The difference in average accuracy of 91.10% vs 98.11% in classification (i.e. 7.01%) is substantial. However, for image resolutions of 64 × 64 and better, there is no significant difference. Thus, there seems to be an inflection point in resolution whereafter there is rapid deterioration in accuracy.

Table 3. Average image downscaling results for colon cancer dataset downscaled from 512 × 512 pixels.

Image Downscaling Permutation	Classes	Accuracy	Precision	Recall	F1-Score	ROC AUC
16 × 16 pixels	2	78.00%	0.85	0.66	0.73	0.88
32 × 32 pixels	2	89.00%	0.93	0.86	0.89	0.96
64 × 64 pixels	2	98.60%	0.95	0.97	0.96	1.00
128 × 128 pixels	2	98.80%	0.98	1.00	0.99	1.00

4.4 Impact of Image Upscaling

Our fourth experiment involved taking the downscaled medical images and upscaling each image to produce a dataset of 1024 × 1024 pixel images that could be tested in the experimental prototype. Table 4 shows these results for colon cancer, along with the measurement metrics used.

At an average accuracy of 92.15%, high-quality images (produced from low-quality) do not significantly diminish the accuracy of the binary classification of cancer-based medical images. However, it is marginally better than low-quality images. There is certainly a case to be made for the effectiveness of super-resolution on the accuracy of binary classification of cancer-based medical images. As with the low-quality images, the margin decrease in accuracy does not warrant increased time and computational resources.

Moreover, the difference in average accuracy of 92.15% vs 98.11% in classification (i.e. 5.96%) is again substantial. However, for image resolutions of 64 × 64 and better, there is again no significant difference. As with the findings of image downscaling, there is an inflection point in resolution whereafter there is rapid deterioration in accuracy.

4.5 Behaviour of the Experimental Pipeline Across Cancer-Based Datasets

The accuracy results, along with the other performance metrics, are consistent across all five cancer datasets. Based on this, we can generalise our results as follows:

Table 4. Average image upscaling results for colon cancer dataset upscaled to 1024 × 1024 pixels.

Image Upscaling Permutation (Downscale Image Resolution)	Classes	Accuracy	Precision	Recall	F1-Score	ROC AUC
16 × 16 pixels	2	79.60%	0.85	0.69	0.74	0.89
32 × 32 pixels	2	91.80%	0.93	0.90	0.91	0.97
64 × 64 pixels	2	98.80%	0.99	0.98	0.99	1.00
128 × 128 pixels	2	98.40%	0.99	0.99	0.99	1.00

- The accuracy of binary classification for cancer-based histopathological images is good provided the images are of good quality and annotated correctly.
- Low-quality and high-quality images do not significantly improve the accuracy of generally good-quality histopathological images.
- The results are consistent across all five cancer-based histopathological images.

4.6 Comparison Against State-of-the-Art

This experiment conducted a comparative analysis between our work and the existing state of the art. The findings in [13] presented fundamental criteria for the evaluation of cancer diagnosis and limited the examination to breast cancer, lung cancer, brain cancer, and skin cancer. This study modelled the performance metrics criteria and agreed with the assertion that previously employed methods for cancer diagnosis are suboptimal, and more advanced and intelligent approaches are needed. Figure 2 depicts sample images from the colon dataset in the image resolutions used in this study.

Table 5 compares current research in cancer diagnosis, and the criteria used to evaluate the model performance. The results confirm that this study employed a more thorough consideration of image resolution techniques and performance metrics to evaluate the binary classification efficacy of cancers from histopathological images.

A limitation of this study is that all performance metrics as per [13] have not been included, and remain an area in which research can be expanded. Notwithstanding this limitation, it remains one of the studies that adheres closely to the principles guiding the accurate evaluation of model performance.

Based on the comprehensive dataset collection in cancer-based medical imaging, the image upscaling and downscaling permutations, the tests that were performed, and the result achieved using the experimental pipeline, the work in the study achieved state-of-the-art in terms of understanding the considerations of image resolution in the image classification for cancer-based histopathological images.

Sufficient research has shown the importance and applicability of super-resolution using deep learning. The use cases of this work have pointed to the application within medical-based histopathological images. Except for [15], little research has clearly articulated the improvement that super-resolution or image resolution has on cancer-based histopathological images. This research has expanded on the current research and presented clear and comprehensive guidance around exploring image resolution for cancer-based histopathological medical images. Due to the comprehensive nature of the image

Table 5. Evaluation of current research related to image resolution and cancer diagnosis.

Cancer Application	Reference	Accuracy	Image Resolution Considered	Performance Metrics Included
Breast Cancer	[32]	98.51%	No	Accuracy, sensitivity, specificity, precision, and F1-score
Breast Cancer	[33]	96.67% to 98.18%	No	Accuracy, sensitivity
Lung Cancer	[34]	82.5% to 90%	No	Accuracy, AUC
Lung Cancer	[35]	97.83%	No	Accuracy, AUC
Kidney Cancer	[3]	89% to 92%	No	Accuracy
Kidney Cancer	[8]	60% to 97%	No	Accuracy, recall, precision, and F1-score
Colon Cancer	[36]	96.33%	No	Accuracy, precision, recall, F-measure, and confusion matrix
Colon Cancer	[37]	99.50%	No	Accuracy, precision, recall, F1-score, SPE, MCC, confusion matrix
ALL	[38]	99%	No	Accuracy, F1-score, kappa, precision, sensitivity, specificity and confusion matrix
ALL	[39]	99.8%	No	Accuracy, precision, recall, and F1-score

Fig. 2. Sample images of colon cancer in the image resolutions of 512 × 512 (original), 2048 × 2048, 16 × 16, 32 × 32, 64 × 64, 128 × 128 and 1024 × 1024 pixels, from left to right

resolution permutations and their applicability across multiple cancer datasets, this study has achieved state-of-the-art results that can guide further understanding around the areas of super-resolution and image resolution for the binary classification of cancer-based histopathological medical images.

5 Conclusion

Super-resolution has proven to be effective in improving the quality of histopathological images. Increased image resolution has improved the binary classification of cancer-based medical images. However, these gains in image quality and accuracy are at a cost. This study aimed to answer whether the gains in image quality and binary classification accuracy were worth the computational and time expense. The work across five widely cited cancer datasets, across numerous image resolution permutations, shows that the increased image quality and binary classification accuracy are not worth the gains cited by certain researchers.

Further work can include comparisons of image classifiers, super-resolution techniques and image downscaling options across a further set of cancer-based medical images. A greater goal can include an understanding of how the deep learning and super-resolution architectures make the decisions and formulate the results, and where the apparent ambivalence to image resolution originates.

References

1. Zhu, J., et al.: MIASSR: an approach for medical image arbitrary scale super-resolution (2021). http://arxiv.org/abs/2105.10738
2. Shahidi, F.: Breast cancer histopathology image super-resolution using wide-attention GAN with improved Wasserstein gradient penalty and perceptual loss. IEEE Access **9**, 32795–32809 (2021)
3. Abdeltawab, H.A., et al.: A deep learning framework for automated classification of histopathological kidney whole-slide images. J. Pathol. Inform. **13**, 100093 (2022)
4. Khamparia, A., Singh, K.M.: A systematic review on deep learning architectures and applications. Expert Syst. **36**(3), e12400 (2019)
5. Voulodimos, A., et al.: Deep learning for computer vision: a brief review. Comput. Intell. Neurosci. **2018**, 1–13 (2018)
6. Gupta, S.: High Accuracy & Faster Deep Learning with High Resolution Images & Large Models. https://sumitgup.medium.com/deep-learning-with-high-resolution-images-large-models-44bfd90482a8
7. Dong, C., et al.: Image super-resolution using deep convolutional networks. IEEE Trans. Pattern Anal. Mach. Intell. **38**(2), 295–307 (2016)
8. Alzu'bi, D., et al.: Kidney tumor detection and classification based on deep learning approaches: a new dataset in CT scans. J. Healthc. Eng. **2022**, 1–22 (2022)
9. Golowczynski, M.: What is image resolution? Everything you need to know. https://smartframe.io/blog/what-is-image-resolution-everything-you-need-to-know/
10. Siipola, J.: What's the best lossless image format? Comparing PNG, WebP, AVIF, and JPEG XL. https://siipo.la/blog/whats-the-best-lossless-image-format-comparing-png-webp-avif-and-jpeg-xl

11. Sabottke, C.F., Spieler, B.M.: The effect of image resolution on deep learning in radiography. Radiol. Artif. Intell. **2**(1), e190015 (2020)
12. Thambawita, V., et al.: Impact of image resolution on deep learning performance in endoscopy image classification: an experimental study using a large dataset of endoscopic images. Diagnostics **11**(12), 2183 (2021)
13. Rukundo, O.: Effects of image size on deep learning. Electronics **12**(4), 985 (2023)
14. Yousef, R., et al.: A holistic overview of deep learning approach in medical imaging. Multimed. Syst. **28**(3), 881–914 (2022)
15. Jiang, X., et al.: Deep learning for medical image-based cancer diagnosis. Cancers **15**(14), 3608 (2023)
16. Haque, I.U., et al.: Effect of image resolution on automated classification of chest X-rays. J. Med. Imaging (2023)
17. Chen, H., et al.: Real-world single image super-resolution: a brief review. Inf. Fusion **79**, 124–145 (2022)
18. Munir, K., et al.: Cancer diagnosis using deep learning: a bibliographic review. Cancers **11**(9), 1235 (2019)
19. Dahl, R., et al.: Pixel Recursive Super Resolution (2017). http://arxiv.org/abs/1702.00783
20. Spanhol, F.A., et al.: A Dataset for Breast Cancer Histopathological Image Classification (2016)
21. Naran, O.S.: Multi Cancer Dataset. https://www.kaggle.com/datasets/obulisainaren/multi-cancer
22. Ghaderzadeh, M., et al.: Acute Lymphoblastic Leukemia (ALL) image dataset (2021)
23. Ghaderzadeh, M., et al.: A fast and efficient CNN model for B-ALL diagnosis and its subtypes classification using peripheral blood smear images. Int. J. Intell. Syst. **37**(8), 5113–5133 (2022)
24. Borkowski, A.A., et al.: Lung and Colon Cancer Histopathological Image Dataset (LC25000)
25. Islam, N., et al.: OPEN vision transformer and explainable transfer learning models for auto detection of kidney cyst, stone and tumor from CT-radiography. Sci. Rep. **2022** (2022)
26. Duchon, C.E.: Lanczos filtering in one and two dimensions. J. Appl. Meteorol. **18**(8), 1016–1022 (1979)
27. Zeng, W., et al.: A comparative study of CNN-based super-resolution methods in MRI reconstruction and its beyond. Signal Process. Image Commun. **81**, 115701 (2020)
28. He, K., et al.: Deep Residual Learning for Image Recognition (2015). http://arxiv.org/abs/1512.03385
29. Gunasekaran, H., et al.: GIT-Net: an ensemble deep learning-based GI tract classification of endoscopic images. Bioengineering **10**(7), 809 (2023)
30. Jiao, L., Zhao, J.: A survey on the new generation of deep learning in image processing. IEEE Access **7**, 172231–172263 (2019)
31. Mukherjee, S.: The Annotated ResNet-50. https://towardsdatascience.com/the-annotated-resnet-50-a6c536034758
32. Toğaçar, M., et al.: BreastNet: a novel convolutional neural network model through histopathological images for the diagnosis of breast cancer. Phys. A Stat. Mech. Appl. **545**, 123592 (2020)
33. Wang, P., et al.: Cross-task extreme learning machine for breast cancer image classification with deep convolutional features. Biomed. Signal Process. Control **57**, 101789 (2020)
34. Paul, R., et al.: Deep feature transfer learning in combination with traditional features predicts survival among patients with lung adenocarcinoma. Tomography **2**(4), 388–395 (2016)
35. Khan, M.A., et al.: VGG19 network assisted joint segmentation and classification of lung nodules in CT images. Diagnostics **11**(12), 2208 (2021)
36. Masud, M., et al.: A machine learning approach to diagnosing lung and colon cancer using a deep learning-based classification framework. Sensors **21**(3), 748 (2021)

37. Sakr, A.S., et al.: An efficient deep learning approach for colon cancer detection. Appl. Sci. **12**(17), 8450 (2022)
38. Ansari, S., et al.: A customized efficient deep learning model for the diagnosis of acute Leukemia cells based on lymphocyte and monocyte images. Electronics **12**(2), 322 (2023)
39. Veeraiah, N., et al.: MayGAN: mayfly optimization with generative adversarial network-based deep learning method to classify Leukemia form blood smear images. Comput. Syst. Sci. Eng. **46**(2), 2039–2058 (2023)

Investigating Frequent Pattern-Based Models for Improving Community Policing in South Africa

Apiwe Macingwane and Omowumni E. IsafIade[✉]

University of the Western Cape, Robert Sobukwe Road, Bellville 7530, South Africa
oisafiade@uwc.ac.za

Abstract. Identifying crime patterns in a region can provide valuable information for developing effective strategies to combat criminal activities. While frequent pattern-based models have proven useful, there is limited research that has considered this within the South African context. Therefore, this research delves into exploring two major frequent pattern-based mining algorithms, which are Frequent Pattern Growth (FP-Growth) and Hyper-Structure Mining (Hmine), and then proposes a Hybrid-Growth algorithm (Hmine_FP-Growth) - an improved FP-Growth algorithm, to generate frequent crime patterns from Crime Stats SA dataset. The study aims to evaluate the algorithms based on factors such as time complexity and memory usage, as well as the patterns' support, lift, and confidence values. The research emphasizes the importance of selecting efficient and cost-effective algorithms for large-scale crime datasets and real-time processing requirements. The results show that FP-Growth exhibits a higher time complexity and memory usage compared to the other two algorithms. Hmine stands out in terms of time complexity, with an average of 7.48 s and a memory usage of 0.03 GB. On the other hand, the Hybrid-Growth algorithm exhibits an average runtime of 79.97 s and utilizes 0.02 GB of memory for pattern generation per province. This distinction highlights Hmine as a fast-processing algorithm, while Hybrid-Growth excels in minimizing memory usage. The outcome of this research has the potential to guide law enforcement agencies in resource allocation and improving crime prevention strategies.

Keywords: Community Policing · Frequent Pattern · Data mining · Machine Learning · Hyper-Structure Mining

1 Introduction

In recent years, South Africa (SA) has witnessed an increasing rate of criminal activities, which has been a major issue regarding the safety and security of its citizens. Figure 1 reveals the trend of crime in the first quarter of the year 2022 [1]. Both local communities and law enforcement agencies are constantly confronted with evolving crime tactics, while law enforcement is mandated to resolve crime issues. Globally, South Africa is known as one of the top 5 most dangerous countries [2] and crime has various negative impacts on people [3], hence the law enforcement agencies are constantly finding ways to deter crime.

A. Pillay et al. (Eds.): SACAIR 2023, CCIS 1976, pp. 203–218, 2023.
https://doi.org/10.1007/978-3-031-49002-6_14

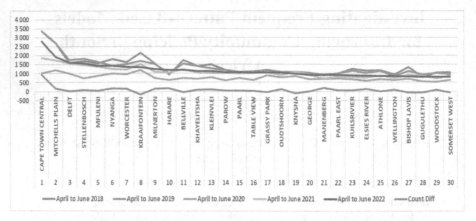

Fig. 1. The First Quarter 2018/2022 Crime Stats of Western Cape, South Africa [1]

Analyzing historical data in SA can reveal crime patterns, enabling effective strategies to mitigate the challenge. In addressing crime issues, Machine Learning (ML) within criminology has been explored, such as crime prediction, patterns, and hotspot detection [4–6]. The introduction of pattern-generating algorithms, an unsupervised learning technique, enables data mining to take charge in revealing crime patterns. Hence, this research delves into exploring two major frequent pattern-based mining algorithms, which are Frequent Pattern Growth (FP-Growth) and Hyper-Structure Mining (Hmine), and then proposes a Hybrid-Growth algorithm (Hmine_FP-Growth), to generate frequent crime patterns from Crime Stats SA dataset, which is a South African crime dataset [7]. Furthermore, the algorithms are evaluated based on factors such as time complexity and memory usage, as well as metrics such as patterns' support, lift, and confidence values. This investigation highlights the importance of selecting efficient and cost-effective algorithms for large-scale crime datasets and real-time processing requirements. While some research has been conducted [4, 8], there has been a lack of comparison of frequent pattern algorithms based on province datasets, using the Hybrid-Growth approach as done in this research.

This paper is organized as follows: Sect. 2 presents the related work and supporting evidence of the importance of this work; Sect. 3 details the methodology; Sect. 4 examines the outcomes of the analysis, and Sect. 5 concludes the paper.

2   Related Work

Several research has been conducted on crime analysis, detection, prediction, and pattern generation in addressing the challenge of increasing crime rate and finding effective crime strategies [5, 9–12]. Manual means of analyzing crime data have shown notable limitations and gaps in terms of functionality, generating dynamic patterns, and lack of real-time data insight [13, 14]. The South African police still rely on manual data entry, which is invasive and error-prone, impeding the implementation of effective crime control strategies [15].

Over the years, numerous studies have aimed to extend the capabilities of ML to encompass a broader spectrum, such as data mining, pattern recognition, and classification [5, 9–12]. Association rule mining is an important aspect of data mining that identifies relationships between items within a large dataset [16]. It allows the identification of frequent patterns by numerical analysis, namely support, lift, confidence, and data pruning. An essential concept within this technique involves identifying patterns within these itemsets. The Apriori algorithm was introduced to generate patterns but faced limitations due to its repetitive data scanning and candidate itemsets generation during the mining process, resulting in high time complexity [11, 16]. The frequent Pattern Growth (FP-Growth) model is an improvement on the Apriori algorithm, to overcome the aforementioned limitations [12, 16–19]. FP-Growth employs an FP-tree-based approach to scan the data and uses a divide-and-conquer approach to create a compressed tree representation of the database [4, 11]. Patterns are generated by performing pattern mining on the generated tree to obtain conditional pattern bases, enabling low-cost pattern mining. However, this algorithm is a bit complex compared to the Apriori and it requires a lot of memory [11, 16].

On the other hand, Hyper-structure mining (H-Mine) of frequent patterns is an algorithm introduced to address the performance bottlenecks of FP-Growth in sparse datasets. It achieves this by using queues instead of a tree data structure [11]. It stores items of transactions in separate queues and connects transactions with the same first item name using a hyperlinks array structure called H-Struct [12]. H-Mine is recognized for outperforming both Apriori and FP-Growth in terms of time complexity and memory usage, even when the minimum support value (MSV) is very small [4]. Selecting an MSV is a crucial stage when mining patterns, it also plays a role in selecting which itemsets are frequent [4, 17]. A lower MSV may result in many irrelevant patterns, which in turn demands more time and memory for processing. Conversely, using an appropriate MSV yields insightful patterns that can provide valuable insights for decision-making.

Research has been carried out on Stats SA datasets such as linear regression analysis, and frequent pattern generation, as shown in Table 1 [4, 8]. However, none of these studies compared frequent pattern algorithms, within a Hybrid-Growth context as done in this study.

Table 1. The frequent pattern-related research conducted on the Crime Stats SA dataset.

Ref	Research Focus	Model used	Data used	Data Pruning	Results
[8]	Proposes a linear regression model to analyze and predict future crime in SA	Linear regression	Stats SA (real)	N/A	A linear regression model was proven to be an efficient model for predicting the crime rate in South Africa

(continued)

Table 1. (*continued*)

Ref	Research Focus	Model used	Data used	Data Pruning	Results
[4]	Developed a crime system (coined CrimeTracker), identify crime hotspots, and patterns using FP-Growth	FP-Growth	Stats SA (real)	Floor and mean function to generate MSV and data filtered to Western Cape	An increase in the dataset tends to increase the runtime, when generating the patterns and requires high memory usage
This work	Compares three pattern-generating algorithms: FP-Growth, H-Mine, and the proposed Hybrid-Growth	FP-Growth, Hmine, Hybrid-Growth	Stats SA (real)	Floor and mean function to generate MSV and data filtered to Western Cape and two stations	Please refer to the outcomes presented in the Results and Discussion section of this paper

3 Methodology

3.1 Data Description

This research used Stats SA datasets, which contains 30862 rows and 14 columns [7]. Detailed information about these 14 columns is as follows: {Province: 9}, {Station: 1143 unique suburbs}, {Category: 27 unique crime types}, and {Years: 11 years (2005–2016)}. The data contained no duplicates, having undergone thorough cleaning, rendering it ready for use in CSV format. Two sample datasets were extracted from the Stats SA dataset: (a) encompassing two stations from each province, and (b) involving all stations within a province. The data underwent preprocessing to derive these two datasets, during which the data is transformed into itemset by excluding the "province" and "station" columns, focusing solely on crime types and their respective counts per year.

3.2 Minimum Support Value (MSV)

The most crucial step in association rule mining is determining the minimum support value (MSV), which plays a significant role in pruning data for the rule generation process [4, 17]. This value helps eliminate infrequent items within an itemset. This research implemented the floor and mean function approach to determine the MSV and the approach is displayed in pseudocode 1 [19].

Pseudocode 1. Generating MSV from itemset
Input: Itemset
Output: MSV
def getMSV (itemset):
    #Count the sum of all the support values of each crime type
    #Find the mean value of the item count
    return floor(mean value)

3.3 Evaluation Metrics

The evaluation metrics is utilized to investigate the patterns generated by each algorithm. The aim is to investigate the mined patterns' quality, and this provides insights into the item's relationships in terms of co-occurrence [16]. The evaluation considers the following key features (where A and B are unique items), namely:

a. **Support value (k):** It measures the number of times an item appears in the transactional database.

$$K(A) = Number\ of\ times\ A\ appears\ in\ a\ transactional\ database \qquad (1)$$

b. **Support:** Calculates the support of itemsets, which is the proportion of transactions in which the item appears.

$$\sup(A \to B) = P(A \cup B) \qquad (2)$$

c. **Lift:** Lift measures the strength of the association between two items, indicating how often they co-occur in comparison to what would be expected if they were independent. If the lift is greater than 1, it indicates a positive correlation, suggesting that the occurrence of one item increases the likelihood of the occurrence of the other item. Otherwise, if less than 1 it indicates a negative correlation, implying that the presence of one item reduces the likelihood of the other item. When it is equal to 1, it implies independence within those items.

$$lift = \frac{P(A \cup B)}{P(A)P(B)} \qquad (3)$$

d. **Pruning:** The process of removing infrequent items from a transactional database using the MSV.

e. **Confidence:** Confidence measures how often the pattern is true. It is calculated as the ratio of the support of the combined items to the support of the antecedent.

$$P(B|A) = \frac{P(A \cup B)}{P(A)} \qquad (4)$$

3.4 Models Used for Experiment

3.4.1 FP-Growth Algorithm

The FP-Growth is an effective data mining model that structures information in a tree-like structure and is mostly used for frequent pattern mining. [4, 18]. Figure 2 summarises the basic steps of the model. The dataset underwent data transformation and pruning using the MSV. The FP-tree was built, where the conditional patterns were generated. This resulted in frequent patterns that were recursively mined.

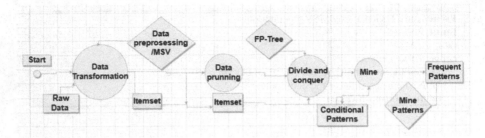

Fig. 2. Process Flow of FP-Growth Algorithm

3.4.2 Hmine Algorithm

Figure 3 illustrates the process flow of the Hmine algorithm utilized in this case for frequent pattern mining in crime data. The algorithm involves transforming the original dataset into a utility dataset to account for repetitions and associate itemsets with support values. The minimum utility threshold is defined for the pruning process, after that the mining process occurs to obtain the patterns.

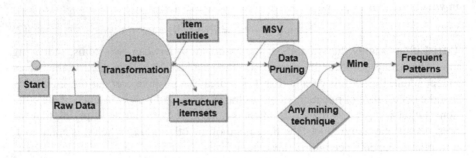

Fig. 3. Process Flow of Hmine Algorithm

3.4.3 Hybrid-Growth Algorithm

This is the proposed algorithm in this research. This memory-based algorithm combines the strengths of Hmine with FP-Growth method for pattern mining.

Figure 4 shows the process flow of the Hybrid-Growth algorithm and how the frequent patterns were mined. This algorithm is subdivided into three main stages: data processing, mining patterns, and evaluation of frequent patterns which result in the extraction of meaningful frequent patterns. The raw data is preprocessed and an MSV value is obtained. The Hmine algorithm receives the transformed transaction with its MSV, and data pruning occurs, where frequent items are generated, containing singletons. Those frequent itemsets are inserted as transactions in the FP-Growth model to mine the frequent patterns using the FP-Tree algorithm. The FP-Tree is generated, from which base patterns are extracted. The conditional patterns are derived from the base

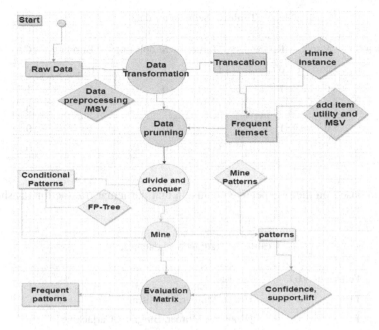

Fig. 4. Process Flow of Hybrid-Growth Algorithm.

pattern through the divide-and-conquer approach. Furthermore, from these conditional patterns, frequent patterns are extracted. Lastly, the evaluation matrix derived from association rule mining is utilized to assess the correlation between items within a frequent pattern. This assessment is based on the support, lift, and confidence values of a pattern.

3.4.4 Algorithm Operation: Hybrid-Growth Approach Combining Hmine and FP-Growth

The functionality of the Hybrid-Growth algorithm involves integrating both the Hmine and FP-Growth algorithms. Hmine is employed after preprocessing as a memory-based algorithm that outperforms FP-Growth in mining patterns with less memory, even with a low minimum support value (MSV). Furthermore, it utilizes the dynamically linked list structure 'H-struct' to enhance mining efficiency by maintaining and updating links between transactions that share the same itemset. On the other hand, FP-Growth employs the divide-and-conquer approach to mine frequent patterns. These mined patterns are evaluated using lift, confidence, and the support of the association rule.

I. Preprocessing

The dataset was presented in a table format, where crime type counts were displayed based on years, as shown in Table 2.

Table 2 illustrates 6 distinct crime types and their corresponding counts associated with Transaction IDs. In the preprocessing stage, the crime types were included in each

Table 2. Sample raw data

Transaction ID	Rape	Robbery	Murder	Shoplifting	Burglary	Carjacking
T1	1	0	1	1	0	0
T2	0	1	1	0	1	1
T3	1	1	1	0	1	0
T4	0	1	0	0	1	0
T5	1	0	1	0	1	0

transaction based on their respective counts on that particular transaction, as shown in Table 3.

Table 3. Transactional database.

Transaction ID	Itemset
T1	{Rape, Murder, Shoplifting}
T2	{Robbery, Murder, Burglary, Carjacking}
T3	{Rape, Robbery, Murder, Burglary}
T4	{Robbery, Burglary}
T5	{Rape, Murder, Burglary}

For each transaction ID, if an item exists, then increment the counter by 1. After going through all the transaction IDs, the counter will represent the support value of tan items, as shown in Table 4. Furthermore, the items were subsequently ordered in descending order based on their support values.

Table 4. Ordered Support values of items.

Item	Support value
Murder	4
Burglary	4
Rape	3
Robbery	3
Shoplifting	1
Carjacking	1

II. **MSV and Data pruning**

MSV serves as the minimum threshold for data pruning, involving the removal of infrequent items from the transactional database using the approach outlined in Pseudocode 1. MSV was determined by the summation support value of each crime type and dividing it by the number of unique crime types in the transactional itemset. The following section presents a step-by-step breakdown of the calculations:

1. **Sum of support values:** support_value (Murder) + support_value (Burglary)
2. + support_value (Rape) + support_value (Robbery) + support_value (Shoplifting) + support_value (Carjacking) = 4 + 4 + 3 + 3 + 1 + 1 = 16
3. **Number of unique crime_type (n):** n = 6
4. **MSV:** MSV = sum of support_value(i)/crime_type (n) = 16/6 ≈ 2.67
5. **Floor Function:** floor (2.67) = 2

Hence, the MSV was set at 2. Crime types were pruned utilizing the MSV of 2, leading to the removal of crime types from the transactional itemset that do not meet the minimum threshold of 2 (see Table 4). Only items greater than or equal to the MSV were kept, as demonstrated in Table 5.

Table 5. Pruned transactional item and frequent item projection.

Transaction ID	itemset	Frequent-item projection
T1	{Murder, Rape}	{Murder, Rape}
T2	{Murder, Burglary, Robbery}	{Murder, Burglary, Robbery}
T3	{Murder, Burglary, Rape, Robbery}	{Murder, Burglary, Rape, Robbery}
T4	{Burglary, Robbery}	{Burglary, Robbery}
T5	{Murder, Burglary, Rape,}	{Murder, Burglary, Rape}

The crime types removed from the transactional itemset were Shoplifting and Carjacking due to their support value being less than the MSV = 2. By scanning the transactional items, {Rape:3, Robbery:3, Murder: 4, Burglary:4, Shoplifting:1, Carjacking:1}, an extraction of the crime type along with its associated support value was performed. Furthermore, the data pruning process (Pseudocode 1) was applied, resulting in the obtained itemset in Table 5, which is considered a frequent-item projection for the Hmine algorithm. The frequent-item projections of the transactional itemset were stored in the main memory, each crime type was stored by two fields (ID, hyperlink). The Frequent set(F-set) was obtained based on the support values in descending order, resulting in an F-list of {Murder, Burglary, Rape, Robbery}.

III. **Hmine-frequent patterns**

The F-set was divided into 4 subsets, which contained the following itemsets as shown in Fig. 5:

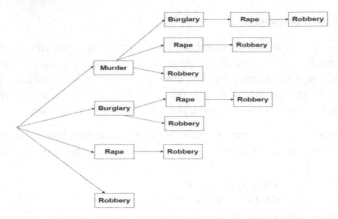

Fig. 5. Frequent itemsets from Hmine

1. Itemset containing Murder: {Murder, Rape}, {Murder, Burglary, Robbery}, {Murder, Burglary, Rape, Robbery}, {Murder, Burglary, Rape}, {Murder, 'Burglary'}, {Murder, Robbery}, {Murder, Rape, Robbery}, {Murder, Burglary, Robbery}
2. Itemset containing Burglary but not Murder: {Burglary, Rape, Robbery}, {Rape, Robbery}, {Burglary, Robbery}
3. Itemset containing Rape, not Murder nor Burglary: {Rape, Robbery}
4. Itemset containing Robbery, not Murder nor Burglary nor Rape. {Robbery}

Table 6. Frequent itemsets from the Hmine algorithm.

Transaction ID	Itemsets
T1	{Murder, Burglary, Rape, Robbery}
T2	{Murder, Burglary, Rape}
T3	{Murder, Burglary}
T4	{Murder, Rape, Robbery}
T5	{Murder, Burglary, Robbery}
T6	{Murder, Rape}
T7	{Burglary, Rape}
T8	{Murder, Robbery}
T9	{Burglary, Rape, Robbery}
T10	{Burglary, Robbery}
T11	{Rape, Robbery}

From the frequent item projection, the following itemsets were obtained as shown in Table 6. The support value for each crime type is: {Murder:7, Burglary:7, Rape:7, Robbery:7}. With the established MSV of 2, no crime types were dropped.

IV. FP-Tree and Frequent patterns mining

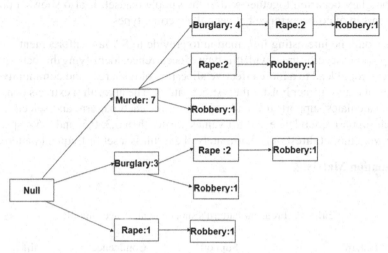

Fig. 6. FP-Tree from Hmine algorithm.

The constructed FP-tree in Fig. 6 resembled the structure displayed in Fig. 5. This FP-tree was formed by tracing transaction paths from the root node to child nodes, with each node representing an item along with its count. The extraction of frequent patterns occurred through the recursive mining of conditional patterns, as shown in Table 7. It is noted that the frequent pattern set(L) = {Murder:7, Burglary:7, Rape:7, Robbery:7}.

Table 7. Base Patterns, Conditional Patterns, and Frequent Patterns.

Item	Base Pattern	Conditional Pattern	Frequent Pattern
Burglary	{Murder}:4	{Murder}:4	{Murder, Burglary}:4
Rape	{Murder, Burglary}:2, {Murder}:2, {Burglary}:2,	{Murder}:4, {Burglary}:4, {Murder, Burglary}:4	{Murder, Burglary, Rape}:4, {Murder, Rape}:4, {Burglary, Rape}:4
Robbery	{Murder, Burglary, Rape}:1, {Murder, Rape}: 1, {Murder}:1, {Burglary, Rape}:1, {Burglary}: 1, {Rape}:1	{Murder, Rape}:2, {Burglary, Rape}:2, {Burglary}:4, {Rape}:4,	{Burglary, Robbery}:4, {Rape, Robbery}:4, {Murder, Robbery}:4, {Murder, Rape, Robbery}:2, {Burglary, Rape, Robbery}:2

For Burglary:

- Base pattern: {Murder}:4, meaning murder occurred 4 times, and was followed by Burglary.

- Conditional pattern: {Murder}:4, extracted from the pattern base.
- Frequent pattern: {Murder, Burglary}:4, indicates 4 instances where both murder and burglary occurred together within the sample dataset. It also shows a potential relationship of co-occurrence between these crime types.

This would be interesting information to provide to SA law enforcement, enabling them to pinpoint specific areas with this type of occurrence. Identifying this pattern within those areas could lead to a more effective allocation of resources and community forum engagement could further bolster their street patrols. The algorithm extracts conditional patterns, calculates support, lift, and confidence for each pattern, and selects patterns with high support, confidence and lift values greater than 0.3, 0.5, and 0.5 respectively, as these were the set thresholds. Ultimately, the result is a set of frequent patterns.

V. Evaluation Matrix

Table 8. Frequent Pattern's support, confidence, and lift.

Frequent Pattern	Support	Confidence	lift
{Murder, Burglary}	0,36	0,57	0,9
{Murder, Burglary, Rape}	0,18	0,5	0,76
{Murder, Rape}	0,36	0,57	0,9
{Burglary, Rape}	0,36	0,57	0,9
{Burglary, Robbery}	0,36	0,57	0,9
{Rape, Robbery}	0,36	0,57	0,9
{Murder, Robbery}	0,36	0,57	0,9
{Murder, Rape, Robbery}	0,18	0,5	0,76
{Burglary, Rape, Robbery}	0,18	0,5	0,76

Based on the minimum support, lift, and confidence, the frequent patterns are {Murder, Burglary}, {Murder, Rape}, {Burglary, Rape}, {Burglary, Robbery}, {Rape, Robbery}, {Murder, Robbery} (Table 8).

4   Results Analysis and Discussion

The experiment was performed using two machines: one equipped with the 12th Gen Intel(R) Core (TM) i7-1255U 1.70 GHz Processor, 16.0 GB (15.7 GB usable) RAM, and the other with an 11th Gen Intel(R) Core (TM) i7-11850H 2.50 GHz Processor, 32.0 GB (31.7 GB usable) RAM. The reason for using these two machines is that the 12th Gen Intel(R) processor had a bottleneck of two stations per province when generating patterns using the FP-Growth algorithm. Furthermore, Python programming through a Jupyter Notebook was used for the experiment.

The two datasets that were extracted from the dataset for the experiment are:

- Two station datasets: {Western cape: 8435}, {Gauteng: 9205}, {Free State: 624}, {Northwest: 833}, {KwaZulu/Natal: 530}, {Mpumalanga: 634}, {Eastern Cape: 2863}, {Northern Cape: 486}, and {Limpopo: 84}.
- Province dataset: {Western cape: 175203}, {Gauteng: 252498}, {Free State: 50417}, {Northwest: 45389}, {KwaZulu/Natal: 139684}, {Mpumalanga: 50077}, {Eastern Cape: 87060}, {Northern Cape: 19535}, and {Limpopo: 43722}.

Table 9 shows a sample of patterns obtained from the two-station dataset from the Limpopo province, which are pruned based on support, lift and confidence values greater than 0.3, 0.5, and 0.5 respectively. The following pattern: {'Malicious damage to property', 'Burglary at residential premises'}, suggest that a high chance of 'Burglary at residential premises' will occur, given that 'Malicious damage to property' has occurred. A potential advice that could be given to South African citizens would be to install security systems and surveillance cameras at their properties. Most patterns generated show a lift value of 1, meaning all the items are independent of each other.

Table 9. Results (from Limpopo province)

Frequent Pattern	Support	Confidence	lift
{'Malicious damage to property, 'Burglary at residential premises'}	1	1	1
{'All theft not mentioned elsewhere', 'Malicious damage to property'}	1	1	1
{'Drug-related crime', 'Malicious damage to property'}	0.91	1	1
{'Assault with the intent to inflict grievous bodily harm', 'Malicious damage to property'}	1	1	1
{'Common assault', 'Malicious damage to property'}	1	1	1

The results were analyzed and discussed for the three algorithms - Hmine, FP-Growth, and Hybrid-Growth - based on Figs. 7, 8, 9 and 10.

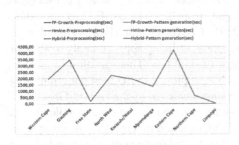

Fig. 7. Two-station dataset (Time).

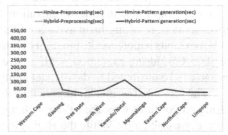

Fig. 8. Province dataset (Time).

Figures 7 and 8 illustrate the analysis of Hmine and the Hybrid-Growth algorithm concerning the time complexity of preprocessing and pattern generation across two distinct datasets. Based on the graphical analysis in Fig. 7 of the two-station dataset, it is evident that FP-Growth exhibits the highest time complexity, while Fig. 8 with the province dataset shows the Hybrid-Growth algorithm incurs the most substantial time complexity during data processing and pattern mining. Hmine displayed an average time of 1.91 s, while Hybrid displayed an average time of 0.12 s for generating patterns, and the FP-Growth algorithm spans from 40.26 to 4219.71 s, with an average time of 1783.47 s on the two-station dataset. However, FP-Growth faced a bottleneck with two stations, rendering it inefficient for generating patterns for all stations within a province. Conversely, Fig. 8, which analyses all stations within a province, shows that the Hybrid-Growth algorithm demonstrates a higher time complexity in generating patterns, ranging from 5.66 to 406.40 s, with an average of 79.97 s. Hmine, in contrast, exhibits a time complexity ranging from 1.58 to 24.77 s with an average of 7.48 s. These results illustrate that Hybrid-Growth and Hmine algorithms perform better compared to FP-Growth. Hybrid-Growth can generate frequent patterns from large-scale datasets at a low time complexity. However, Hmine outperforms both algorithms in time complexity.

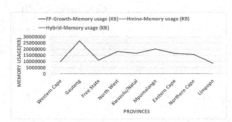

Fig. 9. Two-Station dataset (Memory).   **Fig. 10.** Province dataset (Memory).

Regarding memory usage, Fig. 9 reveals that FP-Growth requires a substantial amount of memory, ranging from 8.54 GB to 26.73 GB, with an average of 16.02 GB. It is noteworthy that the algorithm encounters memory constraints on the 12th Gen Intel(R) with 16 GB RAM, although it was successfully processed on the 32GB RAM-equipped 11th Gen Intel(R) machine. In contrast, Fig. 10 depicts Hmine's relatively conservative memory footprint, with pattern generation ranging from 0.0004 GB to 0.1 GB throughout its execution, resulting in an average of 0.03 GB. Conversely, the Hybrid-Growth algorithm exhibits a more modest memory consumption, utilizing approximately 0.0004 GB to 0.05 GB during execution, with an average of 0.02 GB. This variation in memory usage can be attributed to the distinct approaches employed in frequent pattern mining.

The presented results highlighted the efficiency achieved by the Hybrid-Growth and Hmine algorithms over the FP-Growth algorithm in terms of low time complexity and memory usage gains. While the Hmine algorithm excelled in terms of time complexity for pattern generation, Hybrid-Growth maintained a balanced approach. Similarly, the memory usage analysis emphasized Hybrid-Growth as having the lowest memory usage. Hmine and Hybrid-Growth approaches were suitable choices for pattern mining tasks over the memory-intensive FP-Growth algorithm.

5 Conclusion

Considering the limited frequent pattern-related research on the South African crime dataset, this study aimed to introduce and evaluate the Hybrid-Growth algorithm in comparison to existing pattern-generating algorithms, namely Hmine and FP-Growth, using the Stats SA dataset. Hmine proved to relatively excel in terms of time complexity, while the Hybrid-Growth algorithm demonstrated a good range in memory usage during patterns generation per province, rendering it the most efficient in terms of memory usage. Notably, the Hybrid-Growth approach still outperformed H-Mine when confronted with dense datasets. Conversely, FP-Growth encountered a significant bottleneck, allowing only up to two stations per province due to hardware memory limitations, and required high processing power. FP-Growth resulted in the poorest-performing algorithm, compared to the others. The choice of an algorithm for generating patterns depends on task requirements such as a fast algorithm or a low memory-usage algorithm. This study effectively achieved its objective of comparing the three frequent pattern algorithms and concluded that Hmine and the Hybrid-Growth algorithms are the optimal choices for mining patterns in the Stats SA dataset. Further research should be conducted to confirm the accuracy of the patterns generated.

References

1. South African Police Service (S. A. P. S.), "Crime Statistics: Integrity," Department of Police. https://www.saps.gov.za/services/crimestats.php. Accessed 23 Feb 2023
2. Dlamini, S.: A criminological exploration of community policing forums in Durban, South Africa: a study based on Glenwood Suburb and Cato Manor Township. Doctoral dissertation (2018)
3. Varela, J., Hernández, C., Miranda, R., Barlett, C., Rodríguez-Rivas, M.: Victims of cyberbullying: feeling loneliness and depression among youth and adult Chileans during the pandemic. Int. J. Environ. Res. Public Health 19(10), 5886 (2022)
4. Macingwane, A., Isafiade, O.E.: CrimeTracker: a frequent pattern-based crime knowledge support for strategic community policing. In: Proceedings of IST-Africa Conference (IST-Africa), Tshwane, South Africa, pp. 1–9 (2023). https://doi.org/10.23919/IST-Africa60249. 2023.10187777
5. Adhikary, A., Murad, S.A., Munir, M.S., Hong, C.S.: Edge assisted crime prediction and evaluation framework for machine learning algorithms. In: 2022 International Conference on Information Networking (ICOIN), Jeju-si, Korea, Republic of, pp. 417–422 (2022). https://doi.org/10.1109/ICOIN53446.2022.9687156
6. Agrawal, S., et al.: Federated learning for intrusion detection system: concepts, challenges and future directions. Comput. Commun. (2022)
7. Kaggle "Crime Statistics for South Africa" (2022). https://www.kaggle.com/datasets/slwessels/crime-statistics-for-south-africa. Accessed 28 Apr 2022
8. Obagbuwa, I., Abidoye, A.: South Africa crime visualization, trends analysis, and prediction using machine learning linear regression technique. Appl. Comput. Intell. Soft Comput. 1–14 (2021)
9. Bandekar, S., Vijayalakshmi, C.: Design and analysis of machine learning algorithms for the reduction of crime rates in India. Procedia Comput. Sci. 172, 122–127 (2020)

10. Wibowo, A., Oesman, T.: The comparative analysis on the accuracy of k-NN, Naive Bayes, and Decision Tree Algorithms in predicting crimes and criminal actions in Sleman Regency. In: Journal of Physics: Conference Series, vol. 1450, no. 1, p. 012076. IOP Publishing (2020)

11. Panjaitan, S., Sulindawaty, M.A., Lindawati, S., Watrianthos, R., Sihotang, H.: Implementation of Apriori algorithm for analysis of consumer purchase patterns. In: Journal of Physics: Conference Series, vol. 1255, no. 1, p. 012057 (2019)

12. Pei, J., Han, J., Lu, H., Nishio, S., Tang, S., Yang, D.: H-Mine: fast and space-preserving frequent pattern mining in large database. IIE Trans. **39**(6), 593–605 (2007)

13. Willliam, J.: "Route Fifty," 9 February 2009. https://www.route-fifty.com/cybersecurity/2009/02/static-vs-dynamic-code-analysis-advantages-and-disadvantages/287891/. Accessed 2 Aug 2023

14. Qazi, N., Williams, W.B.: An interactive human-centered data science approach towards crime pattern analysis. Inf. Process. Manag. **56**(6) (2019)

15. Block, K., Kaplan, J.: An analysis of national hockey league playoff games and city-level crime counts. Crime Delinquency **69**(11), 2194–2217 (2023). https://doi.org/10.1177/00111287221118879

16. Zeng, Y., Zeng, Y., Yin, S., Liu, J., Zhang, M.: Research of improved FP-growth algorithm in association rules mining. Sci. Program. 1–6 (2015)

17. Adero, E., Okeyo, G., Mwangi, W.: Using Apriori algorithm technique to analyze crime patterns for Kenyan national crime data: a country perspective. In: 2020 IST-Africa Conference, pp. 1–8 (2020)

18. Heaton, J.: Comparing dataset characteristics that favor the Apriori, Eclat or FP-Growth frequent itemset mining algorithm. In: Southeast Conference 2016, pp. 1–7 (2016)

19. Isafiade, O., Bagula, A., Berman, S.: A revised frequent pattern model for crime situation recognition based on floor-ceil quartile function. Procedia Comput. Sci. **55**, 251–260 (2015)

Financial Inclusion in Sub-Saharan Emerging Markets: The Application of Deep Learning to Improve Determinants

Johnson S. Dlamini[✉], Linda Marshall, and Abiodun Modupe

Department of Computer Science, University of Pretoria, Pretoria 0028, South Africa
u12309355@tuks.co.za, {lmarshall,abiodun.modupe}@cs.up.ac.za

Abstract. Financial inclusion promises to improve worldwide economies by eradicating poverty. In this study, we use high-dimensional household-level survey data from Eswatini, Namibia, Rwanda, and Madagascar to ascertain the main factors that prevent individuals from accessing financial services. This study uses deep learning techniques to perform dimensionality reduction for feature extraction to overcome the common challenge of obscured interpretation in high-dimensional data analysis. To test the effectiveness of reduced features in measuring financial inclusion within and across countries, different algorithms were evaluated on seven performance metrics. From the results, we observed that for Eswatini, Rwanda, and Madagascar, the catboost algorithm was the best predictor of financial inclusion. In each of these three countries, the top three predictors were quality of financial services, spending and income, and remittances; money management and remittances; financial capacity and e-payments and mobile money; and banking and other non-microfinance institutions, remittances, house information, and well-being through farming. For Namibia, the top three features were bank penetration, general and money management, risk, and mitigation with the random forest as the algorithm for financial inclusion prediction. Lastly, it was established that the general cross-region features were banks and non-banks, income/expenditure and money management, and demographics, with the Gaussian Naïve Bayes as the best algorithm for a generalised prediction of financial inclusion.

Keywords: Financial Inclusion · Deep learning · Feature interpretability · Dimension reduction

1 Introduction

Financial inclusion (FI) is a process that ensures that all individuals and businesses have access to and can effectively use suitable financial services that are delivered in a responsible and sustainable manner [1,2]. Primarily, it aims at integrating the economically weaker sections of society into the formal financial system. The FI concept extends from the idea that development should extend to all regions and not just rely on the averaging use of gross domestic product (GDP) to measure inclusive and sustainable economic growth. This is critical because global consensus is growing on the importance of inclusive growth in achieving stable and sustainable economic growth [3]. The Association for FI was established in 2008 [4,5] to advance the development of FI policies in developing and emerging markets [6]. FI quickly became a widely accepted agenda and, in 2010, with the global perspective to promote global FI [7]. Policymakers, researchers, central banks, and other stakeholders consider FI to be critical in achieving inclusive and sustainable growth [6,8]. This is because FI has the potential to reduce poverty, increase economic opportunities and promote financial stability [1,9]. However, challenges such as network infrastructure (road and communication), financial institution penetration, and prerequisites of banking services, among others, still need to be addressed to ensure that everyone has access to the desired and affordable financial services [1,10–12].

An inclusive financial system is very important because it allows people with a wide range of needs to save money, make payments, and manage their risks. This means that many people can access financial services without any economic or non-economic barriers [6]. Hence, FI has the potential to benefit the poor and disadvantaged society groups who, in the absence of an inclusive financial system, rely on their own limited savings and earnings to invest in their education and entrepreneurship, among other ventures. A non-inclusive financial system contributes to persistent inequality and leads to slower economic growth because disadvantaged economic agents rely solely on their limited earnings [6]. As observed from the World Bank statistics, there are approximately 659 million people who live below the extreme poverty line (that is, earn less than $1.90 per day), and there are also those who do not have bank accounts (estimated to be $1.7 billion adults) [13,14]. These people, regardless of their business potential, do not have access to formal financial services, forcing them to live in poverty. Living in poverty implies that paying school fees, medical bills and keeping up with daily household needs becomes a challenge. It is worth noting that the level of poverty varies from region to region within the Sub-Saharan African region (where the countries in this research study are found).

Despite the consensus of key researchers such as [1,6] and [8] on the importance of FI, an understanding of how it is measured, its critical features and how they relate to FI using datasets from Eswatini, Namibia, Rwanda, and Madagascar has not been well established. For example, the requirement to own a bank account for individuals to transcend business and the bank market structure in Africa, amongst other variables, has the unintended consequence of excluding legitimate businesses and thus contributing positively and negatively towards improving FI [6,13,15]. For the positive contribution, people should have access to a safe and convenient bank account for personal finances and the maintenance of the entire family [1,16]. In addition, relying on some of the dimensions of FI while analysing its relationship with other features does not give a full interpretation of the contributing feature that can be used to measure FI in the real sense of it. Finally, including all FI features in a regression analysis at the same time causes high dimensionality (due to the features' high correlation coefficients) and over-parameterisation. This is due to the fact that high-dimensional datasets require the use of complex computer algorithms to analyse in order to uncover facts about the interrelatedness of the features, as well as develop an algorithm to find deep insight features that are generalisable to data from different regions [17–19]. Despite consensus on the importance of FI, an understanding of how it is measured and what constitutes FI using datasets in Eswatini, Namibia, Rwanda, and Madagascar has not been well established. One example is that people need to have a bank account in order to do business. This, along with other factors affecting the bank market structure in Africa, can keep legitimate businesses from participating, which can have both positive and negative effects on the improvement of financial inclusion [6,13,15]. For the positive contribution, people should have access to a safe and convenient bank account for personal finances and the maintenance of the entire family [1,16]. In addition, relying on some of the dimensions of FI while analysing its relationship with other features does not give a full interpretation of the contributing feature that can be used to measure FI in the real sense of it. Finally, including all FI features in a regression analysis at the same time causes high dimensionality (due to the features' high correlation coefficients) and over-parameterisation. This is due to the fact that high-dimensional datasets require the use of complex computer algorithms to analyse in order to uncover facts about the interrelatedness of the features, as well as develop an algorithm to find deep insight features that are generalisable to data from different regions [17–19]. Although these kinds of algorithms require more money to be spent on computer hardware for data storage and processing, the good thing about getting a general representation of features from datasets is that it might be possible to get a subset of features that will help figure out the factors that affect measuring and improving FI.

The globalisation of the Internet and the increasing ease of access to data obscure the ease of developing a measurement tool for FI. The side effects of the ease of data access include increased dataset dimensions and unstructured data. At the same time, capturing as many sub-features as possible is important to accurately measure and interpret the determinants that can be used to promote FI [20,21]. Although there are many methods that have achieved good performance in the literature, such as [22,23], they still lack transparency in interpreting the generated features as they are generalised across multiple datasets to promote FI. To our knowledge, there are no studies in Africa that have developed a measure and reduced the dimensionality of FI data and its nexus with banking structures in Sub-Saharan Africa to provide insightful features to interpret the influence and improve FI. It is against this background that, in this paper, we develop a deep learning (DL) method that reduces the multi-dimensional features in the dataset and then use the compact version to measure FI across the instance of data for African countries and examine the FI nexus with banking structures and other variables.

The main contributions of this paper are summarised as follows:

– The encoder-decoder model is implemented to develop a DL method that drastically reduces the number of dimensions of the features in the input dataset. This is done in order to improve the FI measurement.

– The information value (IV) transformation approach is presented to map the reduced features retrieved from the input dataset into distinct storage bins. This is done to determine how effectively an independent variable predicts the dependent variable based on the text provided for FI.
– We use an interactive system to visualise the results of feature representation using DL methods in a useful evidence-based process for interpreting and determining the impact of the compact features on FI.

The remainder of the paper is organised as follows. Section 2 provides an overview of the related work. In Sect. 3, we describe the structure of the proposed methods and explain the system model. Section 4, presents the results of the proposed methods, and Sect. 5 provides the conclusion of the paper.

2   Related Work

Most of the features used in measuring FI, such as banking market structure and other variables, whether parametric or non-parametric, may be ineffective in Sub-Saharan Africa in terms of providing insightful features and contributing influence factors that could be interpretable for improving FI. This is due to the unique characteristics of the region, such as low literacy rates, inadequate infrastructure, and cultural factors that affect financial behaviour [24–27]. Therefore, there is a need for more context-specific approaches to measuring FI determinants in Sub-Saharan Africa. To improve FI, most studies use a continuous capture of country-level statistics to measure FI by conducting periodic surveys at the company or household level [13,14]. Some of these reports are in public databases such as Global Findex and World Bank Enterprise Survey [28] and FinMark/Finscope [29,30] for researchers to analyse. Thus, FI statistics are often measured using quantitative methods such as surveys, questionnaires, and others [28–32]. These methods are time-consuming as the features that are extracted are often interrelated and do not generalise across various regional datasets to certainly measure FI. The general problem with such datasets is that they incur challenges in the form of difficult interpretation, escalated computational power, and information redundancy that obscures the government decision-making process.

Cámara et al. [22] developed a measure of FI for 82 developed and less-developed countries using 2011 data and applying two-stage principal component analysis (PCA). The authors used access, usage, and barrier dimensions to construct a FI index. Subsequently, the same authors in [33] measured FI for 137 developed and developing countries using 2011 and 2014 data and applied a two-stage PCA. Similar to their previous study, they examined the dimensions of access, usage, and barrier. They proxied the usage dimension by using savings accounts, saving, and borrowing, and the access dimension by the number of bank branches, ATMs, and banking agents per 100,000 adults. They used distance, trust, costs, and document requirements as proxies for the barrier dimension. Park et al. [8] proposed an index of FI for 151 countries for 2011 and 2014, applying a two-stage PCA. The author used availability, accessibility, and usage dimensions.

Based on the literature studied [8,22,31,32], it is clear that there has not been a study with a focus on generalised and interpretable FI features. [8] and [22] tried to address the issue of cross-region financial measurement, but the authors did not give an account of country level and cross-region important features. The author of [32] focused on the Eswatini setting and made use of only payment methods to measure FI without investigating the strength of the other features. In addition, the author did not give an interpretation of the features used in relation to FI. In [31], the author used secondary data without giving an account of how the data reduction was done from the original Finmark/Fincope. Even though the author tries to give an account of feature importance, the work still lacks an explanation of the nature of the features' relationship to FI. With this existing gap, this study proposes a methodology (more details in Sect. 3) to investigate the critical and interpretable features within and across the four chosen countries.

3   Methodology

This research study used an experimental approach to solve the problem of dimensionality in FI data as observed in Finmark/Finscope[1] datasets. The experimental approach was informed by the work of [8,22,31,32] as discussed in Sect. 2. The set of experiments was carried out according to three phases, as highlighted in Fig. 1.

3.1   Input, Preprocessing and Exploratory Data Analysis (EDA) Phase

The Finmark/Finscope datasets from the four countries (Eswatini, Namibia, Rwanda, and Madagascar) were used to conduct this research. This survey data contained sections ranging from A to N. These sections were: *Household Questions(A)*, *Demographics (B)*, *Expenditure and Income (C)*, *Financial Capability (D)*, *Quality of Financial Services (E)*, *Commercial Banks (F)*, *Payments (G)*, *Mobile Money and Technology Connectivity (H)*, Savings (I), Loans and Credit (J), Risk Management and Insurance (K), Remittances (L), Informal Service Providers (ISP) (M), and General (N). Each of these sections contained multiple questions, resulting in various features to analyse per section. DL based feature reduction techniques were used to obtain a compact, interpretable feature set that was used to predict whether or not an individual is financially included based on the status of ownership of a bank account[2]. Although compacting features normally results in obscured interpretation, a guided feature reduction through the renaming of the compact (passed through an *autoencoder*) features ensures that the resultant feature names used in the machine learning (ML) models still resemble the initial features.

---

[1] Data was collected from https://finmark.org.za/ for the years: 2018, 2017, 2016, and 2016 for Eswatini, Namibia, Rwanda, and Madagascar, respectively.

[2] Individuals who own bank accounts are generally regarded as financially included. This has been verified by other authors, such as [1,32] and [31].

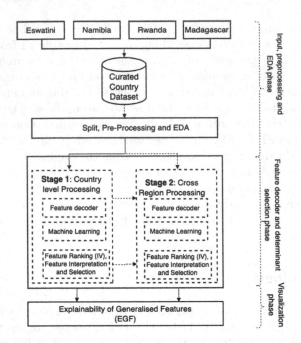

Fig. 1. Proposed architecture for using DL to determine generalised and interpretable features in FI.

First, the separate country data sets were divided into 70% and 30% *training* and *testing* sets, respectively. This was done so that the performance of the ML algorithms could be evaluated on unseen data. The importance of splitting prior to data reprocessing was important as a precautionary measure to avoid information leakage from *training* to *testing* [34].

After data splitting, the Finmark/Finscope datasets were cleaned of data-capturing errors such as missing, duplicate, and irrelevant data (some of the features were not in the initial questionnaire, hence these were deleted). Of these, missing values were the main challenge in this study. To solve this problem, features that had more than 50% missing entries were removed from the input datasets. For the remaining features, the *mean replacement strategy* [35,36] was used to impute missing numeric features while the *mode replacement strategy* [35, 36] was utilised to impute missing categorical feature values. Further, constant features were removed on the basis of their zero variance, which renders them unsuitable for training DL and ML algorithms. Lastly, all numeric features were scaled using the MinMax Scaler as given in Eq. 1.

$$x_{scaled} = \frac{x - x_{min}}{x_{max} - x_{min}} \tag{1}$$

where x_{min} and x_{max} are the minimum and maximum values for feature x. Feature scaling is important when training DL algorithms such as neural networks to ensure that the algorithms training is not influenced by different features recorded on different measuring scales [37].

To adequately capture a high-level view of the state of FI in the four countries, an *exploratory data analysis (EDA)* was carried out. As part of EDA, a high-level exploration of *percentage of bank account holders, mobile money awareness* and *number of rural households* in each of the four countries was done. From Fig. 2, we observe that formally banked individuals are high only for Namibia and Eswatini. Rwanda and Madagascar appear to have most of their populations either informally banked or unbanked. Secondly, the descending mobile money awareness per country follows the following sequence: Eswatini, Madagascar, Rwanda, and Namibia. Lastly, all countries have more than 50% of the population living in rural areas. Rwanda has the highest rural-based population at 85.6%, followed by Eswatini at 73.4%. In the case of Namibia and Madagascar, this proportion is comparable at 53% and 54.8% respectively.

We learn from the above that among the four countries, Eswatini and Namibia appear to be economically sustainable, as more than 50% of the population is formally banked. Rwanda and Madagascar are observed to be economically struggling, as they both have less than a quarter of their population banked.

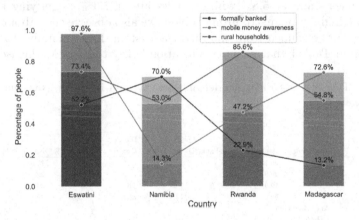

Fig. 2. Finscope mobile money coverage in Eswatini, Namibia, Rwanda and Madagascar.

The state of FI in Eswatini, Namibia, Rwanda, and Madagascar is shown in Tables 1 and 2. For the case of Eswatini, we learn that at least 30.1% (Table 1) of the male population is banked. This is evidence of a gender gap in FI compared to the 22.1% of banked females. *Secondly*, about 14% of the population is fully excluded (neither formally nor informally banked) from the financial system. *Thirdly*, Table 2 shows that in rural places, 33.1% are banked against a 40.2% of unbanked rural population. Lastly, Table 2 also suggests that in Eswatini, people who are married or at least have secondary education are more likely to have a bank account.

For Namibia, Table 2 suggests that the urban population is slightly more banked at 37.9% compared to the rural population which is at 32.2%. *Secondly,* about 20.2% (Table 1) of the population is excluded from the financial system. This suggests that a greater proportion of Namibians do not have access to formal financial services. *Thirdly*, Table 2 suggests that the female gender is more formally banked than the males, and the banking ratio is 40.5% to 29.6% respectively. *Lastly*, the same table suggests that unmarried Namibians are more banked than those of different marriage statuses, and most banked individuals have at least a secondary education.

In Rwanda, we observe that at least 12.5% of the population is formally excluded from the financial system. Here, we also have a peculiar scenario where the rural population enjoys about twice as much FI (15.2% banked from Table 2) than the urban population (7.7% banked). In terms of gender, the female population is more banked than the male at 12% and 10.9%, respectively. Finally, from Table 1, we observe that, given a sample of Rwandans, there is a high probability that they are banked if they are married and have at least primary education.

Lastly, for Madagascar, we observe that Table 2 suggests that this country is predominantly unbanked with at least 13% fully banked population and at least 51.8% of the unbanked stay in rural places. In terms of gender, there is no significant difference in the formal banking ratios of men to women (From Table 2) Banked males stand at 5.8% while females are at 7.3%), signifying that both genders are highly unbanked. From Table 2, we also observe that about 39.4% of the citizens are excluded from the financial system. Lastly, similar to Rwanda, we note from Table 1 that primary education is key to FI in Madagascar.

Table 1. EDA: State of FI by gender and area type in Eswatini, Namibia, Rwanda, and Madagascar

Country/Bank Status	Gender							
	Male				Female			
	Excluded(%)	Banked(%)	Non-bank(%)	Informal(%)	Excluded(%)	Banked(%)	Non-bank(%)	Informal(%)
Eswatini	6.4	30.1	16.5	1	7.2	22.1	15.8	0.9
Namibia	9	29.6	2.4	2.3	11.2	40.5	2.4	2.8
Rwanda	3.9	10.9	18.6	7.2	8.6	12	23.7	15.1
Madagascar	17.3	5.8	8.1	13.4	22.1	7.3	12.1	12.8
	Area Type							
	Urban				Rural			
Eswatini	1.9	19	5.5	0.3	11.9	33.1	26.8	1.6
Namibia	6.1	37.9	1.8	1.2	14.1	32.2	2.9	3.9
Rwanda	1	7.7	4.7	1	11.5	15.2	37.6	21.3
Madagascar	13.5	10.2	13.8	7.8	25.9	3	6.4	19.5

Table 2. EDA: Percentage of banked and unbanked citezens in Eswatini, Namibia, Rwanda, and Madagascar.

Country	Banked Status	Area Type		Gender		Marital Status				Education Level			
		Urban	Rural	Male	Female	Single	Married	Widowed	Divorced	Tertiary	Secondary	Primary	No Education
Eswatini	%Banked	19	33.1	30.1	22.1	22.6	25.5	3.2	0.8	11.6	28.9	8.5	3.2
	%UnBanked	7.6	40.2	23.9	24	24.7	18.8	3.5	0.9	0.6	24.9	15.4	7.0
Namibia	%Banked	37.9	32.2	29.6	40.5	40.8	23.6	3.7	2	8	42.5	12.5	4.7
	%UnBanked	9.1	20.8	13.6	16.3	17.3	9.5	1.8	1.3	0.3	11.1	11.8	6.3
Rwanda	%Banked	7.7	15.2	10.9	12	5.2	14.6	2.2	0.9	1.8	7.7	10.2	2.7
	%UnBanked	6.7	70.4	29.6	47.5	15.8	45.6	11.3	4.3	0.1	9.5	45.5	21.2
Madagascar	%Banked	10.2	3	5.8	7.3	1.6	9.6	1.1	0.8	2.6	7.4	2	0.3
	%UnBanked	35.1	51.8	38.7	48.1	13.2	58	6	9.7	2.2	31.4	38.7	17.2

3.2   Feature Decoder and Determinant Selection Phase

This phase has been divided into two stages, **Country Level processing** and **Cross region processing**. Each of these stages has three further sub-processes, *Feature decoder, Machine Learning* and *Feature ranking, feature interpretation, and selection.* Since the output features of *Stage 1* feed into *Stage 2* (see Fig. 1), we begin a discussion with the former in accordance with the three sub-processes.

Stage 1: Feature Decoder. This phase corresponds to the main process of the study where we do feature reduction for the datasets from separate countries. To achieve this task, the autoencoder (AE) [38] neural network was used per country to get a reduced representation for each of the sections as outlined in Sect. 3.1. In this way, the multiple features per section for all country-level datasets were reduced to the number of sections in the questionnaire. This was done for both the *train* and *test* splits. The AE is an unsupervised neural network that outputs the same number of features as its inputs using two stages. The first stage is called the *encoder stage* (input side) and the second is the *decoder stage* (output side). The *decoder stage* attempts to reconstruct the input from a compressed representation called the *embedding layer*. The *embedding layer* is the critical part of the AE as applied in this study, as this layer was used to predict the representative feature of each section per country dataset. The design of the AE was such that in the *encoder stage*, the input number of neurons (which was initially set to the number of features per country per instantaneous section) used was halved twice, the *embedding layer* was kept at 1 neuron while the *decoder stage* was set to be a reflection of the *encoder stage*. The use of the *embedding layer* to predict the reduced representation for each section resulted in a compact representative feature set equal to the number of sections in the initial questionnaire. Additionally, the final number of compact features for both the *train* and *test* sets was equal.

Stage 1: Machine Learning. Based on the reduced features from the previous process (**Stage 1: Feature Reduction**), fifteen ML algorithms were run to compute the FI at the country level for data sets. At this stage, the most suitable algorithm was selected to calculate FI per specific country.

Stage 1: Feature Ranking (IV), Feature Interpretation and Selection. In this step, the predictive strength of each of the reduced features as used in country-level ML and FI predictions was measured. To achieve this, we computed the weight of evidence (WOE) and information value (IV) scores for each of the features. The WOE specifies how strong an independent numeric feature relates to a binary target variable (bank account ownership status in our case). Hence, to quantify the impact of each of the features, we compute the WOE scores as follows [39,40].

$$WOE_i = ln\left(\frac{N_{noeventi}/NT_{noevent}}{P_{eventi}/PT_{event}}\right) \tag{2}$$

where: WOE_i = the weight of evidence for feature i relative to the target feature, $N_{noeventi}$ = the number of occurrences of the negative class of *target* due to feature i, P_{eventi} = the number of occurrences of the positive class of *target* due to feature i, PT_{event} = total number of the positive target in current bin for feature i, and $NT_{noevent}$ = total number of the negative target in current bin for feature i.

We find the information value (IV) of a feature by adding up the weights of all its WOE categories (bins) [39] and [40].

$$IV_i = \sum_{i=1}^{n} \left(\frac{N_{noeventi}}{NT_{noevent}} - \frac{P_{event}}{PT_{event}} \right) * WOE_i \qquad (3)$$

where: IV_i = the information value score for feature i, n = number of bins (separation categories) of feature i and WOE_i = weight of evidence for feature i. Based on the features' IV scores, only features with Medium, Strong, and Suspiciously Strong were selected as good predictors of country-level FI [41]. These served as candidates for *Stage 2* (Cross-Region Processing) of the study.

Stage 2: Feature Decoder. This step makes use of the output features of **Stage 1: Feature Ranking (IV), Feature Interpretation, and Selection**. Since at this stage, country-level sections had collapsed, these features were then regrouped into the general feature sets: *Money Management, Income, and Expenditure, Banks and Non-Banks, Demographics, Infrastructure Access, Payment Methods*, and *General*. The generalised group names were based on generalised names for the respective original sections across countries for both the *training* and *test* sets. With these feature groups in place, the AE was used to compact each group's features to one representative feature again in a similar procedure explained in *Stage 1: Feature decoder*. The output features of this stage form the input for the next ML stage.

Stage 2: Machine Learning. Similar to **Stage 1: Machine Learning**, we trained fifteen ML algorithms on the data. The main difference to the former stage is that, in this instance, only one dataset remains due to the data regrouping of the previous step of **Stage 2: Feature decoder**.

At this stage, the most suitable ML algorithm for predicting FI across the four countries is selected based on the best performance on the seven evaluation metrics.

Stage 2: Feature Ranking (IV), Feature Interpretation and Selection. This step is similar to **Stage 1: Feature Ranking (IV), Feature Interpretation, and Selection**. The main difference at this stage was that we made use of the generalised dataset. Hence, the feature ranking and computation of WOE and IV values follow Eqs. 2 and 3.

3.3   Visualisation Phase

Adding Shapley values [42] helps us learn more about how features can be used to predict FI, as shown in Eqs. 2 and 3. The Shapley values help to establish whether a particular feature has a positive impact on determining the target variable. The Shapley values for each feature were computed as follows:

$$\phi_i = \sum_{S \subseteq F \setminus \{i\}} \frac{|S|!(|F| - |S| - 1)!}{|F|!} [f_{S \cup \{i\}}(x_{S \cup \{i\}}) - f_S(x_S)] \tag{4}$$

where; $\frac{|S|!(|F|-|S|-1)!}{|F|!}$ is weight contribution of feature i on the Shaply value, and $f_{S \cup \{i\}}(x_{S \cup \{i\}}) - f_S(x_S)$ is the *marginal contribution* term. The *marginal contribution* term measures the performance of the model with and without the feature i. In this study, we used graphs to visually interpret the Shapley values for each feature both at the country and the cross-regional level.

4   Results and Discussions

Following from the Methodology Section and Fig. 1, the results of our research study are twofold, that is, at the country and cross-region level. First, we give country-level results, and later, we discuss cross-regional findings.

4.1   Country Level Results

For Eswatini, Fig. 3 shows the main predictive features, while Table 3 displays the performance of ML algorithms on the Eswatini dataset. In Table 3 we observe that the *catboost* (*catb*) algorithm was the best algorithm in predicting FI in Eswatini. This algorithm performed better than the others in four out of the seven evaluation metrics (accuracy, precision, recall, F1, Balanced Accuracy, Area under the ROC Curve (AUC), and Matthews's correlation coefficient (MCC)). It scored 58.9% MCC, 87% AUC, 79.2% Accuracy and 79.4% Balanced accuracy.

In terms of recall rate, the results of this research study for Eswatini data compare very closely to those obtained by [32]. In his study, the latter used SVC (support vector classifier) and logistic regression (lr) to predict FI and obtained 69.4% and 63.4%. In our case, SVC obtained 65% and lr had 68.3. However, besides not being exhaustive in the number of ML algorithms used, the researcher [32] had utilised only a limited portion of the Finscope dataset.

In Fig. 3, we note that the top three features contributing to predicting whether or not an individual is financially included are *QUALITY OF FINAN-CIAL SERVICES (E), EXPENDITURE AND INCOME (C)* and *REMIT-TANCES (L)*. The plot of Shapley values (Fig. 3) suggests that *EXPENDI-TURE AND INCOME (C)* contributes positively to the Shapley scores. This proposes that improving *EXPENDITURE AND INCOME (C)* enhances FI in Eswatini. The other two features *QUALITY OF FINANCIAL SERVICES (E)* and *REMITTANCES (L)*, are negatively contributing to the Shapley scores. This suggests that an improvement of both of these negatively affects FI in the Eswatini context.

For Namibia, Table 4 shows the performance of fifteen ML algorithms in the Namibian dataset. The *random forest* (*rf*) performed better against all other algorithms on four out of seven evaluation metrics. The best performance scores were on Balance Accuracy, Accuracy, F1, and MCC with values 85.4%, 87.9%, 91.4%, and 71% respectively. Figure 4 shows that the top four features for predicting FI in Namibia are *BANK PENETRATION (J)*, *GENERAL (L)*, *MONEY MANAGEMENT: Saving (F)* and *MONEY MANAGEMENT: RISK AND MITIGATION (H)*. The Shapley values plot shows a positive contribution to FI for *BANK PENETRATION (J)*. The other three features *GENERAL (L)*, *MONEY MANAGEMENT: Saving (F)* and *MONEY MANAGEMENT: RISK AND MITIGATION* appear to have a negative impact on the prediction of Namibian FI.

The benchmark of ML algorithms in the Rwandan data set is shown in Table 5. From this table, we observe that the *catboost (catb)* algorithm performed the best of all the algorithms in four of seven evaluation metrics. The high-performance metrics for *catb* were accuracy, precision, AUC, and MCC with values 85.7%, 79.1%, 85.2%, and 54.8% respectively. In Rwanda, the best features for predicting FI are given in Fig. 5. These include *BANKING/UMURENGE SACCOS AND MFI (K)*, *Money management-Remittances (J)*, *Household Information and Demographics (C)*, *E-Payments and Mobile Money (F)*, and *Household Register (B)*. These results suggest that in Rwanda, FI can be enhanced by improving the banking sector (K), remittances (J), and the location of the household (C).

For Madagascar, the performance of ML algorithms in the dataset according to seven evaluation metrics is shown in Table 6. Similar to Eswatini and Rwanda, the *catboost (catb)* classifier was the best algorithm for predicting FI in Madagascar. Here, the algorithm of choice was also closely contested by the *random forest (rf)* classifier. Therefore, the selection of the best algorithm was based on a direct comparison between the two classifiers. On this basis, the *catboost (catb)* was superior four times on Recall, F1 score, balanced accuracy, and AUC metrics with values 67.8%, 73.1%, 82.5% and 94.5% respectively. Figure 6 shows the best FI features for Madagascar. We note that the top predictive features exclude the individual's financial capacity or income level (C). Instead, *BANKING AND OTHER NON-MFIs (J)*, *REMITTANCES (F)*, and *HOUSE INFO AND WELL-BEING-FARMING (B)* are the top three main features suggested to be relevant to the FI challenge in Madagascar. Hence, contrary to Eswatini and Rwanda, Madagascar needs to not only improve service delivery through banks and improved remittances for better FI, but also through investing in community development and agriculture.

4.2   Cross Region Results

Table 7 shows the performance of fifteen ML algorithms subject to seven evaluation metrics on the unified dataset for the four countries. From this table, we observe the high performances of most of the algorithms across the given metrics. The high performances assert that the algorithms were effective in using the compact feature representation (that is, *Money Management, Income, and Expenditure, Banks, and Non Banks, Demographics, Infrastructure Access, Payment Methods* and *General*) to predict FI across all four countries. In Table 7, we also observe that the Gaussian Naïve Bayes (*gnb*) algorithm is the most effective in predicting FI across all countries for the compact dataset. This algorithm was the best compared to the other algorithms in three metrics: balanced accuracy (0.813), accuracy (0.840), and Matthews correlation coefficient (0.623). Interestingly, the *gnb* algorithm was second best in terms of *precision* (0.891) and *area under the curve* (0.915), losing to AdaBoost (*ada*) and *quadratic discriminant analysis* (*qda*) respectively with good scores.

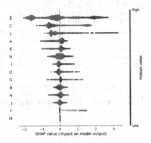

Fig. 3. Eswatini: Contributions of features towards FI.

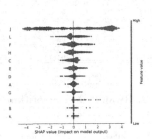

Fig. 4. Namibia: Contributions of features towards FI.

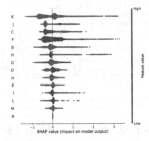

Fig. 5. Rwanda: Contributions of features towards FI.

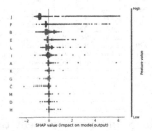

Fig. 6. Madagascar: Contributions of features towards FI.

Note:

- From Fig. 3 A=HOUSEHOLD QUESTIONS/RESPONDENT SELECTION, B=DEMOGRAPHICS, C=EXPENDITURE AND INCOME, D=FINANCIAL CAPABILITY, E=QUALITY OF FINANCIAL SERVICES, F=COMMERCIAL BANKS, G=PAYMENTS, H=Mobile money and Technology connectivity, I=SAVINGS, J=LOANS AND CREDIT, K=RISK MANAGEMENT AND INSURANCE, L=REMITTANCES, M=INFORMAL SERVICE PROVIDERS, and N=General
- From Fig. 4 A=HOUSEHOLD INFORMATION AND DEMOGRAPHICS, B=FARMING, C=INCOME AND EXPENDITURE, D=ACCESS TO INFRASTRUCTURE, E=FINANCIAL CAPABILITY, F=MONEY MANAGEMENT SAVING, G=MONEY MANAGEMENT-BORROWING, H=MONEY MANAGEMENT-RISK AND RISK MITIGATION, I=MONEY MANAGEMENT-REMITTANCES, J=BANK PENETRATION, K=INFORMAL PRODUCTS, and L=GENERAL
- From Fig. 5 A=Localization and Identification of the Household, B=Household Register, C=Household Information and Demographics, D=Access to Infrastructure, E=Financial Capacity, F=E-Payments and Mobile Money, G=Money Management-Saving/Investment, H=Money Management-Borrowing, I=Money Management-Risk and Risk Mitigation, J=Money Management - Remittances, K=BANKING/UMURENGE SACCOS AND MFI, L=Informal Products, M=Farming, and N=Income Sources.
- From Fig. 6 A=HOUSE ID. AND HEAD INCOME AND OCC., B=HOUSE INFO. AND WELL-BEING-FARMING, C=PERSONAL INCOME AND SPENDING, D=COMMUNITY INVOLVEMENT AND INFORMAL GROUPS, E=ACCESS TO AMENITIES AND TECHNOLOGY, F=REMITTANCES, G=RISK PLANNING AND INSURANCE, H=BORROWING (CREDIT/LOANS), I=SAVINGS AND INVESTMENTS, J=BANKING AND OTHER NON-MFIs), K=MFIs, L=Mobile Money, and M=General.

Table 3. Eswatini ML algorithm benchmark

Algorithm	Accuracy	Precision	Recall	F1	B.Accuracy	AUC	MCC
xgb	0.771	0.797	0.753	0.774	0.771	0.854	0.544
lgbm	0.779	0.804	0.765	0.783	0.780	0.858	0.560
mlp	0.706	0.735	0.689	0.710	0.707	0.773	0.415
catb	0.792	0.833	0.753	0.790	0.794	0.870	0.589
knn	0.659	0.681	0.653	0.666	0.660	0.727	0.320
svc	0.688	0.725	0.650	0.685	0.690	0.737	0.381
dt	0.716	0.726	0.735	0.730	0.716	0.716	0.432
ada	0.789	0.815	0.772	0.792	0.790	0.862	0.580
gnb	0.621	0.884	0.320	0.457	0.635	0.777	0.346
lda	0.629	0.778	0.551	0.556	0.633	0.729	0.325
qda	0.592	0.910	0.241	0.381	0.608	0.777	0.311
gbr	0.788	0.831	0.748	0.786	0.790	0.870	0.582
lr	0.675	0.691	0.683	0.687	0.675	0.711	0.349
etc.	0.776	0.824	0.728	0.772	0.779	0.847	0.560
rf	0.787	0.823	0.757	0.787	0.789	0.860	0.579

Table 4. Namibia ML algorithm benchmark

Algorithm	Accuracy	Precision	Recall	F1	B.Accuracy	AUC	MCC
xgb	0.854	0.896	0.895	0.896	0.825	0.926	0.650
lgbm	0.858	0.909	0.888	0.898	0.838	0.930	0.667
mlp	0.843	0.860	0.928	0.892	0.784	0.905	0.608
catb	0.873	0.914	0.905	0.910	0.852	0.937	0.699
knn	0.754	0.775	0.916	0.840	0.643	0.765	0.348
svc	0.757	0.747	0.989	0.851	0.598	0.895	0.353
dt	0.817	0.875	0.863	0.869	0.785	0.785	0.566
ada	0.854	0.893	0.900	0.896	0.822	0.917	0.647
gnb	0.715	0.869	0.713	0.773	0.717	0.800	0.413
lda	0.850	0.880	0.911	0.895	0.808	0.922	0.633
qda	0.649	0.935	0.538	0.677	0.725	0.884	0.422
gbr	0.872	0.911	0.907	0.909	0.848	0.934	0.695
lr	0.734	0.734	0.975	0.837	0.569	0.773	0.253
etc.	0.869	0.889	0.930	0.909	0.827	0.931	0.678
rf	0.879	0.913	0.915	0.914	0.854	0.932	0.710

Table 5. Rwanda ML algorithm benchmark

Algorithm	Accuracy	Precision	Recall	F1	B.Accuracy	AUC	MCC
xgb	0.851	0.743	0.512	0.606	0.731	0.846	0.532
lgbm	0.855	0.771	0.501	0.607	0.729	0.851	0.541
mlp	0.812	0.647	0.354	0.457	0.649	0.784	0.379
catb	0.857	0.791	0.494	0.607	0.728	0.852	0.548
knn	0.807	0.653	0.293	0.404	0.624	0.737	0.344
svc	0.806	0.703	0.231	0.348	0.601	0.744	0.324
dt	0.752	0.450	0.475	0.462	0.654	0.654	0.302
ada	0.833	0.709	0.440	0.542	0.693	0.827	0.466
gnb	0.803	0.616	0.321	0.421	0.631	0.746	0.341
lda	0.798	0.594	0.322	0.417	0.629	0.753	0.329
qda	NaN	NaN	NaN	NaN	NaN	NaN	NaN
gbr	0.841	0.748	0.444	0.556	0.700	0.841	0.491
lr	0.799	0.622	0.268	0.373	0.610	0.752	0.311
etc.	0.821	0.700	0.352	0.468	0.654	0.798	0.406
rf	0.838	0.755	0.412	0.533	0.687	0.824	0.475

Table 6. Madagascar ML algorithm benchmark

Algorithm	Accuracy	Precision	Recall	F1	B.Accuracy	AUC	MCC
xgb	0.934	0.787	0.678	0.728	0.825	0.932	0.693
lgbm	0.933	0.785	0.671	0.724	0.822	0.932	0.688
mlp	0.876	0.615	0.154	0.245	0.569	0.792	0.263
catb	0.935	0.794	0.678	0.731	0.825	0.945	0.697
knn	0.872	0.574	0.087	0.150	0.538	0.737	0.185
svc	0.869	0.000	0.000	0.000	0.500	0.676	-0.003
dt	0.897	0.601	0.621	0.610	0.780	0.780	0.551
ada	0.931	0.768	0.673	0.717	0.821	0.927	0.680
gnb	0.870	0.507	0.388	0.438	0.665	0.863	0.371
lda	0.893	0.678	0.342	0.453	0.659	0.882	0.431
qda	0.875	0.530	0.401	0.455	0.673	0.863	0.391
gbr	0.934	0.789	0.680	0.730	0.826	0.941	0.695
lr	0.869	0.513	0.035	0.064	0.515	0.758	0.103
etc.	0.916	0.797	0.478	0.597	0.730	0.919	0.576
rf	0.936	0.814	0.662	0.730	0.820	0.936	0.699

Table 8 and Fig. 7 present the predictive power of the generalised features across the four countries. It is observed that for all four countries, *Banks and Non-Banks* plays a significant role in predicting FI with an IV score of 3.783. The next generalised features with high predictive power are *Income and Expenditure*, *Demographics*, and *Money Management* with IV scores of 0.179, 0.043 and 0.024 respectively. Based on Table 8, we note that *Banks and Non-Banks* and *Income and Expenditure* have high FI predictive power across the region. While *Demographics*, and *Money Management* may have high predictive power in other countries, they are not entirely effective across the region.

In Fig. 7, we observe that not only do *Banks and Non-Banks* and *Income and Expenditure* have high FI impact, but also that it is suggested that *Banks and Non-Banks* have a positive influence, while *Income and expenses* have negative effects on FI. The same figure also suggests that *Demographics* and *Money Management* have a marginally positive impact on FI.

Table 7. Numeric ML algorithm performance of a generalised dataset for all countries.

Algorithm	Accuracy	Precision	Recall	F1	B.Accuracy	AUC	MCC
xgb	0.810	0.871	0.857	0.863	0.778	0.890	0.551
lgbm	0.814	0.871	0.863	0.867	0.781	0.895	0.560
mlp	0.842	0.877	0.902	0.889	0.802	0.918	0.617
catb	0.830	0.888	0.867	0.877	0.805	0.907	0.601
knn	0.822	0.864	0.886	0.875	0.779	0.904	0.568
svc	0.838	0.879	0.893	0.886	0.801	0.915	0.611
dt	0.791	0.846	0.858	0.852	0.745	0.745	0.495
ada	0.839	0.898	0.870	0.883	0.818	0.906	0.624
gnb	0.839	0.883	0.889	0.886	0.805	0.920	0.615
lda	0.847	0.892	0.890	0.891	0.817	0.919	0.635
qda	0.838	0.886	0.884	0.885	0.807	0.921	0.615
gbr	0.833	0.894	0.864	0.879	0.811	0.907	0.611
lr	0.834	0.852	0.925	0.886	0.772	0.919	0.586
etc.	0.821	0.876	0.867	0.871	0.789	0.899	0.576
rfc	0.831	0.890	0.867	0.878	0.806	0.897	0.604

Note: From Tables 3, 4, 5, 6, and 7 xgb = XGBoost Classifier, catb = CatBoost Classifier, lgbm = Light Gradient Boosting Machine Classifier, mlp = Multilayer Perceptron Classifier, knn= K-Nearest Neighbors Classifier, svc= Support Vector Classifier, dt = Decision Tree Classifier, ada = AdaBoost Classifier, gnb: Gaussian Naive Bayes, lda = Linear Discriminant Analysis, qda = Quadratic Discriminant Analysis, gbr = Gradient Boosting Classifier, lr = Logistic Regression, etc. = Extra Trees Classifier, rfc = Random Forest Classifier

Table 8. Information value (IV) scores of features from the generalised dataset for all countries.

Generalised Feature	IV	meaning
Banks and Non Banks	3.783	Suspicious predictive power
Income and Expenditure	0.179	Medium predictive power
Demographics	0.043	weak predictive power
Money Mngt	0.024	weak predictive power

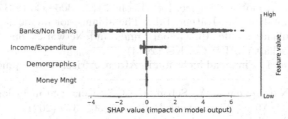

Fig. 7. Cross region generalised features: contributions of features towards FI

5   Conclusion

This paper proposed a DL-based method to deduce the set of generalisable and interpretable features to predict FI across four Sub-Saharan countries. The proposed approach overcomes the challenge of high feature dimension faced when working with survey data. While several ML techniques were used, the application of DL, through the use of the *autoencoder* was central in the reduction of feature dimensions. It was observed that while there exists some country-level variations in critical FI features, all country-level features can be generalised to *Money Management, Income, and Expenditure, Banks and Non-Banks, Demographics, Infrastructure Access, Payment Methods* and *General.* From these generalised features, it was established that the only critical features for cross-region FI computation were *Banks and Non-Banks, Income, and Expenditure, Money Management* and *Demographics.* For the generalised feature set, information value (IV) scores and graphs of feature Shapley scores revealed that the descending predictive power is *Banks and Non-Banks, Income, Expenditure, Demographics,* and *Money Management.* The use of Shapley scores was helpful in determining that the relationship of *Banks and Non-Banks* was generally positive across countries, while that of *Income and Expenditure* was negative towards the improvement of FI. Lastly, it was noted that the impact of *Demographics* and *Money Management* was generally positive. Based on these findings, Sub-Saharan countries are recommended to reform their financial institutional policies for increased accessibility and to enact strategies that promote employment rates, thus facilitating income generation, augmenting expenditure, and fostering economic sustainability.

References

1. Demirgüç-Kunt, A., Klapper, L.: Measuring financial inclusion: explaining variation in use of financial services across and within countries. Brook. Pap. Econ. Act. **2013**(1), 279–340 (2013)
2. World Bank. Development Research Group. Finance and Private Sector Development Team. The little data book on financial inclusion 2012. World Bank Publications (2012)
3. Augustin Kwasi Fosu: Growth, inequality, and poverty reduction in developing countries: recent global evidence. Res. Econ. **71**(2), 306–336 (2017)
4. Hannig, A., Leifheit, M., Rüther, L.L.: The alliance for financial inclusion: bringing smart policies to life. In: Global Funds and Networks, pp. 198–207. Nomos Verlagsgesellschaft mbH & Co. KG (2014)
5. Triki, T., Faye, I.: Financial inclusion in Africa. African Development Bank, p. 146 (2013)
6. Kebede, J., Naranpanawa, A., Selvanathan, S.: Financial inclusion: measures and applications to Africa. Econ. Anal. Policy **70**, 365–379 (2021)
7. Kabakova, O., Plaksenkov, E.: Analysis of factors affecting financial inclusion: ecosystem view. J. Bus. Res. **89**, 198–205 (2018)
8. Park, C.-Y., Jr Mercado, R.V.: Financial inclusion: new measurement and cross-country impact assessment. In: ADB Economics Working Paper Series, p. 539 (2018)
9. Asongu, S.A., Odhiambo, N.M.: Mobile banking usage, quality of growth, inequality and poverty in developing countries. Inf. Dev. **35**(2), 303–318 (2019)
10. Tambunlertchai, K.: Determinants and barriers to financial inclusion in Myanmar: what determines access to financial services and what hinders it? Singap. Econ. Rev. **63**(01), 9–26 (2018)
11. Alamgir, M., et al.: Economic, socio-political and environmental risks of road development in the tropics. Current Biol. **27**(20), R1130–R1140 (2017)
12. Claude, K., Abessolo, H.: Road infrastructure and intra-community trade in the central African economic and monetary community. Int. J. Sci. Res. Manag. (IJSRM) **10**(01), 2905–2913 (2022)

13. Barajas, A., Beck, T., Belhaj, M., Naceur, S.B.: Financial inclusion: what have we learned so far? What do we have to learn? IMF Work. Pap. **2020**(157) (2020)

14. Castaneda Aguilar, R.A., et al.: April 2022 update to the poverty and inequality platform (PIP) (2022)

15. Turner, M.A., Walker, P.D., Chaudhuri, S., Varghese, R.: A new pathway to financial inclusion: alternative data, credit building, and responsible lending (2012)

16. Duryea, S., Schargrodsky, E.: Financial services for the poor: savings, consumption, and welfare (2008)

17. Duda, R.O., Hart, P.E.: Pattern classification and scene analysis. Libr. Q. **44**, 258–259 (1974)

18. Elankavi, R., Kalaiprasath, R., Udayakumar, D.R.: A fast clustering algorithm for high-dimensional data. Int. J. Civ. Eng. Technol. (Ijciet) **8**(5), 1220–1227 (2017)

19. Zhou, L., Pan, S., Wang, J., Vasilakos, A.V.: Machine learning on big data: opportunities and challenges. Neurocomputing **237**, 350–361 (2017)

20. Verleysen, M., François, D.: The curse of dimensionality in data mining and time series prediction. In: Cabestany, J., Prieto, A., Sandoval, F. (eds.) IWANN 2005. LNCS, vol. 3512, pp. 758–770. Springer, Heidelberg (2005). https://doi.org/10.1007/11494669_93

21. Köppen, M.: The curse of dimensionality. In 5th Online World Conference on Soft Computing in Industrial Applications (WSC5), vol. 1, pp. 4–8 (2000)

22. Cámara, N., Tuesta, D.: Measuring financial inclusion: a muldimensional index. BBVA Research Paper (14/26) (2014)

23. Sarma, M., Pais, J.: Financial inclusion and development. J. Int. Dev. **23**(5), 613–628 (2011)

24. Khumalo, Z., Alhassan, A.L.: Read, write, develop: the socio-economic impact of literacy in south Africa. Int. J. Soc. Econ. **48**(8), 1105–1120 (2021)

25. Evans, O.: The effects of economic and financial development on financial inclusion in Africa (2015)

26. Ibragimova, A., Wang, Y., Ivanov, M.: Infrastructure development in Africa's regions: investment trends and challenges. In: E3S Web of Conferences, vol. 295, p. 01029. EDP Sciences (2021)

27. Ekeocha, D.O., Iheonu, C.O.: Household-level poverty, consumption poverty thresholds, income inequality and quality of lives in Sub-Saharan Africa. Afr. Dev. Rev. **33**(2), 234–248 (2021)

28. Demirgüç-Kunt, A., Klapper, L.F.: Financial inclusion in Africa: an overview. World Bank Policy Research Working Paper, p. 6088 (2012)

29. Fanta, A.B., Berkowitz, B., Khumalo, J., Mutsonziwa, K., Maposa, O., Ramsamy, P.: Digitisation of social grant payments and financial inclusion of grant recipients in South Africa-Evidence from Finscope surveys. FinMark Trust, Midrand (2017)

30. Porteous, D.: Just how transformational is m-banking. Commissioned by Finmark (2007)

31. Ismail, Q.F., Al-Sobh, E.S., Al-Omari, S.S., Yaseen, T.M.B., Abdullah, M.A.: Using machine learning algorithms to predict the state of financial inclusion in Africa. In: 2021 12th International Conference on Information and Communication Systems (ICICS), pp. 317–323. IEEE (2021)

32. Akinnuwesi, B.A., Fashoto, S.G., Metfula, A.S., Akinnuwesi, A.N.: Experimental application of machine learning on financial inclusion data for governance in Eswatini. In: Hattingh, M., Matthee, M., Smuts, H., Pappas, I., Dwivedi, Y.K., Mäntymäki, M. (eds.) I3E 2020. LNCS, vol. 12067, pp. 414–425. Springer, Cham (2020). https://doi.org/10.1007/978-3-030-45002-1_36

33. Cámara, N., Tuesta, D., et al.: Measuring financial inclusion: a multidimensional index. IFC Bulletins chapters, p. 47 (2018)
34. Song, C., Ristenpart, T., Shmatikov, V.: Machine learning models that remember too much. In: Proceedings of the 2017 ACM SIGSAC Conference on Computer and Communications Security, Texas, USA, November 2017
35. Jerez, J.M., et al.: Missing data imputation using statistical and machine learning methods in a real breast cancer problem. Artif. Intell. Med. **50**(2), 105–115 (2010)
36. Jäger, S., Allhorn, A., Bießmann, F.: A benchmark for data imputation methods. Front. Big Data **4**, 693674 (2021)
37. Ozsahin, D.U., Mustapha, M.T., Mubarak, A.S., Ameen, Z.S., Uzun, B.: Impact of feature scaling on machine learning models for the diagnosis of diabetes. In: 2022 International Conference on Artificial Intelligence in Everything (AIE), pp. 87–94. IEEE (2022)
38. Hinton, G.E., Salakhutdinov, R.R.: Reducing the dimensionality of data with neural networks. Science **313**(5786), 504–507 (2006)
39. Zdravevski, E., Lameski, P., Kulakov, A.: Weight of evidence as a tool for attribute transformation in the preprocessing stage of supervised learning algorithms. In: The 2011 International Joint Conference on Neural Networks, pp. 181–188. IEEE (2011)
40. Smith, E.P., Lipkovich, I., Ye, K.: Weight-of-evidence (woe): quantitative estimation of probability of impairment for individual and multiple lines of evidence. Hum. Ecol. Risk Assess. **8**(7), 1585–1596 (2002)
41. Siddiqi, N.: Credit Risk Scorecards: Developing and Implementing Intelligent Credit Scoring, vol. 3. John Wiley & Sons, Hoboken (2012)
42. SLundberg, S.M., Lee, S.-I.: A unified approach to interpreting model predictions. Adv. Neural Inf. Process. Syst. **30** (2017)

Viability of Convolutional Variational Autoencoders for Lifelong Class Incremental Similarity Learning

Jiahao Huo[(✉)] and Terence L. van Zyl

University of Johannesburg, Johannesburg 2092, Gauteng, South Africa
216045414@student.uj.ac.za, tvanzyl@uj.ac.za

Abstract. Incremental similarity learning in neural networks poses a challenge due to catastrophic forgetting. To address this, previous research suggests that retaining "image exemplars" can proxy for past learned features. Additionally, it is widely accepted that the output layers acquire task-specific features during later training stages, while the input layers develop general features earlier on. We lock the input layers of a neural network and then explore the feasibility of producing "embedding" models from a VAE that can safeguard the essential knowledge in the intermediate to output layers of the neural network. The VAEs eliminate the necessity of preserving "exemplars". In an incremental similarity learning setup, we tested three metric learning loss functions on CUB-200 (Caltech-UCSD Birds-200-2011) and CARS-196 datasets. Our approach involved training VAEs to produce exemplars from intermediate convolutional and linear output layers to represent the base knowledge. Our study compared our method with a previous technique and evaluated the baseline knowledge (Ω_{base}), new knowledge (Ω_{new}), and average knowledge (Ω_{all}) preservation metrics. The results show that generating exemplars from the linear and convolutional layers is the most effective way to retain base knowledge. It should be noted that embeddings from the linear layers result in better performance when it comes to new knowledge compared to convolutional embeddings. Overall, our methods have shown better average knowledge performance ($\Omega_{all} = [0.7879, 0.7805]$) compared to iCaRL ($\Omega_{all} = [0.7476, 0.7683]$) in the CUB-200 and CARS-196 experiments, respectively. Based on the results, it appears that it is important to focus on embedding exemplars for the intermediate to output layers to prevent catastrophic forgetting during incremental similarity learning in classes. Additionally, our findings suggest that the later linear layers play a greater role in incremental similarity learning for new knowledge than convolutions. Further research is needed to explore the connection between transfer learning and similarity learning and investigate ways to protect the intermediate layer embedding space from catastrophic forgetting.

A. Pillay et al. (Eds.): SACAIR 2023, CCIS 1976, pp. 237–252, 2023.
https://doi.org/10.1007/978-3-031-49002-6_16

1   Introduction

In machine learning, incremental learning is the updating of a model as new data become available or expanded to support further tasks. Ideally, an incrementally trained model should retain previously acquired knowledge while incorporating any new knowledge made available as it trains [22,28]. Many machine learning algorithms cannot retain prior knowledge or do so in an unsatisfactory manner. Models that do not incrementally learn new tasks whilst maintaining prior knowledge suffer from catastrophic forgetting. Catastrophic forgetting usually occurs during training on new data that contains no examples or highly unbalanced examples drawn from previously learned distributions [24]. Catastrophic forgetting in deep neural networks and virtually all of the tasks supported by them remains an open research problem [5–7]. Catastrophic forgetting is when a machine learning model ultimately or abruptly forgets previously learned information. Catastrophic forgetting occurs while learning new information about new data [11]. Historically, analyses have focused almost entirely on incremental supervised classification in neural networks such as those typically encountered in computer vision tasks [2,16,23,25]. However, there is still a lack of evidence detailing the degree to which similarity learning tasks [4,29] and the underpinning pair mining and loss functions are affected by catastrophic forgetting. Currently, majority of methods devised to mitigate catastrophic forgetting in deep neural networks have mainly been designed with classification tasks in mind, shown in research such as [23,25], and [2].

Previous work by Huo and van Zyl [10] compared four state-of-the-art algorithms to reduce catastrophic forgetting during incremental learning. These algorithms included Fully Connected VAEs (FullVAE) [10], Elastic Weight Consolidation [12], Encoder-Based lifelong learning [23], and incremental Classifier and Representation Learning (iCaRL) [25]. Their work analyzed several loss functions on MNIST, EMNIST, Fashion-MNIST, and CIFAR-10. All the techniques considered were effective but unsatisfactory. FullVAE and iCaRL were the most robust in numerous loss functions. However, given that the data sets are somewhat trivial and continuous learning happens on average over fewer than four or five incremental tasks, these results only partially indicate what is expected of real-world data with several tens of incremental tasks.

This study builds on the earlier research mentioned above. The article shows the consequences of catastrophic forgetting in class incremental similarity learning. The methodology approximates the catastrophic forgetting test procedure of Kemker et al. [11], which is described for classification tasks. Our work examines three-loss functions, Contrastive, Center and Triplet loss, using CUB-200 [31] and CAR196 [13] in similarity learning. We contrast the current state of the art for catastrophic forgetting during incremental similarity learning, FullVAE and iCaRL, to our solution ConVAEr. ConVAEr uses convolutional Variational Autoencoders (VAE) to generate representations fed into the convolutional layers to supplement previously seen knowledge. The process eliminates the need to regenerate entire images or keep a collection of previous data. We present results that allow the reader to analyse which techniques retain the most base knowledge, new knowledge, and overall knowledge during incremental metric learning. In summary:

1. We extended the FullVAE method by Huo and van Zyl [10] to the ConVAEr method, which yields better average knowledge retention across all experiments.
2. We expand the work by Huo and van Zyl [10] to include more challenging real-world data sets with several tens of incremental tasks.
3. We reinforce the importance of keeping prior knowledge during incremental similarity learning and demonstrate that image representations can replace exemplars.
4. We show that embeddings from convolutional layers retained base knowledge best, and linear layers retained new knowledge best.

2   Related Work

There has been an increase in the popularity of research related to incremental class learning [16]. Incremental learning solutions have evolved greatly from regularization-based methods to replay-based methods. Regularization methods constrain network parameters when updating by adding a penalty to the loss function or changing the gradient parameters [12,15,16]. Replay-based methods focus on storing exemplars from previous classes to remind the network when learning new classes. These include using generative models to generate past data for replay [27,33]. However, when dealing with complex datasets, using generative models to generate entire images can be challenging. Generative models often require alot of training data and long training times. However, they may still produce lower-quality samples compared to real image data, due to the complexity of the dataset, which can cause more challenges when used [1,14]. Moreover, generative models are prone to mode collapse, a situation in which they fail to capture all modes of the data distribution. Mode collapses causes a lack of variety in generated samples, that often happens where learning to generate images, it is a difficult task for generative models [17]. Current research focuses on incremental learning with an emphasis on classification. As stated by Huo and van Zyl [10], similarity learning is another possible area of focus for research into catastrophic forgetting. This study focuses on the latter, and we investigate methods in a retrieval setting instead of the traditional classification. Due to the complexity and lack of necessity to generate entire images, we extend the work by Huo and van Zyl [10] and explore the viability of generating embedding representations from intermediate convolutional layers on complex real-world datasets.

3   Similarity Learning

This paper focuses on catastrophic forgetting in metric learning methods, and as such, we consider three similarity learning loss functions:

- **Triplet loss** by Wang et al. [30], Schroff et al. [26] has demonstrated the effectiveness of learning representations for the assessment of image and video similarity, as highlighted in a study by Huo and van Zyl [9]. This loss function involves using a triplet, which consists of an anchor image (ground truth), a positive image (same class as the anchor image), and a negative image (a different class compared to the anchor image) [18]. The loss function concurrently optimizes the distance between the anchor and positive examples by reducing their distance apart while also increasing the distance between the anchor and negative images.
- **Contrastive loss** by [3] is a loss function that gets good representations by using pairs of positive images that belongs to the same class and negative images that belongs to different classes. It works by minimizing the distance between positive pairs and maximizing the distance between negative pairs within the embedding space, as discussed by Musgrave et al. [18].
- **Center loss** by Wen et al. [32] is used to learn a unique center for each class representation. The loss penalises the distances between image representations and their respective class centers. This approach aims to maximize the distance between different classes while keeping the representations of the same class closely together. Center loss is normally used in conjunction with the softmax loss.

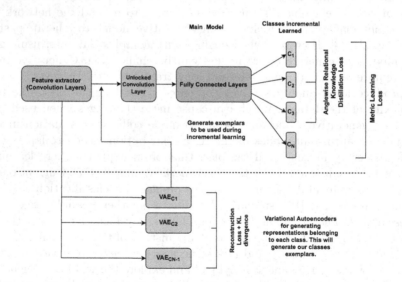

Fig. 1. The feature extractor is the convolution layers of our VGG-11 network. After initial training on the base-training set, the feature extractor is frozen except for the last set of convolutional layers [10].

4   Our Approach (ConVAEr)

Huo and van Zyl [10] takes ideas from the concepts presented in the research by Rannen et al. [23] and Rebuffi et al. [25]. They use Variational Autoencoders (VAEs) for generating individual classes in their approach. The VAEs are trained to generate representations specifically at the end of the convolutional layers. By using the VAEs, the authors can generate examples from classes seen at earlier incremental learning stages. It is necessary to freeze the convolutional layers post-initial training for this approach. The VAEs are then used to generate representations from previously encountered classes and then combined with new classes, thus enabling incremental learning during incremental similarity training. The work of Huo and van Zyl [10] is more closely aligned to overcome catastrophic forgetting. This outcome is achieved using previously learned representations and knowledge for future metric learning tasks. We extended the work by [10] using the author's method and moving the VAE to intercept the input to the convolution layer instead of the input to the fully connected layers shown in Fig. 1.

The autoencoder's reconstruction loss function must vary depending on the network's interception layer's activation function. For example, in our case, the convolutional encoder layers use the ReLU activation function. Therefore, we adopted the Binary Cross-Entropy objective function to compute reconstruction errors, augmented with the Kullback-Leibler divergence, as also employed in the study by Huo and van Zyl [10]. The loss function used to update the VAEs is as follows:

$$L_{VAE} = -\frac{1}{N} \sum_{i=1}^{N} y_i \cdot \log(p(y_i)) + (1 - y_i) \cdot \log(1 - p(y_i)) \tag{1}$$

$$+ \frac{1}{2}(\exp(\sigma^2) + \mu^2 - 1 - \sigma^2),$$

where σ^2 is the variance of the entire data set and μ is the mean. The first term is the Binary Cross-Entropy reconstruction loss, and the second is the Kullback-Leibler divergence.

We also introduce the angle-wise Distillation loss seen in the paper by Huo and van Zyl [10] on the examples generated from the VAEs. Angle-wise Distillation was proposed by Park et al. [20] for transferring mutual relations of data examples between a teacher and student network. Transferring relational knowledge is essential for similarity learning because we want to ensure that similar data remain close in the embedding space in the student network. The angle-wise Distillation loss is defined as:

$$L_A = \sum_{x_i, x_j, x_k \in X^3} \ell_h(V_A(t_i, t_j, t_k), V_A(s_i, s_j, s_k)), \tag{2}$$

where V_A is the angle-wise potential between the triplet images, x_i, x_j, x_k and ℓ_h refers to the Huber loss. t_i, t_j, t_k is the output of the teacher network for the triplet images. s_i, s_j, s_k is the output of the student network. This usage with network update during incremental learning is similar to the approach taken by Huo and van Zyl [10].

The incremental learning step makes use of the following loss function defined as:

$$L_{\text{incremental}} = l_{\text{metric learning}} + \lambda_{\text{distil}} * L_A, \tag{3}$$

where $l_{\text{metric learning}}$ are the three possible metric learning functions, as we state below. L_A is the angle-wise distillation loss for metric learning shown in Eq. 2. λ_{distil} is the importance we place on the angle-wise distillation loss. The student output is the output from the last layer we get from the network that is incrementally trained at each new training step. The teacher output is the last layer from the frozen network before performing the new incremental train step.

For Center loss, we use the distillation loss defined in [8] defined as:

$$L_D(t_o, s_o) = - \sum_{i=1}^{l} t_o^i log(s_o^i) \tag{4}$$

where l is the number of labels, t_o^i and $s_o i$ are the modified versions of the teacher model outputs and the current student model outputs. So, the modified loss function during the incremental step for Center loss is defined as:

$$L_{\text{incremental}} = l_{\text{center loss}} + \lambda_{\text{distil}} * L_D, \tag{5}$$

where L_D is defined above, and the rest remain the same.

Earlier work by Shin et al. [27] used a deep generative model to generate past images in their entirety but did not investigate how generating representations from a network's intermediate layers would perform. The study also did not use knowledge distillation techniques to constrain the update compared to our current approach, which uses angle-wise distillation suitable for similarity learning losses. The need for similarity learning specific modifications was unnecessary in previous studies, as these were predominantly focused on the classification task.

5   Methodology

5.1   Compared Methods

Incremental Classifier and Representation Learning (iCaRL). Incremental Classifier and Representation Learning (iCaRL) is a method proposed by Rebuffi et al. [25] for reducing catastrophic forgetting. We select iCarL as a baseline since it was reported in the article by Mai et al. [16] to be a state-of-the-art method for incremental class learning across multiple datasets when considering classification as a task. iCaRL relies primarily on storing exemplars of previously seen classes. The exemplar set of each class is constructed, using the herding algorithm, from the k closest images to the mean representation of the class. The stored exemplars supplement the incremental learning phase of new classes using knowledge distillation. Since iCaRL does not translate directly to similarity learning tasks, we implemented the modified version described in Huo and van Zyl [10].

Fully Connected VAE (FullVAE). Huo and van Zyl [10] proposed using a VAE to represent images in a representation that can be passed through intermediate layers in the network. The authors focused on preserving knowledge from the fully connected layers (flattened output from the last CNN layer). Their method outperformed iCaRL and other methods using test metrics similar to [11] during incremental similarity learning. Their work experimented with simple network architectures and various simple data sets not representative of real-world requirements.

5.2   Data Sets

All the compared methods are subjected to incremental learning scenarios on well-known data sets to analyze the impact of catastrophic forgetting for metric learning. The data sets used are CUB-200 [31] and CARS-196 [13]. CUB-200 consists of 200 classes of 11,788 images of 200 subcategories belonging to birds. CARS-196 consists of 16,185 images of 196 classes of cars. These are considered two primary datasets for comparison of similarity learning studies [18]. Ordinarily, the first half of the classes are used to train the network, while the second is utilized for testing. For the CUB-200 and CARS-196 data sets, we used the first 120 classes as our base training classes and 40 as incremental classes.

5.3   Architecture of CNN

We utilized a VGG-11 backbone deep neural network architecture. After the convolution layers, the first fully connected layer was changed from the original 4096 to 512 size, and the output layer was modified to 128 for similarity learning purposes. The network serves as an example of a commonly employed deep neural network architecture to assess the real-world performances of the respective methods more accurately. The FullVAEs architectures consisted of 512 input layers (the same size as the output after the convolution layers), followed by 256, 128, and 128 for the bottleneck. These layers are then reconstructed symmetrically in reverse. Our ConVAEs architecture consisted of a 2D Convolution layer with 512 input channels, 2D Convolution with 32 input channels, 2D Convolution with 16 input channels, and a bottleneck linear layer of input size 1600 and 256 output. These layers are reconstructed symmetrically in reverse so that they can perform reconstructions with deConvolutional layers. All kernels for the de/convolution layers are (3, 3).

5.4   Experimental Setup

Exemplars for iCaRL. Rebuffi et al. [25], used 2000 exemplars for CIFAR-100, which makes roughly 3% of the images available to iCaRL. Due to the limited number of images per class for CUB-200 and CARS-196 (60 or fewer images per class) and substantially above previous work. We allow iCaRL to retain 480 exemplars of the total training images of both data sets, approximately 5% for CARS-196 and 6% for CUB-200 of the total training images from 160 total classes.

Test Metrics. The test metrics used in this study vary from the incremental learning test metrics of Kemker et al. [11]. This adaptation is because the models we evaluate output a representation of a vector of features of size 128 per image. We replace the classification precision with the mean precision at R (mAP @ R). Our models do not classify images but use similarity learning for image retrieval. The average precision at R (AP@R) is typically calculated using a single query identity to retrieve the top R related relevant images from the database. The mAP at R is defined by Musgrave et al. [18] as:

$$
mAP@R = \frac{1}{R}\sum_{i=1}^{R} P(i); \quad \text{where} \quad P(i) = \begin{cases} precision \ at \ i & \text{if correct retrieval} \\ 0 & \text{otherwise} \end{cases},
$$
(6)

which is the mean Average Precision with the number of nearest neighbours for each sample to R. In this case, R is the total number of test images that belong to each class.

We want to assess four characteristics of each model: (i) how well a model retains the base knowledge it was initially trained on; (ii) how well a model learns new incremental knowledge; (iii) a combination of retained base and incrementally learned knowledge; and (iv) how well a model will generalize to completely unseen classes during this process.

We train a base model based on base classes to measure these characteristics. We then measure how well new and base knowledge is retained while incrementally learning new classes. We tracked the model's performance on the new classes by measuring the mAP@R. To this end, we utilize the test metrics as defined in the paper by Huo and van Zyl [10], which the authors have adapted from the metrics originally proposed by Kemker et al. [11]. Furthermore, we introduce a novel metric for testing performance in unseen classes, denoted as Ω_{unseen}. These metrics are:

$$
\Omega_{base} = \frac{1}{T-1}\sum_{i=2}^{T} \frac{\alpha_{base,i}}{\alpha_{ideal}}; \quad \Omega_{new} = \frac{1}{T-1}\sum_{i=2}^{T} \alpha_{new,i};
$$
$$
\Omega_{all} = \frac{1}{T-1}\sum_{i=2}^{T} \frac{\alpha_{all,i}}{\alpha_{ideal}}; \quad \Omega_{unseen} = \frac{1}{T-1}\sum_{i=2}^{T} \frac{\alpha_{unseen,i}}{\alpha_{unseen\_base}}
$$
(7)

where T is the total number of training sessions, $\alpha_{new,i}$ is the mAP @ R test for a new incremental class i immediately after learning. $\alpha_{base,i}$ is the test mAP@R on the base classes after i^{th} new class has been learned. $\alpha_{all,i}$ is the test mAP@R of all the classes seen so far. α_{ideal} is the offline model test mAP@R on the test set, which is assumed to be the ideal performance. Note that Ω_{base} and Ω_{all} are normalized with α_{ideal}. The evaluation metrics are between [0,1] unless the results exceed the offline model.

For equation Ω_{unseen}, T is the total number of training sessions, $\alpha_{unseen,i}$ is the mAP@R test for unseen classes immediately after incremental class i has been learned. α_{unseen_base} is the mAP@R test for the unseen classes after training on the base classes to initialize the network. Normalizing by α_{unseen_base} helps us observe any performance gains or decay in unseen classes due to incremental learning steps compared to just training on the base-train data set. If Ω_{unseen} is below one, it means there is a decay in performance. Alternatively, we see an improved unseen class performance if Ω_{unseen} exceeds one during incremental learning.

Training and Testing Setup. For training and testing, we use the data sets CUB-200 and CARS-196 and an adaption of the procedure for incremental learning by Kemker et al. [11]. For our results, we need to assess four characteristics of each model on the data sets: first, the (Ω_{base}) for how well a model retains the base knowledge it was initially trained on; second, the (Ω_{new}) for how well a model obtains new incrementally learned knowledge; third the (Ω_{all}) of the combination of retained and incrementally learned knowledge and forth the (Ω_{unseen}) for how well a model will generalize to completely unseen classes.

Table 1. Summary of data set subsets and splits

Subset	CARS-196 No. of classes	CUB-200 No. of classes	Train/Test Split
Baseline	120	120	80/20
Incremental	40	40	70/30
Unseen	36	40	–

We need a baseline subset to assess the retained base knowledge (Ω_{base}). We use 120 classes from each data set as a baseline subset. To assess the new incremental knowledge obtained (Ω_{new}), we use 40 additional unseen classes from each data set. Combining retained and incremental knowledge (Ω_{all}) does not require separate classes. Finally, CUB-200 and CARS-196 have 200 and 196 classes, respectively. As a result, 40 classes from CUB-200 and 36 from CARS-196 remain. We use these remaining classes to observe whether the performance of unseen classes (Ω_{unseen}) improves during incremental learning. For the retained base knowledge (Ω_{base}) and new incremental knowledge (Ω_{new}), we need a training and test split. These splits are used to evaluate out-of-sample performance[1]. Table 1 gives a detailed breakdown of the data splits. A high-level overview of the steps followed to train a model based on the procedure by Huo and van Zyl [10] is as follows:

1. Split the data set into baseline, incremental, and unseen subsets.
2. Split the baseline subset into base-training and base-test data sets.
3. Split the incremental set into inc-training and inc-test data sets.
4. Use the base training to train our initial base model.

---

[1] These training test splits cut across the classes within each subset (baseline and incremental).

5. Incremental Learning steps.
   (a) Use sequential pairs[2] of classes from the incremental subset to incrementally train the model.
   (b) Use the base-test data set for each incremental training step to assess retention of base knowledge (Ω_{base}).
   (c) Use the inc-test data set to assess new incremental knowledge obtained (Ω_{new}).
   (d) Use base-test plus inc-test data set to assess the combination of retained and incremental knowledge (Ω_{all}).
   (e) Use the unseen data set to assess how well the model generalizes to unseen classes (Ω_{unseen}).
6. Repeat Step 5 until all classes in the incremental subset are exhausted.

The pre-processing performed on the CUB-200 and CARS-196 data sets included normalizing the pixel values with a mean and standard deviation of [0.485, 0.456, 0.406], [0.229, 0.224, 0.225], respectively, for both data sets.

We used semi-hard negative mining on pairs of positive and negative images to generate triplets for the Triplet Loss. Additionally, we used pair-margin mining to create image pairs for the Contrastive Loss. The mining was performed using the Pytorch Metric Learning library [19] with the hyperparameters specified below.

The positive and negative margins for Contrastive loss were [0, 0.3841] for CUB-200 and [0.2652, 0.5409] for CARS-196 from Musgrave *et al.* [18]. Triplet loss margins were [0.0961, 0.1190] for CUB-200 and CARS-196, respectively. Hyperparameters [λ, α] for Center loss were [1.0, 0.5], respectively. We use Adam optimizer with a base set learning rate of 0.001 and 0.0001 incremental learning for both iCaRL and FullVAE. ConvVAEr used the SGD optimiser with the same learning rates.

For our method, we trained one variational autoencoder for each class the network has seen for each incremental training step. The same Adam optimizer was used for training but with a constant learning rate of 0.001. All base models were trained for 200 epochs and additional 100 epochs for each incremental step. Due to the limited training samples in each class, we did not do early stopping with validation. However, we first experimented with validation during "dummy" training to check for the best average epochs needed to reach decent performance before starting our experiments. We settled on 200 and 100 epochs for the base model and the incremental step training, respectively.

The experiments were repeated three times using three random seeds, and the average results of the experiments are reported below.

5.5 Hardware and Software

We used two machines. An Intel(R) Xeon(R) CPU E5-2683 processor and an Intel(R) Core(TM) i7-5820K both have 32 GB of RAM and a GTX 1080 TI 11

---

[2] i.e. First pair [class120,class121], second pair [class121,class122] and so on up till [class139,class140].

GB GPU. Both machines used Linux, Python 3.6, Pytorch 1.7.1 [21], Scikit-learn 0.23.2, and Pytorch Metric Learning 0.9.97.

6   Results and Discussion

6.1   Knowledge Preservation

Figures 2, and 3 highlight how each method implemented reduces catastrophic forgetting during sequential class learning by testing on a base-test set after each new class is introduced. The FullVAE and ConVAEr retain better base knowledge than iCaRL during incremental learning. We further observe from Figs. 2 that as we progress with incremental steps, iCaRL performance drops faster compared to the FullVAE and ConVAERr as the number of samples for each class is reduced.

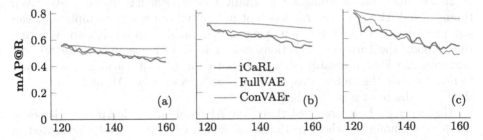

Fig. 2. CARS-196: mean average precision (mAP@R) on base-test for *(a) Triplet*, *(b) Contrastive*, and *(c) Center* loss combined for each incremental learning step.

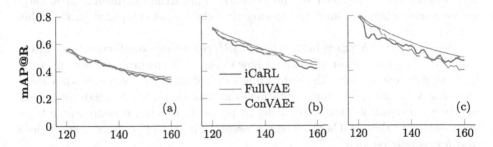

Fig. 3. CARS-196: mean average precision (mAP@R) on base-test plus inc-test for *(a) Triplet*, *(b) Contrastive*, and *(c) Center* loss combined for each incremental learning step.

Table 2 shows results using Eq. 7 for all the models. We tested the methods for their ability to retain previously learned knowledge and learn new knowledge

Table 2. Average mean precision of the incremental class test (mAP @ R) starting from the memorized Ω_{base} with the new classes Ω_{new} added and the overall result Ω_{all}

Data set	Loss	Ω_{base} iCARL	F'VAE	ConVAEr	Ω_{new} iCARL	F'VAE	ConVAEr	Ω_{all} iCARL	F'VAE	ConVAEr
CUB-200	Contrast'	0.8461	0.9272	**0.9920**	0.2102	0.1299	0.1243	0.7380	0.7856	**0.7879**
	Triplet	0.8593	0.8651	0.9710	0.2282	0.2084	0.1300	0.7476	0.7531	0.7784
	Center	0.8008	0.7564	0.8380	0.3020	**0.3052**	0.0985	0.7235	0.6990	0.6761
CARS-196	Contrast'	0.8683	0.8800	**0.9814**	0.4148	0.3843	0.1911	0.7573	**0.7805**	0.7801
	Triplet	0.8952	0.8809	0.9623	0.3467	0.2961	0.1658	0.7683	0.7765	0.7615
	Center	0.7650	0.7499	0.9393	**0.5006**	0.4219	0.1620	0.7090	0.6879	0.7575

by performing tests on both the base-test set (representing old learned classes) and the inc-test set (representing newly learned classes).

The results in Table 2 show ConVAEr as the most robust over a long period of incremental class learning for retaining base knowledge on data sets. Even though ConVAEr is noisy and does not use actual images as exemplars, it can still preserve a class well during incremental learning. However, it's worth noting that among the three loss functions paired with VAE exemplars, Center loss performs the least favorably, struggling with the task of learning new classes. Center loss has the highest drop in base knowledge on FullVAE and iCaRL, but they gain the most new knowledge.

However, we have observed that ConVAEr's ability to learn new classes is the lowest among the three methods, as evidenced by the results presented in the Ω_{new} column of Table 2. We speculate that this may be due to the generated samples from the VAEs in the ConVAEr method being noisier compared to those produced by VAEs of the FullVAE method. This increased noise in the generated samples leads to the occupation of larger regions in the embedding space by old classes, leaving fewer regions available for accommodating new classes as the new classes are introduced to the network. This situation results in a lower performance when it comes to learning new classes, as compared to the other two methods.

We observe ConVAEr is better than iCaRL regarding overall knowledge retention, but iCaRL is better at learning new classes. The results support our argument that we can preserve the embedding space of previously acquired knowledge by using VAE generated representations that can be passed through intermediate layers instead of images, resulting in performance that is comparable to or even better to that of iCaRL. The representation protects old-occupied regions and forms new regions.

Finally, in Table 2, we can observe that the Contrastive loss method retains the most base knowledge, followed by the Triplet loss and Center loss, as indicated by the Ω_{base} column. It is also worth noting that Center loss does not appear to work effectively with the FullVAE and ConvVAEr methods, as supported by the findings of Huo and van Zyl [10].

6.2   Performance on Unseen Classes

The section presents the Ω_{unseen} results on the iCaRL, FullVAE, and ConVAEr on CUB-200 and CARS-196 data set with the VGG-11 architecture using Eq. 7. The Ω_{unseen} results are normalized by the base trained model's unseen class performance. The normalization allows us to observe any decay or improvements from incrementally introducing the models with new classes.

Table 3. Incremental class test's mean average precision (mAP@R) for Ω_{unseen} following Eq. 7

Loss	CUB-200			CARS-196		
	iCARL	F'VAE	ConVAEr	iCARL	F'VAE	ConVAEr
Contrast'	0.7355	0.9525	**1.0040**	0.8247	0.8751	0.9799
Triplet	0.7822	0.8236	0.9853	0.9140	0.9689	**1.0116**
Center	0.5597	0.5528	0.4134	0.5661	0.4785	0.5056

Table 3 shows that ConVAEr's performance decayed the least on unseen performance, followed by FullVAE and iCaRL. The exception is the Center loss function. Furthermore, Triplet loss on CARS-196 and Contrastive loss on CUB-200 using the ConVAEr approach improved the unseen classes' performances over the model that has only trained on the base classes. Taking into account the results from both Table 2 and 3, the decay in base performance, Ω_{base}, are mostly correlated to decay in performances on unseen classes, Ω_{unseen}. The models utilizing the FullVAE and ConVAEr approach were the most robust to unseen class performance decay and even, in some cases, improved. Except for Center loss having the much higher deterioration in unseen performance, Ω_{unseen}, compared to the base performance, Ω_{base}. The results further reinforce that Center loss is not a suitable loss function for class incremental similarity learning. Finally, we can conclude that even though the Ω_{base} and Ω_{new} performance drops as we incrementally train. It does not always result in the decay of unseen class performance. With ConVAEr, we can observe some improvement over not incrementally training the model.

7   Conclusions and Future Work

The results in Table 2 show ConVAEr outperforming FullVAE. This is expected since the additional convolutional layers allow more information to be learned than the only linear layers in FullVAE. From the Ω_{new} results in Table 2, we can observe that iCaRL performed slightly better than FullVAE when learning new classes but vastly outperformed ConVAEr in some cases. In theory, ConVAEr should perform well for new classes, with an additional unlocked convolutional layer in line with FullVAE. We theorize that the VAE representation for the network's simpler intermediate layers (fully connected layers) was more accurate than the VAE representation for more complex intermediate layers (CNN layers), which are more difficult to replicate. The noisier VAE representations could cause difficulty learning new classes, which we can observe from Table 2. Our speculation for the need for better VAE for ConVAEr derived from the Center loss ConVAEr results that performed poorly in the new classes. For Center loss, if the VAE representations are not closely generated around previously optimal centres and VAE representations cover too much space around the centres and cause difficulty learning new classes.

We investigated the robustness ConVAEr method with Contrastive, Center, and Triplet loss functions for catastrophic forgetting during incremental similarity learning, and we have compared it to iCaRL and FullVAE. The results showed the effectiveness of each method, on the three different loss functions, against catastrophic forgetting. We found that Contrastive loss with similarity pair mining performed best at combating catastrophic forgetting on the base knowledge. ConVAEr performed the best average overall for base-test and inc-test data during incremental training. However, iCaRL is more effective when used with Center loss. iCaRL is also more effective at learning new classes.

Further research is necessary to investigate the connection between transfer learning and similarity learning and to explore methods for protecting the intermediate layer embedding space from catastrophic forgetting. Our future work involves designing an incremental learning solution structured around an ensemble-based approach, where each mini-network is responsible for one task. An attention-based model can be used to weight the contributions of the mini-networks during predictions. Furthermore, out-of-distribution techniques can be applied to the mini-networks to help detect classes that it have not been seen before, so the predictions from the mini-networks are either omitted or given much less weight. The predictions of the mini-networks can be concatenated or average before using it for retrievals.

References

1. Brock, A., Donahue, J., Simonyan, K.: Large scale GAN training for high fidelity natural image synthesis. In: International Conference on Learning Representations (2018)
2. Choi, E., Lee, K., Choi, K.: Autoencoder-based incremental class learning without retraining on old data. arXiv preprint arXiv:1907.07872 (2019)
3. Chopra, S., Hadsell, R., LeCun, Y.: Learning a similarity metric discriminatively, with application to face verification. In: 2005 IEEE Computer Society Conference on Computer Vision and Pattern Recognition (CVPR 2005), pp. 539–546. IEEE (2005)
4. Dlamini, N., van Zyl, T.L.: Author identification from handwritten characters using siamese CNN. In: 2019 International Multidisciplinary Information Technology and Engineering Conference (IMITEC), pp. 1–6. IEEE (2019)

5. Draelos, T.J., et al.: Neurogenesis deep learning: extending deep networks to accommodate new classes. In: 2017 International Joint Conference on Neural Networks (IJCNN), pp. 526–533. IEEE (2017)
6. Fernando, C., et al.: PathNet: evolution channels gradient descent in super neural networks. arXiv preprint arXiv:1701.08734 (2017)
7. Goodfellow, I.J., Mirza, M., Xiao, D., Courville, A., Bengio, Y.: An empirical investigation of catastrophic forgetting in gradient-based neural networks. arXiv preprint arXiv:1312.6211 (2013)
8. Hinton, G., Vinyals, O., Dean, J.: Distilling the knowledge in a neural network. arXiv preprint arXiv:1503.02531 (2015)
9. Huo, J., van Zyl, T.L.: Unique faces recognition in videos. In: 2020 23rd International Conference on Information Fusion (FUSION 2020) (2020)
10. Huo, J., van Zyl, T.L.: Incremental class learning using variational autoencoders with similarity learning. Neural Comput. Appl. 1–16 (2023)
11. Kemker, R., McClure, M., Abitino, A., Hayes, T.L., Kanan, C.: Measuring catastrophic forgetting in neural networks. In: Thirty-second AAAI Conference on Artificial Intelligence (2018)
12. Kirkpatrick, J., et al.: Overcoming catastrophic forgetting in neural networks. Proc. Natl. Acad. Sci. **114**, 3521–3526 (2017)
13. Krause, J., Stark, M., Deng, J., Fei-Fei, L.: 3D object representations for fine-grained categorization. In: 4th International IEEE Workshop on 3D Representation and Recognition (3dRR-13), Sydney, Australia (2013)
14. Lesort, T., Caselles-Dupré, H., Garcia-Ortiz, M., Stoian, A., Filliat, D.: Generative models from the perspective of continual learning. In: 2019 International Joint Conference on Neural Networks (IJCNN), pp. 1–8. IEEE (2019)
15. Li, Z., Hoiem, D.: Learning without forgetting. IEEE Trans. Pattern Anal. Mach. Intell. **40**, 2935–2947 (2017)
16. Mai, Z., Li, R., Jeong, J., Quispe, D., Kim, H., Sanner, S.: Online continual learning in image classification: an empirical survey. Neurocomputing **469**, 28–51 (2022)
17. Metz, L., Poole, B., Pfau, D., Sohl-Dickstein, J.: Unrolled generative adversarial networks. In: International Conference on Learning Representations (2016)
18. Musgrave, K., Belongie, S., Lim, S.-N.: A metric learning reality check. In: Vedaldi, A., Bischof, H., Brox, T., Frahm, J.-M. (eds.) ECCV 2020. LNCS, vol. 12370, pp. 681–699. Springer, Cham (2020). https://doi.org/10.1007/978-3-030-58595-2_41
19. Musgrave, K., Belongie, S., Lim, S.N.: Pytorch metric learning (2020)
20. Park, W., Kim, D., Lu, Y., Cho, M.: Relational knowledge distillation. In: Proceedings of the IEEE/CVF Conference on Computer Vision and Pattern Recognition, pp. 3967–3976 (2019)
21. Paszke, A., et al.: Pytorch: An imperative style, highperformance deep learning library. In: Wallach, H., Larochelle, H., Beygelzimer, A., d'Alché-Buc, F., Fox, E., Garnett, R. (eds.) Advances in Neural Information Processing Systems, vol. 32, pp. 8024–8035. Curran Associates Inc. (2019)
22. Polikar, R., Upda, L., Upda, S.S., Honavar, V.: Learn++: an incremental learning algorithm for supervised neural networks. IEEE Trans. Syst. Man Cybern. Part C (Appl. Rev.) **31**, 497–508 (2001)
23. Rannen, A., Aljundi, R., Blaschko, M.B., Tuytelaars, T.: Encoder based lifelong learning. In: Proceedings of the IEEE International Conference on Computer Vision, pp. 1320–1328 (2017)
24. Ratcliff, R.: Connectionist models of recognition memory: constraints imposed by learning and forgetting functions. Psychol. Rev. **97**, 285 (1990)

25. Rebuffi, S.A., Kolesnikov, A., Sperl, G., Lampert, C.H.: iCaRL: incremental classifier and representation learning. In: Proceedings of the IEEE conference on Computer Vision and Pattern Recognition, pp. 2001–2010 (2017)
26. Schroff, F., Kalenichenko, D., Philbin, J.: FaceNet: a unified embedding for face recognition and clustering. In: Proceedings of the IEEE Conference on Computer Vision and Pattern Recognition, pp. 815–823 (2015)
27. Shin, H., Lee, J.K., Kim, J., Kim, J.: Continual learning with deep generative replay. In: Advances in Neural Information Processing Systems, vol. 30 (2017)
28. Syed, N.A., Huan, S., Kah, L., Sung, K.: Incremental learning with support vector machines (1999)
29. Van Zyl, T.L., Woolway, M., Engelbrecht, B.: Unique animal identification using deep transfer learning for data fusion in siamese networks. In: 2020 IEEE 23rd International Conference on Information Fusion (FUSION), pp. 1–6. IEEE (2020)
30. Wang, J., et al.: Learning fine-grained image similarity with deep ranking. In: Proceedings of the IEEE Conference on Computer Vision and Pattern Recognition, pp. 1386–1393 (2014)
31. Welinder, P., et al.: Caltech-UCSD birds 200. Technical report CNS-TR-2010- 001. California Institute of Technology (2010)
32. Wen, Y., Zhang, K., Li, Z., Qiao, Yu.: A discriminative feature learning approach for deep face recognition. In: Leibe, B., Matas, J., Sebe, N., Welling, M. (eds.) ECCV 2016. LNCS, vol. 9911, pp. 499–515. Springer, Cham (2016). https://doi.org/10.1007/978-3-319-46478-7_31
33. Wu, Y., et al.: Incremental classifier learning with generative adversarial networks. arXiv preprint arXiv:1802.00853 (2018)

PuoBERTa: Training and Evaluation of a Curated Language Model for Setswana

Vukosi Marivate[1,2(✉)] (ID), Moseli Mots'Oehli[3], Valencia Wagnerinst[4] (ID),
Richard Lastrucci[1], and Isheanesu Dzingirai[1]

[1] Department of Computer Science, University of Pretoria, Hatfield, South Africa
`vukosi.marivate@cs.up.ac.za`
[2] Lelapa AI, Johannesburg, South Africa
[3] University of Hawaii at Manoa, Honolulu, USA
`moselim@hawaii.edu`
[4] Sol Plaatje University, Kimberley, South Africa
`valencia.wagner@spu.ac.za`

Abstract. Natural language processing (NLP) has made significant progress for well-resourced languages such as English but lagged behind for low-resource languages like Setswana. This paper addresses this gap by presenting PuoBERTa, a customised masked language model trained specifically for Setswana. We cover how we collected, curated, and prepared diverse monolingual texts to generate a high-quality corpus for PuoBERTa's training. Building upon previous efforts in creating monolingual resources for Setswana, we evaluated PuoBERTa across several NLP tasks, including part-of-speech (POS) tagging, named entity recognition (NER), and news categorisation. Additionally, we introduced a new Setswana news categorisation dataset and provided the initial benchmarks using PuoBERTa. Our work demonstrates the efficacy of PuoBERTa in fostering NLP capabilities for understudied languages like Setswana and paves the way for future research directions.

Keywords: Setswana · Natural Language Processing · Language Models

1 Introduction

The development of monolingual models for local languages is essential in better understanding individual languages and further developing tools for those languages [22]. It has been shown that monolingual models can further be improved for downstream tasks later on [8,37] or even allow for cross-lingual transfer [36] and also have a space in improving translation models [9]. Ultimately, the availability of more monolingual models and the curation of data to train them will increase the opportunities for bigger multilingual models [12], as they will have a larger pool of data to learn from, better evaluation for the individual languages and knowledge of the limitations of approaches and knowledge of the various limitations of approaches utilised by monolingual models (e.g. the development

© The Author(s), under exclusive license to Springer Nature Switzerland AG 2023
A. Pillay et al. (Eds.): SACAIR 2023, CCIS 1976, pp. 253–266, 2023.
https://doi.org/10.1007/978-3-031-49002-6_17

of subword segmentation to improve language models for Nguni languages [28]). Additionally, monolingual models will be better suited for tasks that require a deep understanding of a specific language [6].

In this work, we focus our efforts on creating a monolingual language model for Setswana. Setswana is a Bantu languages that is spoken in Botswana as well as several regions of South Africa [32]. Previous work has been done on creating monolingual datasets for Setswana (see NCHLT corpus [15]). This work has been valuable in providing a foundation for further research on Setswana. More recently the NCHLT corpus has been used to train language models that are available online without benchmarks[1] (as per the writing of this paper). The creation of more models provides diversity in approaches and benchmarks, which can help us to better evaluate the quality of resources available for a given language and understand the strengths and weaknesses of different techniques to natural language processing for Setswana. This paper contributes to the Setswana literature and also provides tools for researchers, and identifies areas where more research is needed.

PuoBERTa is a customised masked language model trained specifically for Setswana. It was developed to address the gap in natural language processing (NLP) performance for Setswana. This paper covers how we collected, curated, and prepared diverse multilingual texts to generate a high-quality corpus for PuoBERTa's training. We also built upon previous efforts in creating monolingual resources for Setswana, and evaluated PuoBERTa across several NLP tasks, including part-of-speech (POS) tagging, named entity recognition (NER), and news categorization. Additionally, we introduce a new Setswana news categorization dataset and provide the initial benchmarks using PuoBERTa.

2   Related Work

There have been a few attempts to create a monolingual Setswana language model. TswanaBert[2] [29] is one of the earliest models available for Setswana. It was trained on a combination of datasets from scraped news headlines from social media[3] [27] and the Leipzig Corpus[4] [10,17].

A more recent model based on the NCHLT [15], Autshumato, Leipzig and Webcrawl corpora is the NCHLT RoBERTa Model[5] [14]. For both TswanaBert and NCHLT Setswana RoBERTa language models, at the time of writing, there are no online benchmarks that compare the approaches to other work, such as multilingual models that include the Setswana language. As such we later provide benchmarks for these models on Setswana downstream tasks.

More recently, work on African language models has focused on creating multilingual models. The multilingual approaches aim to exploit the connections

---

[1] http://www.rma.nwu.ac.za/handle/20.500.12185/641.
[2] https://huggingface.co/MoseliMotsoehli/TswanaBert.
[3] https://zenodo.org/record/5674236.
[4] https://corpora.uni-leipzig.de/.
[5] http://www.rma.nwu.ac.za/handle/20.500.12185/641.

between African languages to allow languages with low resources to benefit from those that have more resources. Work by Adelani et.al. [2] focused on creating parallel corpora, for machine translation, for a number of African languages including Setswana. The corpora is then used to fine-tune pre-trained translation models such as MT5 [39], MBART [24], ByT5 [38] and M2M-100 [16]. Other work focused on adapting XLM-RoBERTa into African languages including Setswana [5], training multilingual BERT for African languages [31]. Additionally, with the advent of the age of massive large language models, we have models such as BLOOM [35] which have Setswana within and more Afrocentric models such as Serengeti [1] that are now available.

3   Curating Setswana Corpora PuoData and Pre-training PuoBERTa

3.1   Curating Corpora to Train Word and Sentence Representations

Modern language models require a large amount of data to develop. The challenges of obtaining data for low-resource languages have been well-documented [30,33,34]. In this work, we gather a number of datasets to pre-train Setswana language models. The data was includes data from research organisations, books, official government documents, and online content. The final dataset is referred to as PuoData. Puo means "language" in Setswana.

We believe that PuoData is a valuable resource for the Setswana language community. We hope that PuoData will be used to develop new and innovative applications that benefit the Setswana-speaking community. One of the input corpora that we gather during this project (starting in 2020) was the JW300 monolingual dataset for Setswana from the OPUS Corpora [7]. However, JW300 is no longer an open source dataset available online. As such, we will refer to PuoData+JW300, which is the dataset including a historical JW300 archive we had access to before it was removed from the OPUS. We provide benchmarks with PuoData and PuoData+JW300 in the interest of language development. PuoData+JW300 is a larger dataset than PuoData alone (see Table 1). It contains more text (even if it is text that is religious in nature) than all the other datasets combined.

The data, detailed in Table 1, contains sources such as the NCHLT Setswana [15] corpus, the South African Constitution[6], the Leipzig Setswana BW and ZA corpora. We also include more recent corpora such as the Vuk'zenzele Setswana Corpora [21][7] and South African Cabinet Speeches [21][8]. All of the corpora contain Setswana data only.

PuoData is made available as a single dataset for researchers[12]. We now focus on how we built pre-trained models with PuoData and PuoData+JW300.

---

[6]   https://www.justice.gov.za/constitution/SAConstitution-web-set.pdf.
[7]   https://huggingface.co/datasets/dsfsi/vukuzenzele-monolingual.
[8]   https://huggingface.co/datasets/dsfsi/gov-za-monolingual.
[12]   https://github.com/dsfsi/PuoData, https://huggingface.co/datasets/dsfsi/PuoData.

Table 1. Setswana monolingual curated corpora with relative sizes

Dataset Name	Kind	Num. of Tokens
PuoData contents		
NCHLT Setswana [15]	Government Documents	1,010,147
Nalibali Setswana	Childrens Books	57,654
Setswana Bible	Book(s)	879,630
SA Constitution	Official Document	56,194
Leipzig Setswana Corpus BW	Curated Dataset	219,149
Leipzig Setswana Corpus ZA	Curated Dataset	218,037
SABC Dikgang tsa Setswana FB (Facebook)	News Headlines	167,119
SABC MotswedingFM FB	Online Content	33,092
Leipzig Setswana Wiki	Online Content	230,333
Setswana Wiki	Online Content	183,168
Vukuzenzele Monolingual TSN	Government News	157,798
gov-za Cabinet speeches TSN	Government Speeches	591,920
Department Basic Education TSN	Education Material	708,965
PuoData Total	25 MB on disk	**4,513,206**
PuoData+JW300		
JW300 Setswana [4]	Book(s)	19,782,122
PuoData+JW300 Total	124 MB on disk	**24,295,328**
NCHLT RoBERTa Reported[11]	Mixture	14,518,437

3.2   Training the Masked Language Model: PuoBERTa

We first trained two BPE Tokenizers for the PuoBERTa. One tokenizer with
PuoData and the other with PuoData+JW300 corpora with 52000 tokens each.
The pre-trained masked language models were trained on an NVIDIA Titan
RTX with 24 GB of VRAM on an Intel Core i9-9900K with 32 GB of RAM. The
training information is shown in Table 2.

Table 2. Pre-training setup for PuoBERTa and PuoBERTaJW300 (trained with Puo-
Data+JW300) using the Huggingface library.

	PuoBERTa	PuoBERTaJW300
Epochs	100	40
Steps	525,000	1,585,000
Time Training (approx)	3 days	3 days

The pre-trained models are made available for download for research and development[13],[14] [26]. We next focus on evaluating the new models on a number of downstream tasks.

4 Evaluation of PuoBERTa on Downstream Tasks

The PuoBERTa models were fine-tuned and evaluated on three downstream tasks: Named Entity Recognition (NER), Part of Speech (POS) tagging, and news categorisation/classification. The results showed that the PuoBERTa model achieved state-of-the-art results on all three tasks in terms of F1 score amongst the monolingual models.

4.1 Evaluation on MasakhaNER

We first evaluate PuoBERTa on the MasakhaNER 2.0 [3] benchmark, which is a dataset for named entity recognition in a number of African languages including Setswana. We compare the performance of PuoBERTa to NCHLT RoBERTa and the multilingual models reported in the MasakhaNER 2.0 paper [3], namely Afriberta, AfroXLMR-base and AfroXLMR-large models. The results in Table 3 show that PuoBERTa is very competitive. It does not beat the multilingual models in this case. PuoBERTa+JW300 gets much closer than the rest of the monolingual models.

Table 3. Performance of models on the MasakhaNER datasets

Model	Test Performance (f1 score)
Multilingual Models [3]	
AfriBERTa [31]	83.2
AfroXLMR-base	87.7
AfroXLMR-large	**89.4**
Monolingual Models	
NCHLT TSN RoBERTa	74.2
PuoBERTa	78.2
PuoBERTa+JW300	*80.2*

4.2 Evaluation on MasakhaPOS

For this downstream task we use the same evaluations script as per the Masakha-POS paper [11] to evaluate our PuoBERTa variants plus the NCHLT RoBERTa. The results are shown in Table 4. In this test PuoBERTa almost beats the best multilingual model (AfroLM) but PuoBERTa+JW300 beats all the models available.

---

[13] https://github.com/dsfsi/PuoBERTa.
[14] https://huggingface.co/dsfsi/PuoBERTa.

Table 4. Performance of models on the MasakhaPOS datasets

Model	Test Performance (f1 score)
Multilingual Models [11]	
AfroLM [13]	*83.8*
AfriBERTa	82.5
AfroXLMR-base	82.7
AfroXLMR-large	83.0
Monolingual Models	
NCHLT TSN RoBERTa	82.3
PuoBERTa	83.4
PuoBERTa+JW300	**84.1**

4.3   News Categorisation - Daily News

Dataset Information. We gather our dataset from the official website of the Government of Botswana's Daily News[15]. The website primarily features news content in the English language, while also incorporating some of their materials in Setswana within the designated *Dikgang* segment. Our compiled dataset, denoted as *Daily News - Dikgang*, needed annotation of news categories. Notably, the news items were initially devoid of categorisation information. To rectify this, we embarked on the task of categorising these news items, leveraging the International Press Telecommunications Council (IPTC) News Categories (or codes)[16]. For the categorisation process, we employed the highest hierarchical level of the news code taxonomy. Table 5 provides an overview of these top-level codes, along with their corresponding Setswana translations, which guided our annotation efforts. For this process we involved two expert Setswana annotators who initially annotated the dataset individually and then came together to deal with disagreements on labelling afterwards leading to the final dataset that we use for this paper and also make available. See Appendix A for the Data Statement.

Following the annotation process, it is worth noting that not all annotated data was employed in the subsequent evaluation of models for the classification task. Our strategy involved focusing on the top 10 categories, each comprised of a minimum of 80 labelled samples in the full dataset. This selection criterion was informed by the frequency distribution observed within the collected dataset (see Fig. 1). As such the final full dataset has a total of 4867 samples, 3893, 487 and 487 samples for train, dev and test splits respectively.

In order to delve into the textual data, we present a 2-dimensional visualisation of the news dataset in Fig. 2, where each category is represented by a unique color. This visualisation was constructed by applying a T-distributed Stochas-

---

[15] https://dailynews.gov.bw/.
[16] https://iptc.org/standards/newscodes/.

Table 5. International Press Telecommunications Council (IPTC) Media Codes and Setswana Translations

English Category	Setswana Category
arts, culture, entertainment and media	Botsweretshi, setso, boitapoloso le bobegakgang
conflict, war and peace	Kgotlhang, ntwa le kagiso
crime, law and justice	Bosenyi, molao le bosiamisi
disaster, accident and emergency incident	Masetlapelo, kotsi le tiragalo ya maemo a tshoganyetso
economy, business and finance	Ikonomi, tsa kgwebo le tsa ditšhelete
education	Thuto
environment	Tikologo
health	Boitekanelo
human interest	Tse di amanang le batho
labour	Bodiri
lifestyle and leisure	Mokgwa wa go tshela le boitapoloso
politics	Dipolotiki
religion and belief	Bodumedi le tumelo
science and technology	Saense le thekenoloji
society	Setšhaba
sport	Metshameko
weather	Maemo a bosa

tic Neighbour Embedding (TSNE) [25] model to the TF-IDF representation of the news corpus data. The visualisation gives us an initial view of the distribution of the data and how separable the categories might be. We make the complete dataset available alongside this paper[17] [26], fostering openness and reproducibility.

Training and Performance. For the news categorisation task, we performed 5-fold cross-validation using the different models at hand. We performed the cross validation by combining the train and dev sets and then doing an 80/20 (train/validate) split 5 times randomly. We compared PuoBERTa, PuoBER-TaJW300 to baselines of the NCHLT RoBERTa Model, and also logistic regression with TFIDF. We use the TFIDF+logistic regression to provide an insight on how well the more powerful models perform as well as make sure we have a more efficient baseline. We further provide the performance on the test set for the models we have. For all the RoBERTa-based models, the fine-tuning was at 5 epochs. The results of our classification task experiments are shown in Table 6.

There is a performance improvement by using the RoBERTa based models. The performance improvements by PuoBERTa and PuoBERTaJW300 are even better respectively. To note, the logistic regression TFIDF vectorizer is only

---

[17] https://github.com/dsfsi/PuoBERTa.

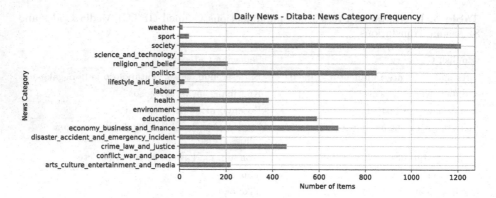

Fig. 1. Frequency of news categories in the *Daily News - Ditaba* dataset

Table 6. Setswana News Categorisation Task Results

Model	5-fold Cross Validation F1	Test F1
Logistic Regression + TFIDF	60.1	56.2
NCHLT TSN RoBERTa	64.7	60.3
PuoBERTa	63.8	62.9
PuoBERTaJW300	**66.2**	**65.4**

trained using the news training data while the RoBERTa tokenizers have not been specifically trained using the Daily News data. It is still an interesting result as development of the news dataset is also useful for less powerful models. To better understand the classification performance, we took the best performing model of the PuoBERTaJW300 and show its confusion matrix in Fig. 3. We observe in this confusion matrix that the models makes more mistakes when trying to predict *economy_business_finance*, *politics*, *society* categories.

4.4   Observations Given the Downstream Tasks

PuoBERTa in general is a competitive model. When we include JW300 data (PuoBERTaJW300), the model improves further and performs as state of the art for the downstream tasks. This brings up a question of what is lost by not being able to use JW300 as a language resource for Setswana and other low resource languages covered by JW300? The performance of PuoBERTa is close to that of PuoBERTaJW300 and is reassuring as it would allow more researchers to focus on building better African language corpora.

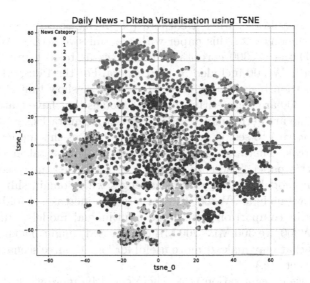

Fig. 2. TSNE Visualisation of TFIDF representation of top 10 category news stories. Categories: 0:*arts_culture_entertainment_and_media*; 1:*crime_law_and_justice*; 2:*disaster_accident_and_emergency_incident*; 3:*economy_business_and_finance*; 4:*education*; 5:*environment*; 6:*health*; 7:*politics*; 8:*religion_and_belief*; 9:*society*.

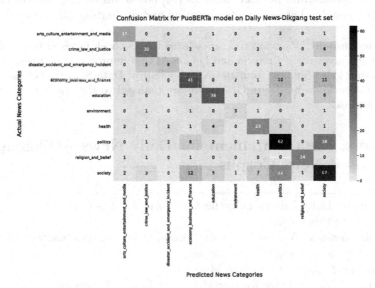

Fig. 3. Confusion Matrix PuoBERTa Model

5   Conclusion and Future Work

In this paper, we presented the creation of a number of artifacts as we pursued creating a state of the art pre-trained Language Model for the Setswana Language. We were able to gather a corpora (of diverse sources) named PuoData that we release along with this paper with a permissive license. We further discuss the PuoData+JW300 corpora which allows us to measure the impact of pretraining with JW300 included in the corpora. For the POS, NER and a new news categorisation task (*Daily News - Dikgang*) we show that PuoBERTa and PuoBERTaJW300 are competitive language models that have state of the art or near state of the art performance in those downstream tasks.

We look forward to seeing how the community uses this model and different application areas. There are a few directions to still be taken that we could not cover in this paper. The corpora needs to be further analysed for bias [18, 19,23] and the models evaluated for challenges with domain shift [2]. We have shown that if we include JW300, the corpora size increases 4 fold, but at what cost? Further in comparison with large multilingual models, PuoBERTa and PuoBERTaJW300 are 300 MB (200 MB compressed) might benefit from further optimisations that may make it more useful for low resource scenarios (compute, access to internet etc.).

For the news categorisation task, there could be more work done to understand how both the simple logistic regression model and other more complex ones can benefit from more feature or label engineering of the dataset to combine labels that might be two similar in hindsight. We also suggest looking into how we can standardise datasets such as [20] etc. to all use a standard categorisation set to enable better transfer learning and evaluation for news.

Acknowledgement. We want to acknowledge the feedback received from colleagues at DSFSI at Univeristy of Pretoria and Lelapa AI that improved this paper. We would like to acknowledge funding from the ABSA Chair of Data Science, Google and the NVIDIA Corporation hardware Grant. We are thankful to the anonymous reviewers for their feedback.

Appendix: Data Statement for Daily News - Dikgang Categorised News Corpus

---

Dataset name: Daily News - Dikgang Categorised News Corpus
Citations: Cite this paper.
Dataset developer(s): Vukosi Marivate (vukosi.marivate@cs.up.ac.za) and Valencia Wagner (valencia.wagner@spu.ac.za)
Data statement author(s): Vukosi Marivate
Organisation: Data Science for Social Impact Research Group
https://dsfsi.github.io),
Department of Computer Science, University of Pretoria, South Africa
and Sol Plaatje University

A. CURATION RATIONALE

The motivation for building this dataset was to provide one of the few annotated news categorisation datasets for Setswana. The task required identifying a high-quality Setswana news dataset, collecting the data, and then annotating leveraging the International Press Telecommunications Council (IPTC) News Categories (or codes)[18]. The identified source was the Daily News[19] (Dikgang Section) from the Botswana Government. All copyright for the news content belongs to Daily News. We collected 5000 Setswana news articles. The distribution of final categories for the dataset are shown in Fig. 1.

B. LANGUAGE VARIETY

The language of this data set is Setswana (primarily from Botswana).

C. SPEAKER DEMOGRAPHIC

Setswana is a Bantu languages that is spoken in Botswana as well as several regions of South Africa [32].

D. ANNOTATOR DEMOGRAPHIC

Two annotators were used to label the news articles based on the *Daily News - Dikgang* news. Their deomographic information is shown in Table 7.

E. PROVENANCE APPENDIX

The original data is from the Daily News news service from the Botswana Government.

Table 7. Annotator demographic

	1	2
Description	Annotator	Annotator
Age	30–35	20–25
Gender	Female	Male
Race/ethnicity	Black/African	Black/African
First Language(s)	Setswana	Setswana and isiXhosa
Linguistics training	Often works as a Setswana - English interpreter	Studied linguistic anthropology and works as a translator/interpreter

---

[18] https://iptc.org/standards/newscodes/.
[19] https://dailynews.gov.bw/.

References

1. Adebara, I., Elmadany, A., Abdul-Mageed, M., Alcoba Inciarte, A.: SERENGETI: massively multilingual language models for Africa. In: Findings of the Association for Computational Linguistics: ACL 2023, pp. 1498–1537. Association for Computational Linguistics, Toronto, Canada (2023). https://doi.org/10.18653/v1/2023.findings-acl.97
2. Adelani, D., et al.: A few thousand translations go a long way! leveraging pre-trained models for African news translation. In: Proceedings of the 2022 Conference of the North American Chapter of the Association for Computational Linguistics: Human Language Technologies, pp. 3053–3070. Association for Computational Linguistics, Seattle, United States (2022). https://doi.org/10.18653/v1/2022.naacl-main.223
3. Adelani, D., et al.: MasakhaNER 2.0: Africa-centric transfer learning for named entity recognition. In: Proceedings of the 2022 Conference on Empirical Methods in Natural Language Processing, pp. 4488–4508 (2022)
4. Agić, Ž., Vulić, I.: JW300: a wide-coverage parallel corpus for low-resource languages. In: Proceedings of the 57th Annual Meeting of the Association for Computational Linguistics, pp. 3204–3210. Association for Computational Linguistics, Florence, Italy (2019). 10.18653/v1/P19-1310
5. Alabi, J.O., Adelani, D.I., Mosbach, M., Klakow, D.: Adapting pre-trained language models to African languages via multilingual adaptive fine-tuning. In: Proceedings of the 29th International Conference on Computational Linguistics, pp. 4336–4349. International Committee on Computational Linguistics, Gyeongju, Republic of Korea (2022)
6. Armengol-Estapé, J., et al.: Are multilingual models the best choice for moderately under-resourced languages? A comprehensive assessment for Catalan. In: Findings of the Association for Computational Linguistics: ACL-IJCNLP 2021, pp. 4933–4946. Association for Computational Linguistics, Online (2021). https://doi.org/10.18653/v1/2021.findings-acl.437
7. Aulamo, M., Sulubacak, U., Virpioja, S., Tiedemann, J.: OpusTools and parallel corpus diagnostics. In: Proceedings of The 12th Language Resources and Evaluation Conference, pp. 3782–3789. European Language Resources Association (2020)
8. Baziotis, C., Zhang, B., Birch, A., Haddow, B.: When does monolingual data help multilingual translation: the role of domain and model scale. arXiv preprint arXiv: 2305.14124 (2023)
9. Burlot, F., Yvon, F.: Using monolingual data in neural machine translation: a systematic study. In: Proceedings of the Third Conference on Machine Translation: Research Papers. pp. 144–155. Association for Computational Linguistics, Brussels, Belgium (Oct 2018). 10.18653/v1/W18-6315
10. Collection, L.C.: Tswana web text corpus (South Africa) based on material from 2019. https://corpora.uni-leipzig.de/en?corpusId=tsn_community_2017. Accessed 22 Aug 2023
11. Dione, C.M.B., et al.: MasakhaPOS: part-of-speech tagging for typologically diverse African languages. In: Proceedings of the 61st Annual Meeting of the Association for Computational Linguistics, vol. 1 Long Papers, pp. 10883–10900. Association for Computational Linguistics, Toronto, Canada (2023). https://doi.org/10.18653/v1/2023.acl-long.609

12. Doddapaneni, S., et al.: Towards leaving no indic language behind: building monolingual corpora, benchmark and models for indic languages. In: Annual Meeting of The Association For Computational Linguistics (2022). https://doi.org/10.18653/v1/2023.acl-long.693

13. Dossou, B.F., et al.: AfroLM: a self-active learning-based multilingual pretrained language model for 23 African languages. SustaiNLP **2022**, 52 (2022)

14. Eiselen, R.: Nchlt Setswana roberta language model (2023). https://hdl.handle.net/20.500.12185/641

15. Eiselen, R., Puttkammer, M.J.: Developing text resources for ten South African languages. In: LREC, pp. 3698–3703 (2014)

16. Fan, A., et al.: Beyond english-centric multilingual machine translation (2020)

17. Goldhahn, D., Eckart, T., Quasthoff, U.: Building large monolingual dictionaries at the Leipzig corpora collection: from 100 to 200 languages. In: Proceedings of the Eighth International Conference on Language Resources and Evaluation (LREC 2012), pp. 759–765. European Language Resources Association (ELRA), Istanbul, Turkey (2012)

18. Gyasi, F., Schlippe, T.: Twi machine translation. Big Data Cogn. Comput. **7**(2), 114 (2023). https://doi.org/10.3390/bdcc7020114

19. Haddow, B., Bawden, R., Miceli Barone, A.V., Helcl, J., Birch, A.: Survey of low-resource machine translation. Comput. Linguist. **48**(3), 673–732 (2022)

20. Ifeoluwa Adelani, D., et al.: MasakhaNEWS: news topic classification for African languages. arXiv e-prints pp. arXiv-2304 (2023)

21. Lastrucci, R., et al.: Preparing the Vuk'uzenzele and ZA-gov-multilingual south african multilingual corpora. In: Fourth workshop on Resources for African Indigenous Languages (RAIL), p. 18 (2023)

22. Limisiewicz, T., Malkin, D., Stanovsky, G.: You can have your data and balance it too: towards balanced and efficient multilingual models. In: Proceedings of the 5th Workshop on Research In Computational Linguistic Typology and Multilingual NLP, pp. 1–11. Association for Computational Linguistics, Dubrovnik, Croatia (2023)

23. Litre, G., et al.: Participatory detection of language barriers towards multilingual sustainability(ies) in Africa. Sustainability **14**(13), 8133 (2022). https://doi.org/10.3390/su14138133

24. Liu, Y., et al.: Multilingual denoising pre-training for neural machine translation. arXiv preprint arXiv: 2001.08210 (2020)

25. Van der Maaten, L., Hinton, G.: Visualizing data using t-SNE. J. Mach. Learn. Res. **9**(11), 2579–2605 (2008)

26. Marivate, V., Mots'Oehli, M., Wagner, V., Lastrucci, R., Dzingirai, I.: Puoberta + puoberta Setswana language models (2023). https://doi.org/10.5281/zenodo.8434795

27. Marivate, V., et al.: Investigating an approach for low resource language dataset creation, curation and classification: Setswana and Sepedi. In: Proceedings of the first workshop on Resources for African Indigenous Languages, pp. 15–20. European Language Resources Association (ELRA), Marseille, France (2020)

28. Meyer, F., Buys, J.: Subword segmental language modelling for Nguni languages. In: Conference On Empirical Methods In Natural Language Processing (2022). https://doi.org/10.48550/arXiv.2210.06525

29. Motsoehli, M.: Tshwanabert (2020). https://huggingface.co/MoseliMotsoehli/TswanaBert

30. Nekoto, W., et al.: Participatory research for low-resourced machine translation: a case study in African languages. In: Findings of the Association for Computational Linguistics: EMNLP 2020, pp. 2144–2160. Association for Computational Linguistics, Online (2020). https://doi.org/10.18653/v1/2020.findings-emnlp.195

31. Ogueji, K., Zhu, Y., Lin, J.: Small data? no problem! exploring the viability of pretrained multilingual language models for low-resourced languages. In: Proceedings of the 1st Workshop on Multilingual Representation Learning, pp. 116–126. Association for Computational Linguistics, Punta Cana, Dominican Republic (2021)

32. Palai, E.B., O'Hanlon, L.: Word and phoneme frequency of occurrence in conversational Setswana: a clinical linguistic application. South. Afr. Linguist. Appl. Lang. Stud. **22**(3–4), 125–142 (2004)

33. Ragni, A., Knill, K.M., Rath, S.P., Gales, M.J.: Data augmentation for low resource languages. In: INTERSPEECH 2014: 15th Annual Conference of the International Speech Communication Association, pp. 810–814. International Speech Communication Association (ISCA) (2014)

34. Ranathunga, S., Lee, E.S.A., Prifti Skenduli, M., Shekhar, R., Alam, M., Kaur, R.: Neural machine translation for low-resource languages: A survey. arXiv e-prints pp. arXiv-2106 (2021)

35. Scao, T.L., et al.: What language model to train if you have one million GPU hours? In: Conference On Empirical Methods in Natural Language Processing (2022). https://doi.org/10.48550/arXiv.2210.15424

36. de Souza, L.R., Nogueira, R., Lotufo, R.: On the ability of monolingual models to learn language-agnostic representations. arXiv preprint arXiv: 2109.01942 (2021)

37. de Vries, W., Bartelds, M., Nissim, M., Wieling, M.: Adapting monolingual models: data can be scarce when language similarity is high. In: Findings of the Association for Computational Linguistics: ACL-IJCNLP 2021, pp. 4901–4907. Association for Computational Linguistics (2021). https://doi.org/10.18653/v1/2021.findings-acl.433

38. Xue, L., et al.: Byt5: towards a token-free future with pre-trained byte-to-byte models. arXiv preprint arXiv: 2105.13626 (2021)

39. Xue, L., et al.: mT5: a massively multilingual pre-trained text-to-text transformer. arXiv preprint arXiv: 2010.11934 (2020)

Hierarchical Text Classification Using Language Models with Global Label-Wise Attention Mechanisms

Jaco du Toit[1,2](✉) [iD] and Marcel Dunaiski[1,2] [iD]

[1] Department of Mathematical Science, Computer Science Division,
Stellenbosch University, Stellenbosch, South Africa
{jacodutoit,marceldunaiski}@sun.ac.za
[2] School for Data Science and Computational Thinking, Stellenbosch University,
Stellenbosch, South Africa

Abstract. Hierarchical text classification (HTC) is a natural language processing task with the objective of categorising text documents into a set of classes from a structured class hierarchy. Recent HTC approaches combine encodings of the class hierarchy with the language understanding capabilities of pre-trained language models (PLMs) to improve classification performance. Furthermore, label-wise attention mechanisms have been shown to improve performance in HTC tasks by placing larger weight on more important parts of the document for each class individually to obtain label-specific representations of the document. However, using label-wise attention mechanisms to fine-tune PLMs for downstream HTC tasks has not been comprehensively investigated in previous work. In this paper, we evaluate the performance of a HTC approach which adds label-wise attention mechanisms, along with a label-wise classification layer, to a PLM which is subsequently fine-tuned on the downstream HTC dataset. We evaluate several existing label-wise attention mechanisms and propose an adaptation to one of the approaches which separates the attention mechanisms for the different levels of the hierarchy. Our proposed approach allows the prediction task at a certain level to leverage the information gained from the predictions performed at all ancestor levels. We compare the different label-wise attention mechanisms on three HTC benchmark datasets and show that our proposed approach generally outperforms the other label-wise attention mechanisms. Furthermore, we show that without using the complex techniques proposed in recent HTC approaches, our relatively simple approach outperforms state-of-the-art approaches on two of the three benchmark datasets.

Keywords: Hierarchical Text Classification · Large Language Models · Attention Mechanisms

1 Introduction

Due to the rapid increase in the availability of digital text documents, many text classification approaches have been proposed to categorise document collections

© The Author(s), under exclusive license to Springer Nature Switzerland AG 2023
A. Pillay et al. (Eds.): SACAIR 2023, CCIS 1976, pp. 267–284, 2023.
https://doi.org/10.1007/978-3-031-49002-6_18

into a set of classes. However, these approaches typically use very general or high-level categories such that users are often required to navigate through a large number of documents when using these categories as filters. Therefore, it is beneficial to categorise these documents into finer-grained classes to improve the accessibility of relevant information.

Hierarchical text classification (HTC) can be used to address this problem by categorising documents into a set of fine-grained classes from a hierarchical class structure. Therefore, HTC systems are suitable approaches for the organisation of a large document collection as they allow users to reduce their search scope from the large collection of documents to a smaller subset of documents with detailed categories. Furthermore, the hierarchical class structure allows users to select the level of detail that they prefer before applying queries or filters on the collection of documents.

Many HTC approaches have been proposed which leverage the hierarchical class structure in various ways to improve classification performance [3,4,7,10, 11,20,22,31–33,36]. However, recent approaches encode the class hierarchy and combine these encodings with pre-trained language models (PLMs) to achieve state-of-the-art performance on HTC benchmark datasets [10,11,32]. Furthermore, several approaches have added label-wise attention mechanisms to deep learning architectures to improve their performance on multi-label and hierarchical text classification tasks [17,18,21,27,30]. These label-wise attention mechanisms obtain label-specific weighted averages of the document features by placing more weight on the most important features for each class separately. Therefore, each class has a representation of the text document which summarises the features of the document that are most important for the classification of that class.

The use of these label-wise attention mechanisms for fine-tuning PLMs on HTC tasks has not been extensively researched in previous work. Therefore, we investigate a HTC approach which adds a label-wise attention mechanism and classification layer to a PLM and fine-tunes the model for a given downstream HTC task. We compare the performance of several label-wise attention mechanisms to determine the most suitable approach for HTC tasks. These attention mechanisms include a label-wise attention layer with a standard dot product alignment function, and an adapted version of the general attention alignment function. Furthermore, we investigate an attention mechanism which splits the attention layers into the different levels of the class hierarchy and uses the information of the parent-level predictions during the prediction task at a certain level. Lastly, we propose an adaptation to this approach which uses the predictions of all ancestor levels during the prediction task at a certain level.

Through comprehensive experiments on three benchmark HTC datasets we show that our proposed label-wise attention mechanism generally outperforms the other approaches. Furthermore, we show that our relatively simple approach obtains state-of-the-art performance on two out of three benchmark HTC datasets. Lastly, we provide further analysis on the behaviour of our approach by contrasting its per hierarchy level performances against the other approaches and investigating the impact that the number of training instances per level has on classification performance.

2 Background

2.1 Hierarchical Text Classification

HTC tasks typically represent the structured class hierarchy as a directed acyclic graph $\mathcal{H} = (C, E)$. $C = \{c_1, \ldots, c_L\}$ is the set of class nodes, where L is the number of classes and E is the set of edges that describe the hierarchical relationships between the class nodes. We consider HTC tasks where each node has a single parent node (apart from the root node which has none) such that the class taxonomy is a tree structure. Therefore, the objective of HTC is to classify a text document comprising T tokens $\mathbf{x} = [x_1, \ldots, x_T]$ into a class set $Y' \subseteq C$ which includes one or more paths in $\mathcal{H}$.

2.2 Attention Mechanisms

Attention mechanisms have been widely used in various deep learning architectures, and most notably, they are the building blocks of transformer models [28]. In the generalised definition of the attention mechanism, we consider a query $\mathbf{Q}$ and a set of key-value pairs $\mathbf{K}$ and $\mathbf{V}$ [28]. Attention weights ($\boldsymbol{\alpha}$) are computed between the query $\mathbf{Q}$ and each of the keys in $\mathbf{K}$ by calculating an alignment score and passing the result through a softmax function as follows:

$$\boldsymbol{\alpha} = \mathrm{softmax}\left(a(\mathbf{Q}, \mathbf{K})\right) \tag{1}$$

where a is an alignment (or compatibility) function which measures the similarity between the given query and key to determine which parts of $\mathbf{V}$ corresponding to that key to focus attention on. The two alignment functions we consider in this paper are referred to as the dot product alignment function:

$$a(\mathbf{Q}, \mathbf{K}) = \mathbf{Q}\mathbf{K}^T \tag{2}$$

and the general alignment function:

$$a(\mathbf{Q}, \mathbf{K}) = \mathbf{Q}\mathbf{W}\mathbf{K}^T \tag{3}$$

where $\mathbf{W}$ is learnt during training.

The output can then be calculated as a weighted sum over the values in $\mathbf{V}$ as $\mathbf{c} = \boldsymbol{\alpha}\mathbf{V}$. The intuition behind this process is that each element of $\boldsymbol{\alpha}$ represents how much weight is placed on each element of $\mathbf{V}$, such that the weighted average captures the most important features of $\mathbf{V}$.

Label-wise attention mechanisms extend this idea by learning a query vector separately for each of the classes in the associated classification task. Therefore, each class obtains separate attention weights which determine how much weight should be placed on each of the values in $\mathbf{V}$ for that class. This allows each class to learn which features are most important for accurately classifying an instance as belonging to that class or not.

2.3   Pre-trained Language Models

In recent years, the field of NLP has undergone a significant paradigm shift due to the introduction of transformer-based [28] language models that are trained through self-supervised tasks on unlabelled text data.

Devlin et al. [5] proposed the Bidirectional Encoder Representations from Transformers (BERT) model which is pre-trained through two unsupervised tasks, mask language modelling (MLM) and next sentence prediction (NSP). BERT obtains deep bidirectional representations of tokens through the MLM task by randomly masking a percentage of the input tokens and instructing the model to predict the masked tokens. NSP is used to pre-train the model by predicting the relationship between two sentences, i.e., whether the two sentences follow each other or not. BERT uses these techniques to extract semantic information by using the surrounding text of a given token to establish context.

Liu et al. [19] proposed the Robustly optimised BERT pre-training Approach (RoBERTa) which improves the original BERT model through various techniques. These techniques include model architecture adaptations such as removing the NSP pre-training task and applying dynamic masking to the training data. Furthermore, RoBERTa uses larger batch sizes, a larger pre-training data volume, and an increased training time.

3   Related Research

Solutions for HTC tasks are typically classified into three categories: flat, local, and global approaches.

Flat approaches ignore the hierarchical class structure by "flattening" the class set to transform the HTC task into a multi-label classification task. These approaches use various deep learning architectures such as convolutional neural networks (CNNs) [17,21], recurrent neural networks [2,25,30], and transformer-based language models [18,27]. Mullenbach et al. [21] introduced the Convolutional Attention for Multi-Label classification (CAML) model which uses a CNN with a dot product label-wise attention mechanism. The label-wise attention mechanism obtains label-specific weighted averages of the features obtained from the convolutional layer by placing attention on the most important features for each class. Li and Yu [17] improved on this by introducing the Multi-Filter Residual Convolutional Neural Network (MultiResCNN) which combines the label-wise attention mechanism from CAML with multi-filter convolutional layers [12] and residual connections [8]. Later, Vu et al. [30] developed the Label Attention Model (LAAT) which uses a bidirectional LSTM [9] followed by a label-wise attention mechanism. However, LAAT uses an adaptation of the general attention function as opposed to the dot product attention used in CAML and MultiResCNN. Furthermore, Vu et al. [30] proposed the JointLAAT model which projects the predictions of the classes at a certain level to a vector and concatenates it to the feature vectors in the lower-level layers in order to leverage the information of the higher-level predictions. Liu et al. [18] proposed the Hierarchical Label-wise Attention Transformer (HiLAT) model which splits the text document into chunks and uses token- and chunk-level label-wise attention mechanisms sequentially. Most recently, Strydom et al. [27] proposed the M-XLNet model which adds the label-wise attention mechanism from CAML on top of a pre-trained XLNet [34] language model and fine-tunes this model on the downstream dataset.

Local approaches attempt to leverage the hierarchical structure of the class taxonomy by creating multiple classifiers and combining their results in various ways to increase the classification accuracy on HTC tasks [1,6,14,15,26].For example, Kowsari et al. [15] proposed the Hierarchical Deep Learning for Text Classification (HDLTex) model which trains a deep neural network classifier for each non-leaf node in the class hierarchy. Other local approaches initialise the weights of classifiers in lower hierarchy levels with the weights of their parent classifiers as a form of transfer learning [1,26].

Unlike local approaches which combine the results from multiple models, global approaches use one model to leverage the hierarchical class structure. Many global methods have been proposed [3,4,7,10,11,20,22,31–33,36] which use various techniques such as recursive regularisation [7], reinforcement learning [20], meta-learning [33], and capsule networks [22]. However, more recent approaches have encoded the structured class hierarchy through a graph encoder [29] and used this information in various ways to improve classification performance [3,4,11,31,32,36]. Chen et al. [3] proposed the hierarchy-aware label semantics matching network (HiMatch) which maps the text embeddings and class hierarchy into the same embedding space and transforms the HTC task into a semantic matching task. Furthermore, Wang et al. [31] proposed Hierarchy Guided Contrastive Learning (HGCLR) which uses contrastive learning techniques to embed the hierarchical class structure into a PLM. Later, Huang et al. [10] introduced the Hierarchy-Aware T5 model with Path-Adaptive Attention Mechanism (PAAMHiA-T5) which fine-tunes a pre-trained T5 model [23] to generate labels from the class hierarchy with a path-adaptive attention mechanism which enables the model to adaptively focus on the path of the currently generated class to avoid noise from the other parts of the hierarchical class structure. Jiang et al. [11] proposed Hierarchy-guided BERT with Global and Local hierarchies (HBGL) which passes hierarchical class embeddings to BERT along with the input text, and a mask token for each level in the hierarchy. Therefore, the objective of the model is to predict the labels at each level by generating the tokens at the positions of the masked tokens. Similarly, Wang et al. [32] proposed the Hierarchy-aware Prompt Tuning (HPT) method which modifies the input sequence that is passed to BERT by adding hierarchy-aware prompts obtained from a graph encoder along with a masked token for each level of the class hierarchy such that the HTC task resembles the MLM pre-training task.

4   Methodology

Figure 1 depicts the high-level architecture of our proposed HTC approach.

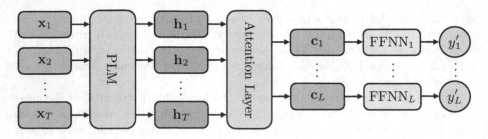

Fig. 1. Our model architecture. We pass the text token sequence (orange) through the PLM to obtain the final hidden states for each token (blue). The final hidden states are used in the label-wise attention mechanism which determines a label-specific representation of the document for each class (green) and passes the representations through associated FFNNs to obtain a confidence score for each class (yellow). (Color figure online)

4.1   PLM

Our approach passes the input token sequence $\mathbf{x} = [x_1, \ldots, x_T]$ to the PLM which converts each token through the embedding layer to obtain the token embeddings $\mathbf{E} = [\mathbf{e}_1, \ldots, \mathbf{e}_T]$. We pass these embeddings through the PLM to obtain $\mathbf{H} = [\mathbf{h}_1, \ldots, \mathbf{h}_T] \in \mathbb{R}^{T \times d_h}$, where $\mathbf{h}_i \forall i \in \{1, 2, \ldots, T\}$ is the final hidden state of the i-th token and d_h is the dimension size of the hidden states. We compare the BERT and RoBERTa PLMs in our experiments.

4.2   Attention Layer

We consider the label-wise attention mechanisms proposed by Mullenbach et al. [21] (CAML) and Vu et al. [30] (LAAT) to convert the concatenated final hidden states ($\mathbf{H}$) to a matrix $\mathbf{C} \in \mathbb{R}^{L \times d_h}$, where L is the number of classes and each row $\mathbf{c}_j \forall j \in \{1, 2, \ldots, L\}$ is a label-specific vector representation of the text document. These two attention mechanisms use similar techniques to obtain the attention weights but differ in that they use the dot product attention (DPA) and general attention (GA) functions respectively. Using the DPA approach the attention weights are calculated as follows:

$$\boldsymbol{\alpha} = \text{softmax}(\mathbf{U}_{\text{DPA}}\mathbf{H}^T) \tag{4}$$

where $\mathbf{U}_{\text{DPA}} \in \mathbb{R}^{L \times d_h}$ is a weight matrix that is learnt during training and $\boldsymbol{\alpha} \in \mathbb{R}^{L \times T}$ is the attention weight matrix which represents the attention weights of each token for each class. The GA approach extends DPA such that the attention weights are calculated as follows:

$$\mathbf{Z} = \tanh(\mathbf{Q}_{\text{GA}}\mathbf{H}^T) \tag{5}$$

$$\boldsymbol{\alpha} = \text{softmax}(\mathbf{U}_{\text{GA}}\mathbf{Z}) \tag{6}$$

where $\mathbf{Q}_{GA} \in \mathbb{R}^{d_p \times d_h}$ and $\mathbf{U}_{GA} \in \mathbb{R}^{L \times d_p}$ are weight matrices that are learnt during training and d_p is a chosen hyperparameter.

The attention weight matrix is used to calculate the label-wise representations of the document as follows:

$$C = \alpha H \tag{7}$$

where the j-th row $(\mathbf{c}_j)$ of the matrix $\mathbf{C} \in \mathbb{R}^{L \times d_h}$ represents the information of the document with regard to class j.

4.3 Output Layer

The output layer uses the label-wise document representations from the $\mathbf{C}$ matrix and passes each row through a separate feed-forward neural network (FFNN) with a sigmoid activation function (σ). Therefore, the confidence score of the document belonging to class j is calculated as follows:

$$y'_j = \sigma(\mathbf{c}_j \mathbf{w}_j^T + b_j) \tag{8}$$

where $\mathbf{c}_j$ is the j-th row in $\mathbf{C}$. We pack the weight vectors $(\mathbf{w}_1, \ldots, \mathbf{w}_L)$ and bias units $(b_1, \ldots, b_L)$ into a weight matrix $(\mathbf{W} \in \mathbb{R}^{L \times d_h})$ and bias vector $(\mathbf{b}_W \in \mathbb{R}^L)$ to compute the predictions of all classes as follows:

$$\mathbf{y}' = \sigma\big((\mathbf{W} \otimes \mathbf{C})\mathbf{1} + \mathbf{b}_W\big) \tag{9}$$

where $\otimes$ is the element-wise multiplication operator and $\mathbf{1} \in R^{d_h}$ is a vector of ones which is used to obtain sums over the hidden dimensions.

These confidence scores are used during training to minimize the binary cross-entropy loss which is calculated as:

$$\mathcal{L} = -\sum_{j=1}^{L} \big(y_j \log(y'_j) + (1 - y_j)\log(1 - y'_j)\big) \tag{10}$$

where $y_j \in \{0, 1\}$ is the ground truth label for class j.

During inference, the confidence scores are used to predict the multi-hot vector $\mathbf{Y}' = [Y'_1, \ldots, Y'_L]$ which represents the class set associated with the text document and is calculated as:

$$Y'_j = \begin{cases} 1, & y'_j \geq \gamma \\ 0, & y'_j < \gamma \end{cases} \tag{11}$$

where γ is a threshold that determines whether the class is assigned to the document or not.

4.4 Hierarchical Attention Mechanisms

We also consider the label-wise attention mechanism proposed by Vu et al. [30] for the JointLAAT model which uses the parent-level predictions and splits the attention layers for the levels of the class hierarchy. Figure 2 illustrates the high-level architecture of this approach which we refer to as the hierarchical label-wise attention (HLA) mechanism. Suppose we have a structured class hierarchy with K levels, and N_k is the number of classes in level $k \in \{1, \ldots, K\}$. HLA obtains the label-wise representation of the document for each class $j \in \{1, \ldots, N_1\}$ in the first level by passing the concatenated hidden states ($\mathbf{H}$) through the attention layer associated with level 1 (Attention$_1$) to obtain $\mathbf{C}_1 = [\mathbf{c}_{1,1}, \ldots, \mathbf{c}_{1,N_1}]$ where $\mathbf{c}_{1,j}$ is the label-specific representation of the document for the j-th class in level 1.

HLA passes the level 1 class representations through their associated FFNNs to obtain the level 1 predictions $\mathbf{y}_1' = [y_{1,1}', \ldots, y_{1,N_1}']$, where $y_{1,j}'$ is the confidence score of the j-th class in level 1. The predictions are passed through a FFNN with a sigmoid activation function to obtain a vector representation of the predictions at level 1 as:

$$\mathbf{s}_1 = \sigma(\mathbf{y}_1' \mathbf{F}_1) \tag{12}$$

where $\mathbf{F}_1 \in \mathbb{R}^{N_1 \times d_k}$ is a weight vector and d_k is a hyperparameter which controls the dimension size of the prediction vector representation. HLA obtains the second level label-specific vectors ($\mathbf{C}_2$) from the associated attention layer (Attention$_2$) and concatenates the vector $\mathbf{s}_1$ to each label-specific vector in level 2 as follows:

$$\mathbf{D}_2 = [Concat(\mathbf{c}_{2,1}, \mathbf{s}_1), \ldots, Concat(\mathbf{c}_{2,N_2}, \mathbf{s}_1)] \tag{13}$$

where $Concat$ concatenates the vectors and $\mathbf{D}_2 \in \mathbb{R}^{N_2 \times (d_h + d_k)}$ contains the label-wise representations of the document for the second level classes. The rows of $\mathbf{D}_2$ are passed through their associated FFNNs to obtain the predictions for the second level $\mathbf{y}_2' = [y_{2,1}', \ldots, y_{2,N_2}']$. These predictions are used to determine the level 2 prediction vector representation ($\mathbf{s}_2$) which is concatenated to the label-wise representations of the third level classes ($\mathbf{C}_3$). This procedure is continued until the final level K is reached and the predictions for each level are combined to form the final prediction scores for the document.

We extend this approach by concatenating all of the ancestor level prediction representations to the current level label-specific representations. Therefore, for the third level of the class hierarchy, the label-wise representations of the document is calculated as:

$$\mathbf{D}_3 = [Concat(\mathbf{c}_{3,1}, \mathbf{s}_1, \mathbf{s}_2), \ldots, Concat(\mathbf{c}_{3,N_3}, \mathbf{s}_1, \mathbf{s}_2)] \tag{14}$$

This allows the current level predictions to leverage the information of all higher-level predictions, as opposed to only the parent-level predictions. We refer to this approach as global hierarchical label-wise attention (GHLA) due to the use of all ancestor level predictions.

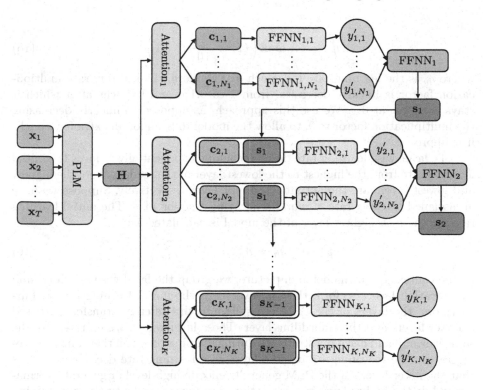

Fig. 2. Our hierarchical model architecture. We pass the input token sequence (orange) through the PLM to obtain the concatenated final hidden states (blue). These are passed to level-specific attention layers to obtain the label-specific representations of the token sequence for each level (green). The predictions of the first level (yellow) are passed through a FFNN to obtain a representation of the output (purple) which is concatenated to the label-specific vectors of level two. We repeat this process of using the previous level predictions until the final level of the hierarchy is reached. (Color figure online)

4.5   Learning Rate Adaptions

To effectively fine-tune the PLM with the added label-wise attention and classification layers, we use three learning rate adaption techniques which include: learning rate warmup, learning rate decay, and layer-wise learning rate decay.

The learning rate warmup linearly increases the learning rate from approximately 0 to the assigned learning rate (η) within a number of steps during the first epoch of training. This prevents the model from overfitting on the first few steps of the training process by gradually increasing the learning rate. The number of steps used to linearly increase the learning rate is calculated by multiplying the total number of steps in an epoch with a warmup fraction λ.

The learning rate decay is applied after the learning rate warmup steps and linearly decreases the learning rate based on the current epoch number (v) as follows:

$$\eta_v = \eta \times \max\left((1 - \frac{v-1}{10}), 0.1\right) \tag{15}$$

where η_v is the learning rate for epoch v. Therefore, the learning rate multiplication factor is linearly decreased from 1 to 0.1 over 10 epochs after which it stays constant at 0.1. We use this approach, as opposed to linearly decreasing the multiplication factor to 0, to allow the model to train for an extra few epochs if it improves.

The layer-wise learning rate decay technique exponentially reduces the initial learning rate from the highest to the lowest layer of the model, where the highest and lowest layers are the output and input layer respectively. Suppose we have an assigned learning rate of η and a decaying fraction of β. The initial learning rate for the r-th highest layer of the model is calculated as:

$$\eta_r = \eta \times \beta^{r-1} \tag{16}$$

In the context of our model architecture, we group the label-wise attention and classification layers to form the top layer with the initial learning rate η. Furthermore, the layers of the PLM are split into the different transformer-based encoder layers and the embedding layer. These layer-wise learning rates use the same warmup and decaying strategies during training based on their initial learning rates. The reasoning behind the layer-wise learning rate decay approach is that the lower layers of the PLM generally encode high-level or general information while the higher layer representations are more specific to the pre-training task of the PLM [35]. Therefore, by decreasing the learning rates from the highest to the lowest layers of our model, we are able to leverage the language understanding capabilities of the PLM while allowing the task-specific layers to be learnt more effectively.

5   Experiments

5.1   Datasets and Evaluation Metrics

We evaluate the performance of the different label-wise attention methods using the three commonly used HTC benchmark datasets: Web of Science (WOS) [15], RCV1-V2 [16], and NYTimes (NYT) [24]. The WOS dataset comprises research publications from the Web of Science publication database while RCV1-V2 and NYT comprise news articles from Reuters and The New York Times respectively. We use the same preprocessing steps and dataset splits as in previous studies [3,4,10,11,31,32,36]. Tables 1 and 2 present summary statistics and hierarchical properties of these datasets.

We use the standard Micro-F1 and Macro-F1 evaluation metrics which are the most common metrics used for HTC tasks [3,4,10,11,31,32,36]. Micro-F1 considers the overall precision and recall of all instances and may be biased by class imbalances while Macro-F1 assigns equal importance to each class in the hierarchy by averaging the per class F1-scores over all classes.

Table 1. Characteristics of the benchmark datasets. The columns "Levels" and "Classes" give the number of levels and classes in the class structure. "Avg. Classes" is the average number of classes per document, while "Train", "Dev", and "Test" are the number of instances in each of the dataset splits.

Dataset	Levels	Classes	Avg. Classes	Train	Dev	Test
WOS [15]	2	141	2.0	30,070	7,518	9,397
RCV1-V2 [16]	4	103	3.24	20,833	2,316	781,265
NYT [24]	8	166	7.6	23,345	5,834	7,292

Table 2. The average per-level branching factor of the hierarchy in each benchmark dataset, which is calculated as the average number of child nodes for the nodes at a particular level. The node counts per level are given in parentheses.

Dataset	Level 1	Level 2	Level 3	Level 4	Level 5	Level 6	Level 7	Level 8
WOS	19.14 (7)	0.0 (134)	–	–	–	–	–	–
RCV1-V2	13.75 (4)	0.78 (55)	0.02 (43)	0.0 (1)	–	–	–	–
NYT	6.75 (4)	1.89 (27)	0.92 (51)	0.36 (47)	0.71 (17)	0.5 (12)	0.33 (6)	0.0 (2)

5.2 Implementation Details

We implemented our approaches with the PyTorch, PyTorch Lightning, and Hugging Face libraries. We compared BERT and RoBERTa as the PLM in our models and used the `bert-base-uncased` and `roberta-base` models from Hugging Face.

We used the Adam optimiser [13] and performed hyperparameter tuning on the initial learning rate (η) and batch size with possible values of $\{7.5e-5, 1e-4, 2e-4, 3e-4\}$ and $\{16, 32\}$ respectively. We chose the learning rate warmup and layer-wise learning rate decay fractions as $\lambda = 0.1$ and $\beta = 0.8$ respectively. Furthermore, we tuned the value of d_p with possible values of $\{256, 512\}$ and chose $d_k = 128$ along with a threshold $\gamma = 0.5$. We used a maximum of 12 epochs and stopped training if the harmonic mean between the Micro-F1 and Macro-F1 on the development set did not increase for 5 epochs. For the hierarchical attention mechanisms we use GA as the attention layers associated with each level.

We trained each model on the training set and chose the hyperparameter combination which obtained the highest harmonic mean between the Micro-F1 and Macro-F1 scores on the development set. The chosen models were retrained with three random seeds and evaluated on the test set. Therefore, for all experiments in the following subsections, we report the average over three runs with different random seeds.

5.3 Main Results

Table 3 presents the results of our approaches compared to the recent HTC approaches described in Sect. 3. We compare the different attention mechanisms with a BERT PLM and select the best performing approach and substitute RoBERTa as the PLM for comparison purposes.

Table 3. Performance comparisons of the various approaches discussed in this paper using the three commonly used benchmark datasets.

Model	WOS		RCV1-V2		NYT	
	Micro-F1	Macro-F1	Micro-F1	Macro-F1	Micro-F1	Macro-F1
HiMatch [3]	86.20	80.53	84.73	64.11	–	–
HGCLR [31]	87.11	81.20	86.49	68.31	78.86	67.96
PAAMHiA-T5[a] [10]	**90.36**	81.64	87.22	70.02	77.52	65.97
HBGL [11]	**87.36**	**82.00**	87.23	**71.07**	80.47	70.19
HPT [32]	87.16	81.93	87.26	69.53	80.42	70.42
DPA$_{BERT}$	87.13	81.48	87.07	68.45	79.67	68.27
GA$_{BERT}$	87.05	81.46	86.88	69.11	80.06	68.56
HLA$_{BERT}$	87.17	81.55	86.71	68.45	79.60	68.06
GHLA$_{BERT}$	87.17	81.55	87.19	68.62	79.67	68.67
GHLA$_{RoBERTa}$	87.00	81.44	**87.78**	70.21	**81.41**	**72.27**

[a] Results obtained using twice the number of model parameters as the other approaches.

Our results show that the DPA and GA approaches generally perform similarly across the three benchmark datasets. However, GA shows improvements over DPA in terms of Macro-F1 on the RCV1-V2 (0.66) and NYT (0.29) datasets. Our proposed GHLA approach outperforms the DPA and HLA approaches on all three datasets. It should be noted that the results on the WOS dataset for HLA and GHLA approaches are the same because the WOS class hierarchy only comprises two levels. Furthermore, GHLA outperforms the GA approach on all of the performance measures except for the RCV1-V2 Macro-F1 score and the NYT Micro-F1 score. We hypothesise that GHLA generally outperforms the other label-wise attention mechanisms because it is able to leverage the information of all the ancestor level predictions by splitting up the attention layers at each level. Therefore, GHLA is able to focus on the predictions at a particular level while ignoring classes in lower levels and utilising the predictions of previous levels to improve performance on HTC tasks.

We select GHLA since it is the best performing label-wise attention mechanism and use RoBERTa as the PLM to compare the difference in performance when using the two PLMs. The RoBERTa PLM significantly improves the results on the RCV1-V2 and NYT datasets in terms of Micro-F1 (0.59 and 1.74) and Macro-F1 (1.59 and 3.60). Furthermore, GHLA$_{RoBERTa}$ improves on state-of-the-art approaches on RCV1-V2 in terms of Micro-F1 (0.52) and NYT in terms of Micro-F1 (0.94) and Macro-F1 (1.85). The NYT dataset has the most complex class hierarchy while the RCV1-V2 class hierarchy is also more complex than WOS. Therefore, due to the significant performance improvements on the NYT dataset particularly, we hypothesise that using RoBERTa as the PLM allows our GHLA approach to more effectively handle complex class hierarchies in general.

5.4   Level-Wise Results

Figure 3 presents the Micro-F1 and Macro-F1 scores for the classes at each level for the three benchmark datasets. Our results show that the average number of training instances per class at a certain level generally has a direct correlation with the performance at that level. On the WOS dataset we see that the Micro-F1 and Macro-F1 scores are significantly higher for the first level classes than the second level classes. We also see that the average number of training instances are much larger for level 1 classes (4295) than level 2 (224). Similarly, the Micro-F1 and Macro-F1 scores on the NYT dataset show how the average number of training instances per level relates to the performance differences for the classes at each level. Furthermore, the performance metrics on the RCV1-V2 and NYT datasets show that the GHLA$_{\text{RoBERTa}}$ model is generally able to leverage the class hierarchy more effectively at each level to outperform the other approaches particularly on the levels with fewer average training instances per class in terms of Macro-F1.

5.5   Stability Analysis

Table 4 presents the results of our proposed approaches with standard deviations over the three runs with different seeds. The results show that the DPA and GHLA approaches produce the most stable results on the WOS and RCV1-V2 datasets respectively. Furthermore, the HLA approach has a significantly higher standard deviation on the RCV1-V2 dataset compared to the other label-wise attention mechanisms. Lastly, the results show that the Macro-F1 metric generally produces higher deviation across the multiple runs than the Micro-F1 scores.

Table 4. The average performance results of evaluating the different models over three independent runs. The values in parentheses show the corresponding standard deviations.

Model	WOS		RCV1-V2		NYT	
	Micro-F1	Macro-F1	Micro-F1	Macro-F1	Micro-F1	Macro-F1
DPA$_{\text{BERT}}$	87.13 **(0.09)**	81.48 **(0.13)**	87.07 (0.18)	68.45 (0.22)	79.67 (0.11)	68.27 (0.26)
GA$_{\text{BERT}}$	87.05 (0.17)	81.46 (0.23)	86.88 (0.07)	69.11 (0.04)	80.06 (0.09)	68.52 (0.25)
HLA$_{\text{BERT}}$	87.17 (0.13)	81.55 (0.14)	86.71 (0.46)	68.45 (1.74)	79.60 **(0.04)**	68.06 (0.30)
GHLA$_{\text{BERT}}$	87.17 (0.13)	81.55 (0.14)	87.19 **(0.04)**	68.62 **(0.03)**	79.67 (0.22)	68.67 (0.33)
GHLA$_{\text{RoBERTa}}$	87.00 (0.16)	81.44 (0.23)	87.78 (0.05)	70.21 (0.39)	81.41 (0.09)	72.27 **(0.23)**

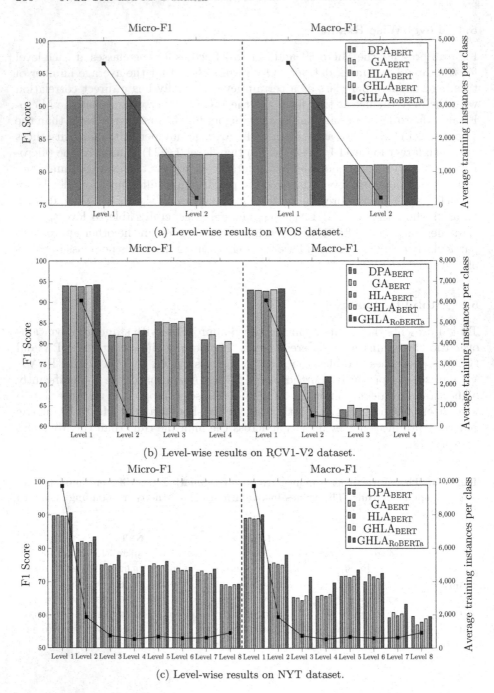

(a) Level-wise results on WOS dataset.

(b) Level-wise results on RCV1-V2 dataset.

(c) Level-wise results on NYT dataset.

Fig. 3. Level-wise performance results of our approaches on the three benchmark datasets. The bar plots present the F1 scores (left y-axis) for the different approaches at each level of the hierarchy while the line plot shows the average training instances for the classes at a particular level (right y-axis). The Micro-F1 and Macro-F1 scores are presented in the left and right panels respectively.

6   Conclusion

In this paper we proposed a new hierarchical text classification approach which adds label-wise attention and classification layers to a pre-trained language model (PLM) which is then fine-tuned on a given downstream classification task. It is based on an approach which splits the attention layers for each of the levels of the class hierarchy. We suggested an improvement to allow the predictions at a certain level to leverage the information of all ancestor level predictions. We evaluate our suggested improvement by comparing it to different label-wise attention mechanisms approaches by performing comprehensive experiments on three benchmark datasets and showed that our proposed label-wise attention mechanism generally outperforms the other approaches. In addition, we also compared BERT against RoBERTa as the underlying PLM and found that using the RoBERTa model significantly improved performance on benchmark datasets that comprise more complex class hierarchies. Finally, we showed that our label-wise attention mechanism with the RoBERTa PLM obtains state-of-the-art performance on two of the three benchmark datasets.

References

1. Banerjee, S., Akkaya, C., Perez-Sorrosal, F., Tsioutsiouliklis, K.: Hierarchical transfer learning for multi-label text classification. In: Proceedings of the 57th Annual Meeting of the Association for Computational Linguistics, pp. 6295–6300. Association for Computational Linguistics, Florence, Italy (2019). https://doi.org/10.18653/v1/P19-163
2. Baumel, T., Nassour-Kassis, J., Cohen, R., Elhadad, M., Elhadad, N.: Multi-label classification of patient notes: study on ICD code assignment. In: AAAI Workshops (2018)
3. Chen, H., Ma, Q., Lin, Z., Yan, J.: Hierarchy-aware label semantics matching network for hierarchical text classification. In: Proceedings of the 59th Annual Meeting of the Association for Computational Linguistics and the 11th International Joint Conference on Natural Language Processing, pp. 4370–4379. Association for Computational Linguistics (2021). https://doi.org/10.18653/v1/2021.acl-long.337
4. Deng, Z., Peng, H., He, D., Li, J., Yu, P.: HTCInfoMax: a global model for hierarchical text classification via information maximization. In: Proceedings of the 2021 Conference of the North American Chapter of the Association for Computational Linguistics: Human Language Technologies, pp. 3259–3265. Association for Computational Linguistics (2021). https://doi.org/10.18653/v1/2021.naacl-main.260
5. Devlin, J., Chang, M.W., Lee, K., Toutanova, K.: BERT: pre-training of deep bidirectional transformers for language understanding. In: Proceedings of the 2019 Conference of the North American Chapter of the Association for Computational Linguistics: Human Language Technologies, Volume 1 (Long and Short Papers), pp. 4171–4186. Association for Computational Linguistics, Minneapolis (2019). https://doi.org/10.18653/v1/N19-1423
6. Dumais, S., Chen, H.: Hierarchical classification of web content. In: Proceedings of the 23rd Annual International ACM SIGIR Conference on Research and Development in Information Retrieval, SIGIR 2000, pp. 256–263. Association for Computing Machinery, New York (2000). https://doi.org/10.1145/345508.345593

7. Gopal, S., Yang, Y.: Recursive regularization for large-scale classification with hierarchical and graphical dependencies. In: Proceedings of the 19th ACM SIGKDD International Conference on Knowledge Discovery and Data Mining, KDD 2013, pp. 257–265. Association for Computing Machinery, New York (2013). https://doi.org/10.1145/2487575.2487644

8. He, K., Zhang, X., Ren, S., Sun, J.: Deep residual learning for image recognition. In: 2016 IEEE Conference on Computer Vision and Pattern Recognition (CVPR), pp. 770–778 (2016). https://doi.org/10.1109/CVPR.2016.90

9. Hochreiter, S., Schmidhuber, J.: Long short-term memory. Neural Comput. **9**, 1735–1780 (1997). https://doi.org/10.1162/neco.1997.9.8.1735

10. Huang, W., et al.: Exploring label hierarchy in a generative way for hierarchical text classification. In: Proceedings of the 29th International Conference on Computational Linguistics, pp. 1116–1127. International Committee on Computational Linguistics, Gyeongju, Republic of Korea (2022)

11. Jiang, T., Wang, D., Sun, L., Chen, Z., Zhuang, F., Yang, Q.: Exploiting global and local hierarchies for hierarchical text classification. In: Proceedings of the 2022 Conference on Empirical Methods in Natural Language Processing, pp. 4030–4039. Association for Computational Linguistics, Abu Dhabi (2022). https://doi.org/10.18653/v1/2022.emnlp-main.268

12. Kim, Y.: Convolutional neural networks for sentence classification. In: Proceedings of the 2014 Conference on Empirical Methods in Natural Language Processing (EMNLP), pp. 1746–1751. Association for Computational Linguistics, Doha (2014). https://doi.org/10.3115/v1/D14-1181

13. Kingma, D., Ba, J.: Adam: a method for stochastic optimization. In: 3rd International Conference on Learning Representations, San Diego (2015)

14. Koller, D., Sahami, M.: Hierarchically classifying documents using very few words. In: Proceedings of the Fourteenth International Conference on Machine Learning, ICML 1997, pp. 170–178. Morgan Kaufmann Publishers Inc., San Francisco (1997)

15. Kowsari, K., Brown, D.E., Heidarysafa, M., Jafari Meimandi, K., Gerber, M.S., Barnes, L.E.: HDLTex: hierarchical deep learning for text classification. In: 2017 16th IEEE International Conference on Machine Learning and Applications (ICMLA), pp. 364–371 (2017). https://doi.org/10.1109/ICMLA.2017.0-134

16. Lewis, D.D., Yang, Y., Rose, T.G., Li, F.: RCV1: a new benchmark collection for text categorization research. J. Mach. Learn. Res. **5**, 361–397 (2004)

17. Li, F., Yu, H.: ICD coding from clinical text using multi-filter residual convolutional neural network. In: proceedings of the AAAI Conference on Artificial Intelligence, vol. 34, pp. 8180–8187 (2020). https://doi.org/10.1609/aaai.v34i05.6331

18. Liu, L., Perez-Concha, O., Nguyen, A., Bennett, V., Jorm, L.: Hierarchical label-wise attention transformer model for explainable ICD coding. J. Biomed. Inf. **133**, 104–161 (2022). https://doi.org/10.1016/j.jbi.2022.104161

19. Liu, Y., et al.: RoBERTa: a robustly optimized BERT pretraining approach. ArXiv (2019)

20. Mao, Y., Tian, J., Han, J., Ren, X.: Hierarchical text classification with reinforced label assignment. In: Proceedings of the 2019 Conference on Empirical Methods in Natural Language Processing and the 9th International Joint Conference on Natural Language Processing (EMNLP-IJCNLP), pp. 445–455. Association for Computational Linguistics, Hong Kong (2019). https://doi.org/10.18653/v1/D19-1042

21. Mullenbach, J., Wiegreffe, S., Duke, J., Sun, J., Eisenstein, J.: Explainable prediction of medical codes from clinical text. In: Proceedings of the 2018 Conference

of the North American Chapter of the Association for Computational Linguistics: Human Language Technologies, Volume 1 (Long Papers), pp. 1101–1111. Association for Computational Linguistics, New Orleans (2018). https://doi.org/10.18653/v1/N18-1100

22. Peng, H., et al.: Hierarchical taxonomy-aware and attentional graph capsule RCNNs for large-scale multi-label text classification. IEEE Trans. Knowl. Data Eng. **33**(6), 2505–2519 (2021). https://doi.org/10.1109/TKDE.2019.2959991

23. Raffel, C., et al.: Exploring the limits of transfer learning with a unified text-to-text transformer. J. Mach. Learn. Res. **21**(1), 5485–5551 (2020)

24. Sandhaus, E.: The New York Times annotated corpus. Technical report, Linguistic Data Consortium, Philadelphia (2008)

25. Shi, H., Xie, P., Hu, Z., Zhang, M., Xing, E.P.: A neural architecture for automated ICD coding. In: Annual Meeting of the Association for Computational Linguistics (2017). https://doi.org/10.18653/v1/P18-1098

26. Shimura, K., Li, J., Fukumoto, F.: HFT-CNN: learning hierarchical category structure for multi-label short text categorization. In: Proceedings of the 2018 Conference on Empirical Methods in Natural Language Processing, pp. 811–816. Association for Computational Linguistics, Brussels, Belgium (2018). https://doi.org/10.18653/v1/D18-1093

27. Strydom, S., Dreyer, A.M., van der Merwe, B.: Automatic assignment of diagnosis codes to free-form text medical note. J. Univ. Comput. Sci. **29**(4), 349–373 (2023). https://doi.org/10.3897/jucs.89923

28. Vaswani, A., et al.: Attention is all you need. In: Guyon, I., et al. (eds.) Advances in Neural Information Processing Systems, vol. 30. Curran Associates, Inc. (2017)

29. Veličković, P., Cucurull, G., Casanova, A., Romero, A., Liò, P., Bengio, Y.: Graph attention networks. In: International Conference on Learning Representations (2018)

30. Vu, T., Nguyen, D.Q., Nguyen, A.: A label attention model for ICD coding from clinical text. In: Proceedings of the Twenty-Ninth International Joint Conference on Artificial Intelligence, IJCAI-20, pp. 3335–3341 (2020). https://doi.org/10.24963/ijcai.2020/461

31. Wang, Z., Wang, P., Huang, L., Sun, X., Wang, H.: Incorporating hierarchy into text encoder: a contrastive learning approach for hierarchical text classification. In: Proceedings of the 60th Annual Meeting of the Association for Computational Linguistics (Volume 1: Long Papers), pp. 7109–7119. Association for Computational Linguistics, Dublin (2022). https://doi.org/10.18653/v1/2022.acl-long.491

32. Wang, Z., et al.: HPT: hierarchy-aware prompt tuning for hierarchical text classification. In: Proceedings of the 2022 Conference on Empirical Methods in Natural Language Processing, pp. 3740–3751. Association for Computational Linguistics, Abu Dhabi (2022). https://doi.org/10.18653/v1/2022.emnlp-main.246

33. Wu, J., Xiong, W., Wang, W.Y.: Learning to learn and predict: a meta-learning approach for multi-label classification. In: Proceedings of the 2019 Conference on Empirical Methods in Natural Language Processing and the 9th International Joint Conference on Natural Language Processing, pp. 4354–4364. Association for Computational Linguistics, Hong Kong (2019). https://doi.org/10.18653/v1/D19-1444

34. Yang, Z., Dai, Z., Yang, Y., Carbonell, J., Salakhutdinov, R.R., Le, Q.V.: XLNet: generalized autoregressive pretraining for language understanding. In: Wallach, H., Larochelle, H., Beygelzimer, A., d'Alché-Buc, F., Fox, E., Garnett, R. (eds.) Advances in Neural Information Processing Systems, vol. 32. Curran Associates, Inc. (2019)

35. Zhang, T., Wu, F., Katiyar, A., Weinberger, K.Q., Artzi, Y.: Revisiting few-sample BERT fine-tuning. arXiv preprint arXiv:2006.05987 (2020)
36. Zhou, J., et al.: Hierarchy-aware global model for hierarchical text classification. In: Proceedings of the 58th Annual Meeting of the Association for Computational Linguistics, pp. 1106–1117. Association for Computational Linguistics (2020). https://doi.org/10.18653/v1/2020.acl-main.104

Multimodal Misinformation Detection in a South African Social Media Environment

Amica De Jager$^{(\boxtimes)}$, Vukosi Marivate , and Abiodun Modupe

Department of Computer Science, University of Pretoria, Pretoria 0028, South Africa
amicadejager@gmail.com, {vukosi.marivate,abiodun.modupe}@up.ac.za

Abstract. The world is witnessing a growing epidemic of misinformation. Misinformation can have severe impacts on society across multiple domains: including health, politics, security, the environment, the economy and education. With the constant spread of misinformation on social media networks, a need has arisen to continuously assess the veracity of digital content. This need has inspired numerous research efforts on the development of misinformation detection (MD) models. However, many models do not use all information available to them and existing research contains a lack of relevant datasets to train the models, specifically within the South African social media environment. The aim of this paper is to investigate the transferability of knowledge of a MD model between different contextual environments. This research contributes a multimodal MD model capable of functioning in the South African social media environment, as well as introduces a South African misinformation dataset. The model makes use of multiple sources of information for misinformation detection, namely: textual and visual elements. It uses bidirectional encoder representations from transformers (BERT) as the textual encoder and a residual network (ResNet) as the visual encoder. The model is trained and evaluated on the Fakeddit dataset and a South African misinformation dataset. Results show that using South African samples in the training of the model increases model performance, in a South African contextual environment, and that a multimodal model retains significantly more knowledge than both the textual and visual unimodal models. Our study suggests that the performance of a misinformation detection model is influenced by the cultural nuances of its operating environment and multimodal models assist in the transferability of knowledge between different contextual environments. Therefore, local data should be incorporated into the training process of a misinformation detection model in order to optimize model performance.

Keywords: Misinformation · Disinformation · Fake News Detection · Multimodal · Natural Language Processing · Transfer Learning · Deep Learning · Social Media · South African

A. Pillay et al. (Eds.): SACAIR 2023, CCIS 1976, pp. 285–299, 2023.
https://doi.org/10.1007/978-3-031-49002-6_19

1   Introduction

Misinformation can have severe impacts on society across multiple domains, including health, politics, security, the environment, the economy and education. These impacts are significantly amplified on social media networks due the effortless content creation capabilities, easy accessibility to global audiences, and the speed and scope at which information can be diffused across networks in real-time. Due to the aforementioned characteristics of social media networks, misinformation is granted an enormous potential to cause real harm, within minutes, for millions of users [5]. For example, the sharing of misinformation on social media networks was used in South Africa to encourage civil unrest, resulting in city-wide looting and riots, known as the Durban Riots during 2021 [7]. Another example includes how misinformation was used by domestic and foreign actors to influence public opinion, undermining democracy, in the 2016 American presidential elections [2]. Misinformation was also used extensively during the COVID-19 pandemic to cause confusion, encourage risk-taking behaviours and public mistrust in health authorities, thereby undermining the public health response [15]. Overall, misinformation is a global issue that warrants the need to monitor, assess and regulate the veracity of digital content. Thus, inspiring research in the pursuit of mitigating this crisis through the development of misinformation detection models.

In the subsections to follow, misinformation and the subtypes of misinformation will be defined. Additionally, different groups of misinformation detection models will be outlined, followed by a discussion on concept drift and its applicability to misinformation detection models for the South African context.

1.1   Misinformation

Misinformation refers to false or inaccurate information. It should be noted that there are several terms relating to misinformation which may be confusing, such as: fake news and disinformation. Fake news refers to misinformation that is distributed in the form of news articles. Both misinformation and disinformation refer to false or inaccurate information, however the distinction between these terms is whether the information was created with the intent to deceive. Misinformation often refers to the unintentional instances, while disinformation, the intentional instances where the information has ill intent and is often a deliberate attack to a particular group [16]. However, since it is difficult for researchers to determine intent, misinformation is used throughout this article as an umbrella term to include all false or inaccurate information.

Misinformation can be classified into different categories according to the type of content being created or shared, the motivations of the content creator and the methods used to share the content. Wardle [14] defined seven categories of misinformation, based on a scale measuring the content creator's intent to deceive, namely: satire or parody, false connection, misleading content, false context, imposter content, manipulated content and fabricated content. These categories are defined in Table 1. These same categories were later used by Naka-mura, Levy and Wang [8] in creating a misinformation dataset, Fakeddit, to support in the development of misinformation detection models. Misinforma-tion detection is often treated as a binary classification task (i.e., true or false), grouping all types of misinformation into a single parent class. However, mis-information detection can be extended into a multi-class classification problem, thus accounting for the different types of misinformation, as described in Table 1.

Table 1. Wardle's seven categories of misinformation [14].

Measure of Intent to Deceive	Misinformation Category	Definition
1	Satire or Parody	The meaning of the content is twisted or misinterpreted in a satirical or humorous way
2	False Connection	Headlines, visuals or captions do not support the content
3	Misleading Content	Misleading use of information to frame an issue or individual
4	False Context	Genuine content is shared with false contextual information
5	Imposter Content	Genuine sources are impersonated
6	Manipulated Content	Genuine information or imagery is manipulated to deceive (e.g., photo editing)
7	Fabricated Content	New content is created that is completely false, designed to deceive and cause harm

1.2 Misinformation Detection Models

Before the development of misinformation detection (MD) models, misinforma-tion detection was performed solely by professional fact-checkers. Professional fact-checking as a process involves manually reading through content, researching and comparing knowledge extracted from the content with factual information in that particular domain, and making a decision on the content's authenticity based on the subsequent findings. Although manual fact-checking plays a critical role in misinformation detection, it is not sufficient when analysing the immense volume and variety of digital content that is being rapidly diffused on social media networks [4,18]. For example, text created on high frequency social media platforms can be upwards of one million sentences every few minutes, making it impossible for professional fact-checkers to read through and validate all the information. Therefore, MD models, which involves leveraging the benefits of machine learning, are developed to automate the process, assisting fact-checkers in the task of misinformation detection.

Misinformation detection models can be categorized based on the type of information the model uses for detection. These categories, displayed in Fig. 1, include: content-based, context-based, propagation-based and early detection. Content-based MD models detect misinformation directly on the content of a sample, which forms part of a training dataset. This involves using text, audio, image, video, or any combination thereof to generate the training data used in a model. Context-based MD models, on the other hand, detect misinformation based on the contextual information surrounding a sample on a social media network. These types of models work by using information such as a sample's geolocation, the time that a sample is created, the account, profile or source responsible for a sample and the reaction of network users to a sample. Propagation-based MD models detect misinformation based on the propagation patterns of a sample over a network, also referred to as information diffusion. Propagation-based MD models also consider the users who share a sample and the speed at which a sample is diffused. Early detection models focus on detecting misinformation at an early stage before the content becomes viral. Models in this category need to be robust in order to handle the two major challenges that arise in the task of early misinformation detection, namely: a lack of data and a lack of labels [16]. Early detection models are either one of the aforementioned models (content-based, context-based or propagation-based), or a hybrid between these models. This paper focuses on content-based methods, as these models consider the veracity of the actual content, as opposed to the circumstances surrounding the content on social media networks.

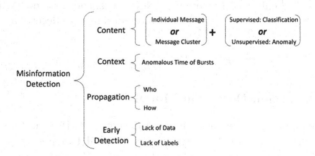

Fig. 1. An overview of the categorization of misinformation detection models [16].

Misinformation detection models can also be categorized according to the quantity of information sources that are utilized. If a single source of information is used for detection (i.e., text), then the model is referred to as a unimodal model. In contrast, if multiple sources of information are utilized (i.e., text and image), then the model is referred to as a multimodal model. Since the majority of content shared on social media networks include both text and images, unimodal models tend to be insufficient for misinformation detection on these platforms. Thus, multimodal models, which are capable of processing both textual and visual inputs, are the preferred choice for improving the classification of misinformation on social media networks [9].

1.3 Concept Drift

Misinformation detection models are vulnerable to concept drift. Concept drift occurs when the underlying relationship between the input data and the target variable changes over time [3]. In context of misinformation detection, this can occur when many labelled samples from verified misinformation content becomes outdated, with the introduction of newly developed content. For example, a MD model that is trained on misinformation data before the COVID-19 pandemic may struggle to classify misinformation during the pandemic [10]. Another example is a MD model that is trained on a misinformation dataset, originating from a specific country or cultural context, may struggle to classify misinformation when applied to a different country or cultural context because the nuance of that setting might not be applicable for another region or culture. Therefore, it is important to consider the contextual environment in which the MD model is deployed and to ensure the model is trained on relevant, updated datasets.

1.4 Misinformation Detection in South Africa

Due to the potential influence of concept drift on the performance of misinformation detection models, it is important to discuss misinformation in the South African context. There are, currently, no South African misinformation datasets available to support in the development of MD models, specifically for use in South Africa. However, there is a manual fact-checking program, called Real411 [11]. Real411 hosts a website on which members of the South African community can submit complaints flagging potential cases of misinformation, hate speech, incitement to violence or harassment. Complaints typically include both textual and visual elements. The most common complaints are displayed, by tag, in Fig. 2. Once a complaint is submitted, the Real411 Digital Complaints Committee (DCC) assesses the veracity of the flagged content and if the complaint is proven valid, the DCC initiates appropriate action, which may include: seeking assistance from the relevant online platforms, being referred to the South African Human Rights Commission (SAHRC) or reported to the South African Police Service (SAPS). For the purposes of this research, Real411 has made their data available, to become South Africa's first misinformation dataset, in order to assist in developing MD models capable of functioning in the South African social media environment.

2   Related Works

Yuan, et al., [17] developed an Event Adversarial Neural Network (EANN) for multimodal misinformation detection. The model consists of a modified convolu-

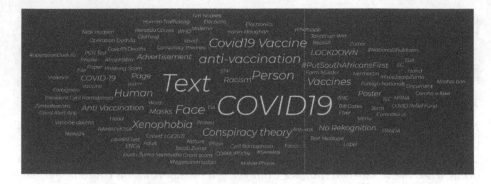

Fig. 2. Word cloud of most common Real411 complaints by tag [11].

tional neural network (CNN) for textual feature extraction and a pre-trained, 19-layer CNN from Visual Geometry Group (VGG19) for visual feature extraction. An event discriminator is also used in the framework to remove all event-specific features, maintaining the shared features among events. The EANN model is trained and evaluated on a Twitter and Weibo misinformation dataset. Khattar, Goud, Varma and Gupta [6] developed a Multimodal Variational Autoencoder (MVAE) which makes use of a bidirectional long-short term memory (Bi-LSTM) encoder for textual feature extraction and VGG19 for visual feature extraction. The MVAE model learns a joint latent vector by optimizing the process of encoding a text-image sample and decoding the result to reconstruct the original sample. This learnt representation is then used for binary classification. Like EANN, MVAE is also trained and evaluated on a Twitter and Weibo misinformation dataset. Singhal, et al., [13] developed a multimodal model, SpotFake+, which was one of the first misinformation detection models to leverage the benefits of transfer learning by making use of the pre-trained large language model (LLM), XLNet, for textual feature extraction and VGG19 for visual feature extraction. The SpotFake+ model is trained and evaluated on the Politifact and Gossipcop datasets from the FakeNewsNet repository [12]. Politifact incorporates misinformation from the political domain, while Gossipcop incorporates misinformation from the entertainment domain. Alonso-Bartolome and Segura-Bedmar [1] developed a multimodal model which makes use of CNN architecture for both textual and visual feature extraction. The model is trained and evaluated using the Fakeddit dataset which is a cross-domain, multi-class misinformation dataset [8]. The Fakeddit dataset is one of the largest misinformation datasets available with over one million samples. The model achieves state-of-the-art results with an overall accuracy of 87% at detecting misinformation. Palani, Elango and Viswanathan [9] developed a multimodal model, CB-Fake, which makes use of bidirectional encoder representations from transformers (BERT), a pre-trained LLM, for textual feature extraction and a capsule neural network (CapsNet) for visual feature extraction. The CB-Fake model is also trained and evaluated on the Politifact and Gossipcop datasets, achieving current state-of-the-art results in misinformation detection. Zhou, Ying, Qian, Li, and Zhang [19] developed a multimodal misinformation detection network (FND-CLIP) based on contrastive language-image pretraining (CLIP). The FND-CLIP framework uses pre-trained BERT for textual feature extraction, pre-trained residual neural network (ResNet) for visual feature extraction and CLIP encoders to allow for finer-grained feature fusion between the textual and visual features. The framework also incorporates a modality-wise attention module to adaptively reweight and aggregate the features. FND-CLIP is trained and evaluated on Weibo, Politifact and Gossipcop datasets, outperforming state-of-the-art multimodal misinformation detection models. A review of literature highlights a lack of research of misinformation detection within a South African context and all existing datasets originate from America, Europe or Asia. Therefore, the purpose of this paper is to contribute a multimodal MD model capable of detecting misinformation in the South African social media environment, as well as introduces the first South African misinformation dataset.

3   Methodology

This section formulates the problem and discusses the proposed framework of a multimodal misinformation detection model used in a South African social media environment. The section concludes with a description of the experimental setup.

3.1   Problem Formulation

This article presents two tasks. The aim of Task 1 is to evaluate the proposed model on a well-known dataset in three different classification settings: binary, 3-class and 6-class classification. The purpose of classification is to identify whether a sample, sourced from a social media network, is misinformation or not. Task 2 extends Task 1 by comparing the model's performance on a local South African dataset, when trained on non-local samples, a mix of non-local and local samples and local samples. The second task investigates the effect of knowledge transferability of the MD model between different contextual environments. It is also important to note that during both tasks, the proposed multimodal MD model is compared with both textual and visual unimodal models. This is performed in order to investigate the contribution of multimodality to the task of misinformation detection. For the purpose of this article, misinformation detection is modelled as a binary classification problem, as presented in Eq. 1.

$$F(S) = \begin{cases} 0 & \text{if } S \text{ is not misinformation} \\ 1 & \text{if } S \text{ is misinformation} \end{cases} \tag{1}$$

Let $S = \{s_1, s_2, \ldots, s_n\}$ represent a misinformation dataset of n samples, with $s_i = t_i \cup v_i$, where t_i and v_i are the text and image extracts associated with the ith sample, respectively. Let $Y = \{y_1, y_2, \ldots, y_n\}$ represent the set of ground-truth labels for the misinformation dataset, such that the ith label, for the ith sample, is given by: $y_i \in \{0, 1\}$. Then, a multimodal MD model $F : S \rightarrow Y$, is represented in Eq. 1.

3.2   Proposed MMiC Model

This article proposes a multimodal misinformation detection model that is capable of detecting misinformation in the South African social media environment. The Multimodal Misinformation-in-Context (MMiC) model uses pre-trained bidirectional encoder representations from transformers (BERT) model for textual feature extraction and pre-trained residual neural network (ResNet) model for visual feature extraction, as these encoders are used by the current state-of-the-art FND-CLIP model [19]. The MMiC model is outlined in Fig. 3.

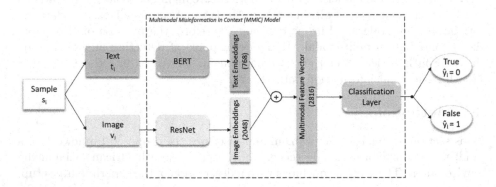

Fig. 3. Diagram of proposed MMiC model for misinformation detection.

A misinformation dataset is pre-processed before entering the MMiC model framework. After pre-processing is complete, the multimodal sample enters the framework, as displayed in Fig. 3. The sample is split into its separate modalities: the textual component of the sample is fed into the BERT textual encoder, while the visual component of the sample is fed into the ResNet50 visual encoder. In the next stage of the process, BERT outputs a textual feature vector of length 768 and ResNet50 outputs a visual feature vector of length 2048. The two feature vectors are concatenated and the joint feature vector is then fed into the classification layer. The classification layer is a two-layer fully connected network which predicts the label $\hat{y}$, an estimate of the ground-truth label y, as formulated in Sect. 3.1. Further architectural details are outlined below in Sect. 3.3 (under Parameter Setup).

3.3   Experimental Setup

This subsection discusses the experimental setup used for the investigation. This includes describing the datasets, baseline models, environment, parameters and evaluation metrics used.

Datasets. Two datasets are used for training and evaluation. The first dataset is the Fakeddit dataset by Nakamura, Levy and Wang [8] and represents the non-local dataset. This dataset is based on Reddit and was chosen because it is large, contains multi-domain misinformation and contains five different types of misinformation (a subset of the types outlined in Sect. 1.1). The different types of misinformation included in the Fakeddit dataset are manipulated content, false connection, satire/parody, misleading content and imposter content. For the purposes of this paper, the five types of misinformation present in the Fakeddit dataset are considered as a single misinformation class. The Fakeddit dataset used contains 3523 true samples and 2474 misinformation samples. The second dataset is a new South African specific multimodal misinformation dataset sourced from Real411[1] [11] and represents the local dataset. This dataset contains samples from various social media networks, namely: WhatsApp, Instagram, Facebook and Twitter. The Real411 dataset also contains different categories, however these categories do not align with those used in the Fakeddit dataset. The categories present in the Real411 dataset are misinformation, hate speech, incitement to violence and harassment. Since hate speech, incitement to violence and harassment do not fall under the label of misinformation, samples from these classes are excluded. The Real411 dataset used contains 378 true samples and 341 misinformation samples.

Table 2. Combinations of samples, from different datasets, used for training in Task 2.

Combination	Number of Samples from Fakeddit Dataset	Number of Samples from Real411 Dataset	Total
Non-Local	5997	0	5997
Mixed	5997	646	6643
Local	0	646	646

For the purpose of Task 1 (as discussed in Sect. 3.1), the multimodal Fakeddit dataset is used for both training and evaluation. For Task 2, three combinations of the Fakeddit and Real411 datasets are used for training, while a separate subset of the Real411 dataset is used for evaluation. The training combinations are outlined in Table 2.

Baseline Models. The proposed MMiC model is tested on a benchmark dataset, Fakeddit, and compared with two base modalities, namely: unimodal textual and visual models. The baseline unimodal textual models include two classical machine learning models: Gaussian Naive Bayes (NB) and Logistic

---

[1] https://www.real411.org/.

Regression (LR). Although considered a popular algorithm for textual classification tasks, the Support Vector Machine (SVM) model is not chosen as a baseline because results from existing misinformation detection research suggest that the NB and LR models outperform the SVM model [1,9,13]. SVM is also computationally expensive to implement in comparison to NB and LR, which require less computation. In addition, SVM requires a careful selection of hyperparameters which can be time-consuming. Also included as a baseline unimodal textual model is a simple BERT classifier, consisting of the textual encoder, BERT, and a classification layer. Likewise, the baseline unimodal visual model is a simple ResNet classifier consisting of the visual encoder, ResNet50, and a classification layer.

Parameter Setup. The experiments are carried out on a server with the specifications described in Table 3. This setup is used for building, training and evaluating the models. Python is used to implement all code and a virtual environment isolates all the packages necessary for running the project. All models are implemented using PyTorch and Scikit-Learn libraries. The datasets are split into 80% training and 20% test samples. Textual pre-processing includes: converting all characters to lowercase, removing new line characters, leading and trailing spaces, URLs and punctuation. Once the text is cleaned, the textual samples are vectorized using TD-IDF (for the NB and LR models) and BERT (for the BERT and MMiC models). The pre-trained BERT model that is used is the 'bert-base-uncased' model. The maximum length of the input text is set to 300 words, in order to accommodate the length of the largest textual sample. The pre-trained ResNet model that is used us the 'microsoft/resnet-50' model. The input images are normalized and resized to a dimension of 560×560. All models are trained for a maximum of 10 epochs, however an Early Stopping optimization technique is used to select the actual number of epochs in order to prevent over-fitting. Thus, the model from the epoch which achieves the best validation accuracy is retained. The cross-entropy loss function and AdamW optimizer are used to train the models. The learning rates for the BERT, ResNet and MMiC classifiers are set to 1×10^{-5}, 1×10^{-3} and 5×10^{-6}, respectively.

Table 3. Server Specifications

Component	Specification
OS	Ubuntu Server 20.04
CPU	12th Gen Intel(R) Core(TM) i9-12900
RAM (CPU)	125GB DDR4
GPU	2 x NVIDIA RTX A6000
RAM (GPU)	2×50 GB
Disk Size	2TB after RAID

Evaluation Metrics. Traditional machine learning performance metrics are used for evaluation. These metrics originate from the confusion matrix and include the terms (TP, TN, FP and FN) which stand for true positive, true negative, false positive and false negative, respectively. The first metric used is a measurement of the model's overall accuracy when classifying misinformation, as shown in Eq. 2:

$$Accuracy = \frac{TP + TN}{TP + TF + FP + FN} \tag{2}$$

The second metric used is precision, which is a measurement of the model's ability to correctly predict misinformation, as shown in Eq. 3:

$$Precision = \frac{TP}{TP + FP} \tag{3}$$

The third metric used is recall, which is a measurement of the model's ability to identify all cases of misinformation, as shown in Eq. 4:

$$Recall = \frac{TP}{TP + FN} \tag{4}$$

The final metric used is the F1-score, which takes into account both the model's precision and recall, and is shown in Eq. 5:

$$F1\text{-}Score = \frac{2 \times Precision \times Recall}{Precision + Recall} \tag{5}$$

4   Results

The results of the proposed MMiC model, as well as the results of the unimodal baseline models, for Task 1 and Task 2 are displayed in Table 4 and Table 5, respectively. For each task, the experiment is repeated five times and the mean result recorded. Furthermore, there are two phases involved during the evaluation on each task, namely: cross validation (CV) and test. The Hold Out CV approach is used, as this is computationally inexpensive when compared to other CV techniques. During CV, the models are trained and tested on random portions of the datasets (i.e., for each repetition of the experiment, the train and test sets are different). During the test phase, the models are trained and tested on the same portions of the datasets (i.e., for each repetition of the experiment, the train and test sets remain unchanged). The CV phase is used to validate the results collected during the test phase.

The results in Table 4 highlight that the MMiC model matches the performance of existing state-of-the-art misinformation detection models on the benchmark dataset, Fakeddit. The MMiC model also outperforms all other tested models across the different classification settings for both the CV and test phases. MMiC achieves top performance during the test phase: 88% for binary, 88% for 3-class and 83% for 6-class classification settings. The BERT model is second to

Table 4. F1-Scores of models trained and evaluated on Fakeddit dataset. Standard deviations are shown in parenthesis, for CV.

	Model	Binary	Multi-3	Multi-6
CV	NB	0.68 (0.02)	0.58 (0.01)	0.45 (0.01)
	LR	0.74 (0.01)	0.75 (0.02)	0.62 (0.01)
	BERT	0.83 (0.01)	0.81 (0.02)	0.70 (0.02)
	ResNet	0.74 (0.01)	0.73 (0.02)	0.70 (0.03)
	MMiC	**0.85** (0.01)	**0.85** (0.02)	**0.83** (0.01)
Test	NB	0.69	0.57	0.43
	LR	0.76	0.75	0.62
	BERT	0.86	0.85	0.72
	ResNet	0.75	0.73	0.71
	MMiC	**0.88**	**0.88**	**0.83**

the MMiC model, followed by the ResNet model. The LR model shows similar performance to the ResNet model for binary (76%) and 3-class (75%) classification settings but drops off significantly on the 6-class classification setting (62%). The NB model performs the worst out of all the models in this task: 69% for binary, 57% for 3-class and 43% for 6-class classification. The fact that BERT outperforms the ResNet classifier may suggest that the textual component of the misinformation samples provide more information (and therefore, more evidence) to the classification process, than the visual components of the samples. In Fig. 4, it can be seen that as the number of classification classes increases, the performance of the models decrease. However, as the number of classification classes increases, the MMiC model performs significantly better than the other models. This may suggest that multimodal models are less vulnerable to performance decay as the classes in the misinformation classification task increases.

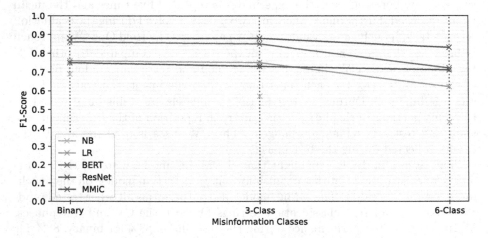

Fig. 4. Line graph of results from Task 1.

Table 5. F1-Score of models trained on combination of local and non-local samples. Standard deviations are shown in parenthesis, for CV.

	Model	Non-Local	Mixed	Local
CV	NB	0.53 (0.02)	0.55 (0.02)	0.54 (0.03)
	LR	0.50 (0.04)	0.59 (0.04)	0.63 (0.05)
	BERT	**0.60** (0.08)	0.83 (0.02)	0.84 (0.02)
	ResNet	0.54 (0.05)	0.57 (0.02)	0.57 (0.02)
	MMiC	0.54 (0.03)	**0.84** (0.02)	**0.85** (0.03)
Test	NB	0.55	0.60	0.59
	LR	0.57	0.59	0.64
	BERT	**0.63**	**0.87**	0.87
	ResNet	0.54	0.59	0.61
	MMiC	0.53	0.86	**0.88**

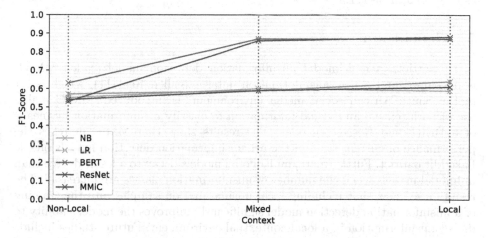

Fig. 5. Line graph of results from Task 2.

The results in Table 5 illustrate that the MMiC model performs the best when trained on local samples to classify local misinformation (88%). It is also observed that the MMiC and BERT show similar performance, outperforming all other models across all different training environments, during the CV and test phases. The fact that the MMiC and BERT models show similar performance on this task may be a result of the images in the Real411 dataset not adding significant information to the classification process. This may be a consequence of many Real411 images being screenshots of social media posts (i.e., textual content).

From the results in Table 5, it is observed that all models appear to increase in performance when local misinformation samples are introduced into the training dataset. This trend can also be observed in Fig. 5 and highlights the importance of incorporating local samples into the training process of a MD model. Furthermore, most models (except NB) perform the best when trained in a local contextual environment. However, the mean performance increase of the models between a mixed and local training environment is 1%. Whereas, the mean performance increase of the models between a non-local and mixed training environment is 14%. The total mean performance increase of the models between a non-local and mixed or local training environment is 15%. Therefore, the results show that it is important to include local misinformation samples when training a misinformation detection model to detect misinformation in a local environment. The effect of including local samples is especially influential for the BERT and MMiC models (as observed in Fig. 5) resulting in a mean performance increase of 29% between a non-local and mixed or local training environment. Overall, it should be noted that the training dataset does not necessarily need to only include local samples in order to significantly improve performance. The results observed provides evidence supporting the potential influence of concept drift, as discussed in Sect. 1.3.

5   Conclusion

In this article, a multimodal misinformation detection model (called MMiC) is developed and a new misinformation dataset (called Real411), which is set in the South African social media environment, is introduced. MMiC makes use of both textual and visual information to classify misinformation by means of a BERT and ResNet encoder. The results show that MMiC matches the performance of current state-of-the-art misinformation detection models on the Fakeddit dataset. Furthermore, multimodal models decay to a lesser degree than unimodal models when the number of misinformation classes are increased. The results also show that including local misinformation samples into the training of a misinformation detection model significantly improves the model's ability to detect misinformation in a local contextual environment. Future studies include investigating the performance of different deep learning models within the MMiC framework on the Real411 dataset. Although Real411 is a predominately English dataset, South Africa has 11 national languages, which may influence misinformation detection in the South African social media environment.

Acknowledgements. We want to acknowledge Real411, a program of Media Monitoring Africa, for the dataset. We would like to acknowledge funding from the ABSA Chair of Data Science, Google and the NVIDIA Corporation hardware Grant.

References

1. Alonso-Bartolome, S., Segura-Bedmar, I.: Multimodal fake news detection. arXiv (2021)
2. Bentzen, N.: Trump's disinformation 'magaphone': consequences, first lessons and outlook (2021). https://www.europarl.europa.eu/. Accessed from European Parliament: 20 Aug 2023

3. Brownlee, J.: A gentle introduction to concept drift in machine learning (2017). https://machinelearningmastery.com/gentle-introduction-concept-drift-machine-learning/. Accessed from Machine Learning Mastery: 20 Aug 2023

4. Capuano, N., Fenza, G., Loia, V., Nota, F.D.: Content-based fake news detection with machine and deep learning: a systematic review. Neurocomputing **530**, 91–103 (2023)

5. Figueira, A., Oliveira, L.: The current state of fake news: challenges and opportunities. Procedia Comput. Sci. **121**, 817–825 (2017)

6. Khattar, D., Goud, J.S., Gupta, M., Varma, V.: MVAE: multimodal variational autoencoder for fake news detection. In: The World Wide Web Conference, pp. 2915–2921. Association for Computing Machinery, New York (2019)

7. Lapping, G.: The July riots: an inflection point for digital media (2021). https://www.businesslive.co.za/redzone/news-insights/2021-09-01-the-july-riots-an-inflection-point-for-digital-media/. Accessed from Business Live: 20 Aug 2023

8. Nakamura, K., Levy, S., Wang, W.Y.: Fakeddit: a new multimodal benchmark dataset for fine-grained fake news detection. In: Proceedings of the Twelfth Language Resources and Evaluation Conference, Marseille, France, pp. 6149–6157. European Language Resources Association (2020)

9. Palani, B., Elango, S., Viswanathan, K.V.: CB-fake: a multimodal deep learning framework for automatic fake news detection using capsule neural network and BERT. Multimed. Tools Appl. **81**(4), 5587–5620 (2022)

10. Raza, S., Ding, C.: Fake news detection based on news content and social contexts: a transformer-based approach. Int. J. Data Sci. Anal. **13**(4), 335–362 (2022)

11. Real411: Home (2023). https://www.real411.org/. Accessed from Real 411: 20 Aug 2023

12. Shu, K., Mahudeswaran, D., Wang, S., Lee, D., Liu, H.: FakeNewsNet: a data repository with news content, social context and spatialtemporal information for studying fake news on social media. arXiv (2019)

13. Singhal, S., Kabra, A., Sharma, M., Shah, R.R., Chakraborty, T., Kumaraguru, P.: SpotFake+: a multimodal framework for fake news detection via transfer learning. In: Proceedings of the AAAI Conference on Artificial Intelligence, vol. 34, no. 10, pp. 13915–13916 (2020)

14. Wardle, C.: Fake news. it's complicated (2017). https://firstdraftnews.org/articles/fake-news-complicated/. Accessed from First Draft News: 20 Aug 2023

15. WHO: Infodemic (2023). https://www.who.int/health-topics/infodemic/understanding-the-infodemic-and-misinformation-in-the-fight-against-covid-19. Accessed from World Health Organization: 20 Aug 2023

16. Wu, L., Morstatter, F., Carley, K.M., Liu, H.: Misinformation in social media: definition, manipulation, and detection. SIGKDD Explor. Newsl. **21**(2), 80–90 (2019)

17. Yuan, Y., et al.: EANN: event adversarial neural networks for multi-modal fake news detection. In: Proceedings of the 24th ACM SIGKDD International Conference on Knowledge Discovery and Data Mining, KDD 2018, pp. 849–857. Association for Computing Machinery, New York (2018)

18. Zhang, X., Ghorbani, A.A.: An overview of online fake news: characterization, detection, and discussion. Inf. Process. Manag. **57**(2), 102025 (2020)

19. Zhou, Y., Yang, Y., Ying, Q., Qian, Z., Zhang, X.: Multimodal fake news detection via clip-guided learning. In: 2023 IEEE International Conference on Multimedia and Expo (ICME), Los Alamitos, CA, USA, pp. 2825–2830. IEEE Computer Society (2023)

Improving Semi-supervised Learning in Generative Adversarial Networks Using Variational AutoEncoders

Faheem Moolla[1] , Terence L. van Zyl[2(✉)] , and Hairong Wang[1]

[1] University of the Witwatersrand, Johannesburg 2000, GT, South Africa
1234085@students.wits.ac.za
[2] University of Johannesburg, Johannesburg 2092, GT, South Africa
tvanzyl@uj.ac.za

Abstract. Semi-supervised learning is a deep learning paradigm that has shown significant value for general machine learning and generative modelling. To date, Generative Adversarial Networks (GANs) still suffer from challenges related to mode collapse and other sources of instability. Further, little research has been done to investigate how incorporating semi-supervised learning (using SGAN) and pre-training (using VAE) into GAN training might alleviate some of these challenges. To this end, this study proposes SSGAN, a combination of VAE and SGAN, to tackle some of these challenges. Our extensive qualitative and quantitative analysis shows that the proposed approach significantly improves the stability of GAN training and the quality of generated images. Further, the results indicate that this can be done with relatively few additional labelled examples. In conclusion, continued research and exploring foundation models and other semi- and self-supervised learning mechanisms will likely lead to further improvements.

Keywords: GAN · VAE · Semi-supervised GAN

1 Introduction

Beyond supervised learning, machine learning encompasses both unsupervised and semi-supervised paradigms [14]. An example of unsupervised learning is generative modelling, which enables the sampling of novel outputs that are representative of the underlying probability distribution of the source dataset [4,6]. Specifically, Generative adversarial networks (GANs) have found utility in numerous tasks, including image generation, image manipulation, 3D object synthesis, video synthesis, and natural language generation [3,5].

Semi-supervised learning has gained attention due to its ability to learn from labelled and unlabelled data. Here, the number of labelled samples is often far less than that of unlabelled samples. It is especially useful for training models with limited labelled samples and abundant unlabelled samples [13].

1.1 Related Work

Semi-supervised GAN (SGAN) is an example of using labelled data during GAN training [11]. Unlike traditional GANs, where the discriminator makes a binary classification of 'real' and 'fake' images, SGAN is designed as a multiclass classifier, predicting which of N classes an input belongs to. The number of classes is extended to $N + 1$ during training, with the additional class corresponding to 'fake' outputs. This setup differs from regular (or normal) GANs as labels can now be assigned to the generated samples. Training times for the Generator of the network also decreased. Further, the Discriminator provided better classification accuracy with limited labelled data than Convolutional Neural Networks (CNNs) [11]. Salimans et al. [15] extended the work of Odena [11] by simplifying the architecture of the discriminator while improving the visual quality of the images produced [15]. Kumar et al. [8] proposed a further improvement by including fake samples in the training process [8].

A unification of Variational AutoEncoder (VAE) and GAN has been considered previously. Yu et al. [18] proposed VAEGAN, which uses a VAE as the generator and the discriminator from the traditional GAN model. Both SGAN and VAEGAN produce higher-quality images than vanilla GAN. However, VAEGAN makes it difficult for the model to learn the latent distribution and for the discriminator to be well-trained. Further, VAEGAN does not take advantage of existing labelled data like SGAN.

1.2 Contributions

Despite the success of GAN, issues relating to the convergence of models still hamper practitioners. One factor preventing convergence is mode collapse, which results in a lack of diversity in the samples produced by the generator [3]. Further, GANs may also experience difficulties in training due to vanishing gradients [10]. In addressing these challenges, GANs, being inherently unsupervised, often fail to exploit labelled data that may be available like SGAN does.

This study aims to enhance GANs by incorporating pre-training and semi-supervision into the training process. In particular, the generator of the GAN was pre-trained using a VAE. The advantage of this setting is to provide the discriminator with higher-quality images at earlier training times while using semi-supervised learning later on to exploit existing labels. The proposed setup, dubbed SSGAN is then trained to generate class-specific samples that can be effectively used in downstream semi-supervised learning tasks. In so doing, the setup can exploit labelled data that may be available. The resulting architecture consists of a pre-trained VAE as the generator and a discriminator trained in a semi-supervised learning framework.

The study involves creating and training three GAN setups. First, a typical semi-supervised GAN (SGAN) was trained. Second, the generator of the SGAN was pre-trained using a VAE. Finally, these are compared against a regular GAN. All GANs were trained on three datasets: MNIST, Fashion MNIST (FMNIST), and CIFAR-10. In the semi-supervised learning GAN, only a small subset of the samples had labels to facilitate the training process.

To assess the GANS, the samples generated were compared qualitative and quantitative. To support the qualitative evaluation of the samples generated, the quality of generated samples is evaluated by training a CNN classifier using generated samples and then using real images as test data. Statistical analysis is also performed to understand the significance of the results.

In summary, the main contribution of this study lies in the proposal of a semi-supervised GAN architecture where the decoder of a pre-trained VAE is used as the generator, and the discriminator is trained in a semi-supervised learning setting. The proposed SSGAN model achieves the same asymptotic performance as SGAN, generating high-quality images whilst demonstrating both jumpstart and improved area under the learning curve. The test results show significant improvements over normal GANs in classification accuracies on the benchmark datasets.

The rest of the paper is organized as follows. Section 2 presents the proposed method together with implementation details; Sect. 3 gives the experimental results and discussions; and Sect. 4 concludes the paper and points out potential future work.

2   Methodology

The proposed methodology can be summarised in two steps. The first step facilitates the training of a VAE, and prepares the decoder of the VAE as the pre-trained generator of the proposed semi-supervised GAN (termed SSGAN for this study). The second step involves implementing the semi-supervised GAN model with a stacked architecture for the discriminator and training using mainly unlabelled data (unsupervised learning) and a small number of labelled data (supervised learning). Hence, the GAN was trained in a semi-supervised fashion overall.

The following subsections introduce the implemented VAE and the proposed semi-supervised GAN, SSGAN, and other implementation details.

2.1   VAE and Pre-trained Generator

To improve the architecture of SGAN, the generator of the GAN was pre-trained as the decoder from a VAE [7]. VAEs are based on two key ideas: encoding the input data into a latent representation and decoding the latent representation back into data space. Neural networks model the encoding and decoding processes, and the VAE objective function is optimised using variational inference [7].

The VAE objective function consists of the reconstruction loss and the KL-divergence regularisation terms:

$$\mathcal{L}_{VAE} = \mathbb{E}_{z \sim q_\phi(z|x)}[\log p_\theta(x|z)] - D_{KL}(q_\phi(z|x)||p(z)) \tag{1}$$

where $\mathcal{L}_{VAE}$ is the VAE loss function, $q_\phi(z|x)$ is the encoder network that maps the input data x to the latent representation z, $p_\theta(x|z)$ is the decoder network that maps the latent representation z to the reconstructed data $\hat{x}$, and $p(z)$ is the standard normal distribution. Given the latent representation z, the reconstruction loss is modelled as the negative log-likelihood of the data as

$$\log p_\theta(x|z) = -\frac{1}{2}\sum_{i=1}^{D}(x_i - \hat{x}_i)^2 - \frac{D}{2}\log(2\pi\sigma^2) \tag{2}$$

where D is the number of samples, x_i and $\hat{x}_i$ are the i-th elements of the input and reconstructed data, respectively, and σ^2 is the variance of the reconstruction distribution. The second term in Eq. 1 is the KL divergence regularisation term, which ensures that the learned latent representation learns a distribution. This term is computed as

$$D_{KL}(q_\phi(z|x)||p(z)) = -\frac{1}{2}\sum_{i=1}^{K}(1 + \log\sigma_i^2 - \mu_i^2 - \sigma_i^2) \tag{3}$$

where K is the dimensionality of the latent space, and μ_i and σ_i are the mean and standard deviation of the i-th dimension of the latent representation z, respectively.

The VAE used the architecture in Fig. 1 for this study. A similar architecture was used for each of the selected datasets, with modifications to the input and output layers and the latent dimensions to cater to the different image sizes of the datasets. The latent dimension that resulted in the lowest average validation loss was selected for each dataset. The optimal latent dimensions for MNIST, FMNIST and CIFAR-10 are 2, 5 and 10, respectively, from the search space of $\{2, 5, 10, 20, 100\}$. In the experiment, we used Adam optimiser to train the VAE model parameters with learning rates of 0.0002, 0.0002 and 0.0001 for MNIST, FMNIST and CIFAR-10 datasets.

2.2   GAN

The architecture of a GAN consists of two neural networks, namely a generator network and a discriminator network, that play an adversarial game against each other [5]. The generator network transforms a random input vector into a data instance that resembles the training data. The discriminator network takes a data instance from the training data or the generator and classifies it as real or fake. The generator is trained to create data instances that fool the discriminator, and the discriminator is trained to classify real and fake data instances correctly [5].

Formally, given the generator G, $G(z; \theta_g) = X$ where z is the input vector to the generator, X is the output of the generator, and θ_g are the parameters of the generator, and discriminator D, given by $D(X; \theta_d) = p_{data}(X)$, where $p_{data}(X)$ is the probability of X being a real data instance, and θ_d are the parameters of

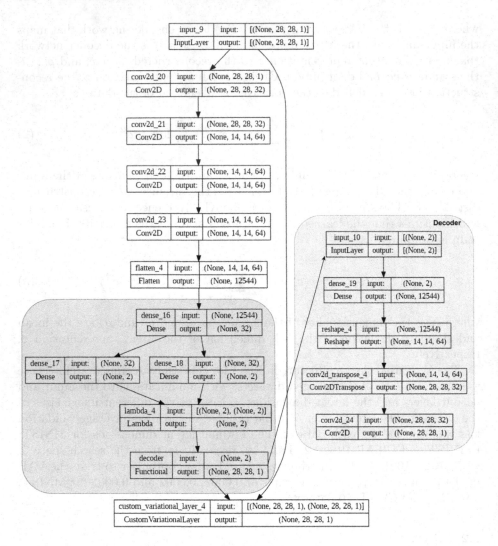

Fig. 1. Architecture used for the VAE model. The blue block depicts where the latent dimensions are specified, and the decoder for the MNIST dataset is illustrated in a yellow block. (Color figure online)

the discriminator, a GAN network is trained using the objective function,

$$\min_{G} \max_{D} V(D, G) = \mathbb{E}_{x \sim p_{data}(x)}[\log D(x)] + \mathbb{E}_{z \sim p_z(z)}[\log(1 - D(G(z)))] \quad (4)$$

where $p_{data}(x)$ is the distribution of real data instances, $p_z(z)$ is the distribution of random input vectors, $D(x)$ is the discriminator's output when given a data instance x, and $G(z)$ is the generator's output when given a random input vector z. The objective of the discriminator is to maximise this function, while the generator's objective is to minimise it.

After training, the generator can generate realistic new data instances by sampling from the distribution of random input vectors. The generated data instances can be used for various applications, such as image synthesis, data augmentation, and anomaly detection [3].

The optimisation of GANs is a challenging task due to the min-max nature of the objective function in Eq. 4. The generator and discriminator are trained alternately, which can lead to instability and oscillations in the training process. One common problem is mode collapse, where the generator produces limited variations of the same data instances. Various techniques have been proposed to address this issue, such as adding noise to the input of the discriminator, using different architectures for the generator and discriminator, and using regularisation techniques to prevent overfitting [1,2,5]. Further, mode collapse and overfitting are closely related because the generator and discriminator are adversarial. Therefore, balancing the two is critical to avoiding mode collapse and overfitting in GANs [1,15].

To address some of these issues, for the architecture of our model, the VAE decoder was used as a pre-trained generator for the GAN. Figure 2 illustrates the architecture for an SGAN modified with the VAEGAN. The yellow block indicates the additional parts of the architecture for a semi-supervised GAN compared to a regular GAN. The purple block indicates the included VAE. The architecture remains mostly the same, except for the labelled data input to the discriminator to help train it for semi-supervised learning.

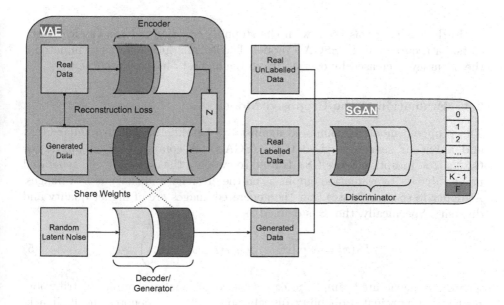

Fig. 2. SSGAN with the modifications to GAN for SGAN are shown in the yellow block and for VAE in the purple block. (Color figure online)

The discriminator model consists of three convolutional layers with 14×14, 7×7 and 4×4 filters, respectively. The convolutions are followed by dropout, a dense layer of width 2048, and a dense layer of width $N + 1$.

2.3 Hyperparameter Tuning

A GAN consists of multiple components, including the generator and discriminator, each with its own set of hyperparameters. These hyperparameters may significantly affect the quality of the generated output. Similarly, the fine-tuning of hyperparameters of the CNN is crucial for optimal classification performance. To tune the GAN and CNN's learning rate, batch size and dropout rate, the Keras Bayesian Optimisation[1] was used [12]. The loss functions, noise dimensions and the number of layers remained constant. Table 1 shows the hyperparameter ranges specified.

Table 1. Search range for hyperparameters

	Start Point	End Point	Step Size
Batch Size	32	256	16
Dropout Rate	0.1	0.5	0.1
Learning Rate	0.00001	0.001	0.00001

Early Stopping was used, with the stopping criterion being a change in the classifier accuracy of the SSGAN model. The learning process was terminated if the accuracy decreased by over 3% from the highest obtained accuracy.

2.4 Evaluation of the Proposed Model

The proposed model is compared against two baseline models: a regular GAN and a regular semi-supervised GAN (SGAN). A commonly used metric for GANs [16], Inception Score (IS), is used to evaluate the quality of the generated images due to its evaluation emphasis on the diversity and sharpness of images. A higher IS score indicates that the generated images are of higher quality and diversity. Specifically, the IS is defined as:

$$IS(x) = \exp\left(\mathbb{E}_{x \sim p_g} D_{KL}(p(y|x)\|p(y))\right), \tag{5}$$

where x is a generated sample, $p(y|x)$ is the conditional probability distribution, $p(y)$ is the marginal probability distribution, and D_{KL} denotes the Kullback-Leibler divergence.

---

[1] The following parameters were passed to the Bayesian Optimiser for each of the datasets: *num_initial_points*: 10; *max_trials*: 100 and *seed*: 42.

Compared to the Fréchet Inception Distance score, which is another commonly used metric for evaluating the quality of generated images in GANs, the IS is a better choice, especially for smaller datasets, as it is less sensitive to the size of the dataset and can provide a reasonable estimate of the quality of the generated images [17].

3 Results and Discussion

3.1 Hyperparameter Optimisation Results

GAN. After using Bayesian Optimisation and Early Stopping, the Generative Adversarial Network (GAN) hyperparameters were obtained for each of the three datasets. The obtained hyperparameter values are provided in Table 2.

Table 2. Hyperparameter values used in GAN

	MNIST	FMNIST	CIFAR-10
Epochs	50	80	145
Batch Size	256	160	128
Dropout Rate	0.1	0.2	0.4
Learning Rate	0.0002	0.0002	0.0001

VAE. The VAE architecture that was used is illustrated in Fig. 1. Early Stopping was also used for the VAE to determine the number of epochs used for each dataset. In the case of this model, the stopping criterion used was a change in the validation loss of the learned data. The model would stop training if the loss did not improve after five epochs. These values are 55, 80 and 135 for MNIST, FMNIST and CIFAR-10, respectively.

3.2 Results for GAN Models

In the following subsections, we present results from various experiments. These include visual comparisons of generated images by a regular GAN and SSGAN, classification results from using the discriminator component of SSGAN as a classifier and training CNN classifiers using the generated images from the three GAN models.

Image Quality. For qualitative analysis, the images generated by a regular GAN and the SSGAN are shown in Figs. 3 and 4 for MNIST, FMNIST, and CIFAR-10 datasets. As illustrated in the figures, the quality of images generated for each dataset by SSGAN is visually better than that generated by a regular

GAN, especially for the earlier epochs. Figure 3 shows that SSGAN produces distinguishable numbers and clothing shapes for the MNIST and FMNIST datasets in early epochs. In contrast, a regular GAN model generates blurry shapes. Figure 4 presents the images from CIFAR-10. At early epochs, the images by our model are clear but do not explicitly indicate animals. In contrast, the regular GAN model only presents blurry images. This validates the superiority of using the decoder of a pre-trained VAE as the generator in the proposed model. At later epochs, the CIFAR-10 images slightly increase in quality, but it is still difficult to distinguish which type of animal is presented. This suggests that our model works well for simpler datasets, such as MNIST or FMNIST, but not for more complicated datasets, such as RGB images. However, the model produces better-quality images than a regular GAN model. Figure 4 contains images from the later epoch (Epoch number 130) to show whether mode collapse occurred.

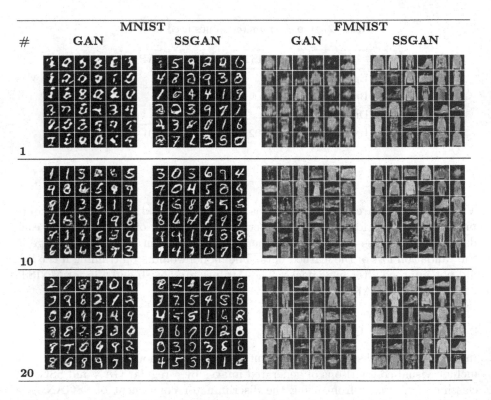

Fig. 3. Generated MNIST and FMNIST images after varying numbers of epochs.

Classification Accuracy of the SSGAN Discriminator as a Classifier. In our semi-supervised GAN setting, the discriminator also functions as a classifier to classify the generated samples into $N + 1$ different classes, one being the

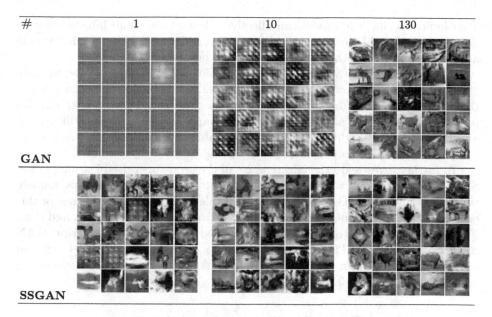

#	1	10	130

GAN

SSGAN

Fig. 4. Generated CIFAR-10 images after varying numbers of epochs.

'fake' class. To evaluate the performance of this discriminator, i.e., classifier, we performed classification tasks using the discriminator part of our model. To do this, we trained the SSGAN for each dataset using a limited number of labelled samples. We use 100 labelled samples for the MNIST, 1000 for the FMNIST, and 4000 for the CIFAR-10 dataset. These values are chosen based on the existing literature's state-of-the-art models for semi-supervised GAN. The chosen numbers of labelled samples also indicate that more labelled samples are required as the datasets become more complicated. The accuracy scores, along with their standard deviations from the classifier of the SSGAN, are shown in Table 3 for each dataset.

Table 3. Classification results of the classifier from the stacked discriminator in our SSGAN model (n represents the number of labelled samples used during training of the GAN).

	MNIST (n = 100)	FMNIST (n = 1000)	CIFAR-10 (n = 4000)
Accuracy	99.32% ± 0.21	92.78% ± 0.56	83.22% ± 1.08

The results in Table 3 demonstrate that the model's classification component successfully learned the images' latent features and accurately predicted their corresponding outputs. The model's performance on MNIST, which only had 100 labelled images, was good, achieving a classification accuracy of 99.32%.

This indicates that the model can effectively learn from small labelled sample sizes and yield highly accurate results. Moreover, the model loss graphs provide additional support for the robustness of the model.

When applied to FMNIST, which had 1000 labelled images, the model's classification accuracy was slightly lower at 92.78%, but still impressive given the complexity of the dataset. In comparison, the CIFAR-10 dataset, with its intricate features and a greater variety of shapes within images, achieved an accuracy of 83.22%, which is also commendable since it contains many more complexities than the other datasets.

To further investigate the effectiveness of the semi-supervised learning in the SSGAN, a comparison was performed with two state-of-the-art models, namely the SGAN and Triple GAN models [8,9]. Table 4 displays the outcomes of this comparison. These results were taken directly from the original papers, and thus, FMNIST is omitted as no results were provided for the SGAN and Triple GAN models on FMNIST. The results obtained in Table 4 for our model were an average of results taken over five separate runs of the model, and they are the same as those in Table 3.

Table 4. Classifier scores for each dataset

	MNIST ($n = 100$)	CIFAR-10 ($n = 4000$)
Triple GAN	90.90%	91.52%
SGAN	99.11%	82.74%
SSGAN	99.32%	83.22%

As illustrated, our model outperforms the accuracy scores obtained for the MNIST and CIFAR-10 datasets in the SGAN model. Our model's accuracy is 0.21% better than SGAN for MNIST and 0.48% better for CIFAR-10. Our model's accuracy on the MNIST dataset is also better than that of Triple GAN, with an improved accuracy of 8.42%. However, the Triple GAN accuracy for CIFAR-10 is significantly higher than our model by 5.3%.

The overall promising performance of SSGAN suggests that utilising the decoder of a VAE as the discriminator enhances image generation and classification accuracy. The higher image quality produced during the early training epochs likely facilitated the discriminator's learning process, resulting in a better-performing classification component.

Loss Graphs. The loss graphs for the discriminator during training are analysed to investigate how well the proposed SSGAN can avoid fast convergence of the discriminator and model collapse. Figure 5 shows the discriminator and generator losses for the MNIST and CIFAR-10 datasets and comparisons to a regular GAN model. It is important to note that the x-axis measures steps taken, not epochs. As depicted in the figure, the losses of normal GANs exhibit oscillatory behaviour during the initial stages of the training process, gradually decreasing as the training progresses. On the other hand, our model's losses show smaller oscillations from the start. In addition, the losses at the early and subsequent epochs have smaller differences and remain almost constant. Our model also shows smaller standard deviations compared to the normal GANs.

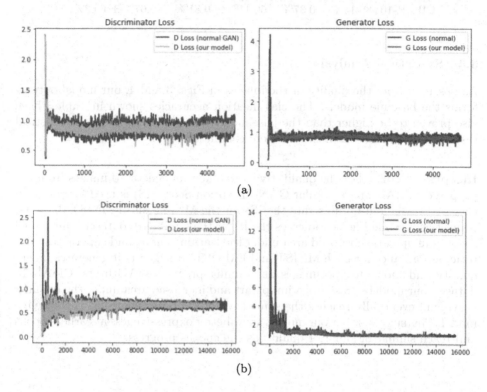

Fig. 5. (a) MNIST GAN loss functions, (b) CIFAR-10 GAN loss functions.

Classification Results. To evaluate the quality of the images generated by SSGAN, we trained several CNN classifiers using real images and images generated from a regular semi-supervised GAN (SGAN) and SSGAN, respectively. Each classifier was run five times for each dataset to ensure completeness, and the mean accuracy and standard deviation were recorded. The results obtained for each classifier on the various datasets are illustrated in Table 5.

The classification accuracies of SSGAN compared to the SGAN model are 5.17% higher for MNIST, 4.72% higher for FMNIST, and 9.56% higher for the CIFAR-10. This means that the proposed SSGAN model outperforms the SGAN for all datasets. The results further support that pre-training the generator as the decoder of a VAE improves the training of semi-supervised GANs.

Table 5. Average classifier scores and standard deviations for each model

Model	CNN using real images	CNN using SGAN images	CNN using SSGAN images
MNIST	99.31% ± 0.15%	92.87% ± 0.29%	98.04% ± 0.21%
FMNIST	93.17% ± 0.21%	86.41% ± 0.44%	91.13% ± 0.33%
CIFAR-10	89.14% ± 0.37%	60.11% ± 0.81%	69.67% ± 0.63%

3.3   Statistical Analysis

As observed from the quality of the images in Figs. 3 and 4, our model outperforms the baseline models. The classification accuracies shown in Table 3 have also proven to be higher than the baseline models. For completeness, statistical tests and statistical classifications were conducted.

Inception Score. For the qualitative assessment of generated images from the proposed SSGAN and a regular GAN, Inception Score (IS) is used for comparison. Figures 6a and 6b illustrate the ISs for the MNIST and FMNIST datasets, respectively, while Fig. 6c indicates the IS for the CIFAR-10 dataset. Our model shows a jumpstart, improved area under the learning curve, and superior asymptotic performance for both MNIST and FMNIST. It suggests it generates higher quality and more diverse images as training progresses. With the CIFAR-10 dataset, our model's IS shows a jumpstart and increased area under the learning curve and eventually reaches the same asymptotic performance as the baseline model. This suggests that our models show better expressiveness at earlier epochs and reach similar or superior quality as the epochs progress.

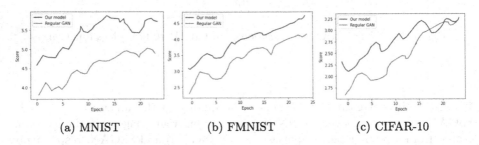

(a) MNIST   (b) FMNIST   (c) CIFAR-10

Fig. 6. Inception score for the dataset. Our model is in blue. A regular GAN is in red. (Color figure online)

Statistical Significance. To evaluate the statistical significance of the observed differences in accuracy scores among the different datasets (MNIST, FMNIST, and CIFAR-10), an analysis of variance (ANOVA) followed by Tukey's Honestly Significant Difference (HSD) post-hoc test was performed.

The ANOVA test was conducted to assess whether there were statistically significant differences in the means of accuracy scores among the datasets. The null hypothesis (H_0) assumed that the means were equal across all datasets, while the alternative hypothesis (H_a) suggested that at least one dataset's mean accuracy score was significantly different. The obtained F-statistic was 47.85 with a corresponding p-value of 0.0001. Since the p-value is below the chosen significance level of 0.05, the null hypothesis was rejected, indicating significant differences in accuracy scores among the datasets.

Following the significant result from the ANOVA, Tukey's HSD post-hoc test was conducted to identify pairwise differences in accuracy scores. The post-hoc test results indicated that the accuracy scores for all three datasets were significantly different from each other $(p < 0.05)$. Specifically, the accuracy scores for MNIST were significantly higher than both FMNIST and CIFAR-10 $(p < 0.05)$ datasets. The accuracy score for FMNIST was significantly higher than that of the CIFAR-10 dataset $(p < 0.05)$.

These statistical analyses provide robust evidence that the observed differences in classification accuracy scores among the datasets are statistically significant, supporting the conclusion that the dataset characteristics impact the performance of our GAN-based classifier.

4   Conclusion

Despite the advancements made in the training of GANs, the potential of this class of generative models is still hampered by issues, including unstable training and mode collapse. This study attempts to provide partial solutions to some of these issues. We present a modified architecture of semi-supervised GAN, named SSGAN, by replacing the generator of the regular SGAN model with the decoder of a pre-trained VAE. Experimental results show that the proposed semi-supervised GAN model effectively generates high-quality images at early epochs, implying more stable GAN training. When the generated images are used to train a CNN classifier, the proposed SSGAN model led to higher accuracy scores on the test dataset that consists of real images, outperforming the baseline model, a normal semi-supervised GAN, by 5.17% for MNIST, 4.72% for FMNIST, and 9.56% for CIFAR-10.

While this study uses the decoder from a vanilla VAE model, more complex VAE models, such as a Conditional Variational Autoencoder (CVAE) or an InfoVAE, can be considered in the proposed framework. In future work, we will investigate whether decoders from more complex VAE models in the SSGAN training can further improve the quality of generated images. Adapting different GAN architectures, such as DCGAN or CGAN models, to include the stacked discriminator is also an interesting future research direction.

References

1. Arjovsky, M., Bottou, L.: Towards principled methods for training generative adversarial networks (2017)
2. Arjovsky, M., Chintala, S., Bottou, L.: Wasserstein generative adversarial networks. In: International Conference on Machine Learning, pp. 214–223. PMLR (2017)
3. Creswell, A., White, T., Dumoulin, V., Arulkumaran, K., Sengupta, B., Bharath, A.: Generative adversarial networks: an overview. IEEE Signal Process. Mag. **35**(1), 53–65 (2018). https://doi.org/10.1109/MSP.2017.2765202
4. Harshvardhan, G.M., Gourisaria, M.K., Pandey, M., Rautaray, S.S.: A comprehensive survey and analysis of generative models in machine learning. Comput. Sci. Rev. **38**, 100285 (2020). https://doi.org/10.1016/j.cosrev.2020.100285
5. Goodfellow, I., et al.: Generative adversarial nets. In: Advances in Neural Information Processing Systems, vol. 27 (2014). https://doi.org/10.1145/3422622
6. Huo, J., van Zyl, T.L.: Incremental class learning using variational autoencoders with similarity learning. Neural Comput. Appl. 1–16 (2023)
7. Kingma, D.P., Welling, M.: An introduction to variational autoencoders. Found. Trends® Mach. Learn. **12**(4), 307–392 (2019). https://doi.org/10.1561/2200000056
8. Kumar, A., Sattigeri, P., Fletcher, T.: Semi-supervised learning with GANs: manifold invariance with improved inference. In: Advances in Neural Information Processing Systems, vol. 30 (2017)
9. Li, C., Xu, T., Zhu, J., Zhang, B.: Triple generative adversarial nets. In: Advances in Neural Information Processing Systems, vol. 30 (2017)
10. Mao, X., Li, Q., Xie, H., Lau, R., Zhen, W.: Multi-class generative adversarial networks with the L2 loss function (2016). arXiv preprint arXiv:1611.04076
11. Odena, A.: Semi-supervised learning with generative adversarial networks. arXiv preprint arXiv:1606.01583 (2016)
12. O'Malley, T., et al.: Kerastuner (2019). https://github.com/keras-team/keras-tuner
13. Reddy, Y., Viswanath, P., Reddy, B.E.: Semi-supervised learning: a brief review. Int. J. Eng. Technol. **7**, 81 (2018). https://doi.org/10.14419/ijet.v7i1.8.9977
14. Sah, S.: Machine learning: a review of learning types (2020). https://www.preprints.org/manuscript/202007.0230/v1
15. Salimans, T., Goodfellow, I., Zaremba, W., Cheung, V., Radford, A., Chen, X.: Improved techniques for training GANs. In: Advances in Neural Information Processing Systems, vol. 29 (2016)
16. Szegedy, C., et al.: Going deeper with convolutions. In: Proceedings of the IEEE Conference on Computer Vision and Pattern Recognition, pp. 1–9 (2015). https://doi.org/10.1109/CVPR.2015.7298594
17. Wu, Y., Burda, Y., Salakhutdinov, R., Grosse, R.: On the quantitative analysis of decoder-based generative models (2016). https://doi.org/10.48550/ARXIV.1611.04273
18. Yu, X., Zhang, X., Cao, Y., Xia, M.: VAEGAN: a collaborative filtering framework based on adversarial variational autoencoders. In: IJCAI, pp. 4206–4212 (2019). https://doi.org/10.24963/ijcai.2019/584

Impacts of Architectural Enhancements on Sequential Recommendation Models

Mufhumudzi Muthivhi[1]([✉])(ID), Terence L. van Zyl[2](ID), and Hairong Wang[1](ID)

[1] University of the Witwatersrand, Johannesburg 2000, Gauteng, South Africa
1599695@students.wits.ac.za
[2] University of Johannesburg, Johannesburg 2092, Gauteng, South Africa
tvanzyl@uj.ac.za

Abstract. Improving the architecture of sequential recommendation models has long been associated with enhanced accuracy metrics. However, new evaluation methods reveal that these improvements often favour a specific user cohort. The study aims to show that their popularity bias limits the performance of sequential recommendation models despite the numerous architectural enhancements adopted from NLP and Computer Vision. We propose a novel evaluation methodology to reflect users' preferences for popular and unpopular items accurately. We vary the threshold across the power law distribution to obtain two item subsets. This process sheds light on the extent of bias and performance discrepancies across the user spectrum. Our analysis of the experimental results reveals sequential recommendation models are limited to performing only as well as the epsilon-greedy algorithm. Consequently, the enhanced accuracy metrics such as Hit Rate and Normalized Discounted Cumulative Gain, frequently highlighted in the research, tend to stem primarily from the user group positioned at the short head of a power-law distribution.

Keywords: Sequential recommendation model · SASRec · Popularity bias

1 Introduction

Recommendation systems enhance user experiences by suggesting relevant items based on user preferences. Learning user behaviour is needed for developing effective recommender systems, as it allows for more personalized and pertinent recommendations [12,28].

Sequential recommendation models predict the next item a user would prefer based on the user's historical preference behaviour. Recent advancements in sequential recommendation involve incorporating architectural advances motivated by recent successes in Natural Language Processing (NLP) and Computer Vision. Moreover, there is a growing emphasis on modifying attention mechanisms within sequential recommendation methods to enhance their predictive

© The Author(s), under exclusive license to Springer Nature Switzerland AG 2023
A. Pillay et al. (Eds.): SACAIR 2023, CCIS 1976, pp. 315–330, 2023.
https://doi.org/10.1007/978-3-031-49002-6_21

capabilities [21,22,26,35]. These modifications improve the model's ability to capture relevant patterns and temporal dependencies in the user's interaction history. Although these algorithmic developments in recommendation systems have generated substantial improvements in a wide range of applications, these systems continue to face the challenges of being biased, including a popularity bias. Popularity bias is manifested as a user being exposed to popular items more frequently. The excess exposure of users to popular items has a compounding effect that makes popular items more popular. When considering the probability distribution of the frequency of item recommendation, this compounding can lead to recommendation systems with the distribution heavily skewed towards a small subset of popular items, with the majority of items ending up in a long tail with little probability of being recommended.

A common goal in the sequential recommendation domain is to enhance the model's accuracy by re-imagining its architecture, drawing inspiration from advancements in NLP and Computer Vision. However, a prevalent popularity bias endures when evaluating the models across the entire item collection. The study aims to show that even a simple epsilon-greedy algorithm-based recommendation approach outperforms the commonly used sequential recommendation models regardless of architectural improvements. In summary, we contribute to the existing literature by:

- Proposal of a novel evaluation methodology accurately reflecting users' preferences for popular and unpopular items.
- Implementation of several architectural enhancements on the unidirectional transformer model (SASRec).
- Demonstration of the existence of a performance ceiling for sequential recommendation models despite their architectural advancements.
- Identification of the enduring superiority of the Epsilon Greedy algorithm over proposed state-of-the-art sequential recommendation algorithms.
- Enrichment of the unidirectional transformer model by incorporating context data using an auxiliary-information embedding as the item embedding.

The rest of the paper is structured as follows. Section 2 provides an overview of the related work. In Sect. 3, we detail the methodology employed to develop several enhanced sequential recommendation models. Section 4 proposes a novel evaluation method and outlines the datasets and baseline models used. Our findings are presented in Sect. 5, and the paper is concluded in Sect. 6.

2 Related Work

Sequential recommendation models predict a user's next preferred item based on their historical behaviour [12,28]. A recommendation model can provide more relevant and personalized recommendations by understanding users' preferences. Early works focused on embedding the transition information between adjacent behaviours into the item latent factor using matrix factorization and Markov chains [28].

Due to the success of Deep Learning (DL) models for feature learning in Natural Language Processing (NLP) and Computer Vision, many DL-based sequential models were developed [9]. DL-based sequential recommendation models focus on capturing intricate user preferences and item dependencies by incorporating temporal context. Earlier DL-based models consisted of Recurrent Neural Networks that capture global sequential patterns and learn dynamic user interest representations [14,37].

Recently, attention mechanisms have shown success in modelling sequential data for machine translation and text classification [5,32,33]. As a result, there has been increasing interest in integrating fully attention-based neural networks for sequential recommendation. Notably, Kang and McAuley [17] introduce a unidirectional transformer model called SASRec, which employs adaptive item-specific weighting and attains state-of-the-art results. Building upon SASRec, Sun et al. [30] enhance the framework by implementing a bidirectional transformer architecture and a Cloze task, yielding their BERT4Rec model.

A commonly adopted strategy in NLP and Computer Vision communities is to improve the transformer models by redesigning the model architecture [7]. Researchers in sequential recommendation have taken a similar approach by redesigning the architecture of SASRec and Bert4Rec. Methods such as learning to rank, contrastive learning, pseudo labels, sequence-to-sequence training, and diffusion, are adapted for sequential recommendation tasks in different application domains [21,22,26,35].

Recent studies have found that recommender systems that perform well on accuracy metrics focus their recommendations on a tiny fraction of the item spectrum [13]. Often, these items are the most popular and sit at the short head of a power law probability distribution. This is formally known as popularity bias and occurs when already popular items are recommended even more frequently than their popularity would warrant [2]. This is caused by the experimental nature of the recommendation process leading to the probability distribution of the frequency of item recommendation following a power law or Pareto distribution [25]. As a result, we have the data divided into roughly two groups: one with those at the short head of the distribution (popular items) and the other with those found at the long tail of the distribution (unpopular items).

Several investigations have focused on addressing popularity bias in recommendation systems [1,4]. Nonetheless, only a few studies have incorporated bias mitigation techniques within the context of sequential recommendations [36]. Most papers persist in reporting biased results, often due to their reliance on the "leave-one-out" strategy to evaluate their recommendation models [16]. The "leave-one-out" strategy involves selecting a relevant item from the user's preference profile and ranking it against a randomly chosen subset of items. This evaluation approach, known as sampling metrics, accelerates the computation of metrics [16]. The sampling metrics approach has attracted significant scrutiny in recent years [20]. Krichene and Rendle [20] argue that the routine use of sampling metrics produces high-bias and low-variance estimators of the exact metrics. They propose that researchers evaluate their models by ranking the relevant item against the entire collection.

By evaluating the accuracy metrics against the entire set we found that even a basic recommendation approach based on the epsilon-greedy algorithm surpasses the performance of state-of-the-art sequential recommendation models. We argue that popularity bias persists in sequential recommendation models, irrespective of their algorithmic or architectural enhancements.

3   Methodology

The study uses Self-Attentive Sequential Recommendation (SASRec) as the base model [17]. We redesigned the model by adopting several NLP and Computer Vision architectural designs. We also incorporate auxiliary information to improve the model's context-awareness [8]. Several other works have reported improved results using a combination of sequential recommendation and auxiliary information [10,23,29,38]. Fischer et al. [10] concatenates the item attributes (such as category, brand and genre keywords) to the item embedding and then train on a bidirectional transformer. Zhang et al. [38] extends the item attributes to text descriptions and employs two separate self-attention blocks for item IDs and attributes. Singer et al. [29] track the changes of item attributes over time through a transformer that handles 2D input sequences. Unlike their work, we utilized all available auxiliary information using a multi-modal sequential recommendation model. We extracted learned embeddings from popular image, text, tabular and audio transformer models on HuggingFace [23,34].

3.1   Notations

In the following subsections, the following notations are frequently used. A set of users is denoted by $\mathcal{U} = \{u_1, u_2, \ldots, u_{|\mathcal{U}|}\}$, a set of items by $\mathcal{V} = \{v_1, v_2, \ldots, v_{|\mathcal{V}|}\}$, each item's auxiliary information embedding by $\mathcal{K} = \{\mathbf{k}_1, \mathbf{k}_2, \ldots, \mathbf{k}_{|\mathcal{V}|}\}$ where $\mathbf{k}_v$ is a joint vector embedding of text, tabular and image data for item v, and the sequential item data by $\mathcal{S}_u = \{v_1^u, v_2^u, \ldots, v_{n_u}^u\}$. Each term, v_t^u $(1 \leq t \leq n_u)$, in $\mathcal{S}_u$ represents an interaction of user u with item v at relative time t, where n_u is the number of items rated by user u. Finally, given $\mathcal{S}_u$, the next preferred item to be predicted for a user u is denoted by $v_{n_u+1}^u$.

3.2   Embedding

The embedding layer consists of two different learned embeddings: an auxiliary information embedding $\mathcal{K} \in \mathbb{R}^{|\mathcal{V}| \times d}$ and a positional embedding $\mathcal{P}$. The positional embedding encodes the positions of the items within a user's sequential item data $\mathcal{S}_u$. The vector is scaled to the embedding size d using a linear layer. For each item in the sequence $\mathcal{S}_u$, we have the auxiliary information embedding $\mathbf{k}_u \in \mathbb{R}^{n \times d}$, positional embedding $\mathbf{p}_u \in \mathbb{R}^{n \times d}$, and the summation of the two, $\mathbf{h}_u^0 = \mathbf{k}_u + \mathbf{p}_u$. The input to the transformer is then $\mathbf{H}^0 = \{\mathbf{h}_1^0, \mathbf{h}_2^0, \ldots, \mathbf{h}_{|\mathcal{U}|-1}^0, \mathbf{h}_{|\mathcal{U}|}^0\}$.

3.3   Models

SASRec serves as our base model upon which we construct several architectural modifications discussed in the following subsections. SASRec uses a single-head self-attention mechanism:

$$\mathbf{H}^{l+1} = \text{Attention}(\mathbf{H}^l \mathbf{W}^Q, \mathbf{H}^l \mathbf{W}^K, \mathbf{H}^l \mathbf{W}^V) \tag{1}$$

and a point-wise feed-forward network:

$$\mathbf{F}_i = \text{FFN}(\mathbf{h}_i^l) = \text{ReLU}(\mathbf{h}_i^l \mathbf{W}^{(1)} + \mathbf{b}^{(1)})\mathbf{W}^{(2)} + \mathbf{b}^{(2)} \tag{2}$$

where $\mathbf{W}$ and $\mathbf{b}$ are the learnable parameters. SASRec predicts the next item i given the first t items by calculating a relevance score $p_{i,t}$:

$$p_{i,t} = \mathbf{F}_t \mathcal{K}_i \tag{3}$$

We introduce several architectural enhancements aimed at regularizing the original SASRec Objective. The sequence-to-item SASRec objective aligns user sequences and items in vector space. These additional loss functions serve as regularization terms, addressing various limitations. The Pairwise loss encourages proper item ranking in pairs, and LambdaRank optimizes the change in NDCG for item importance assessment. Contrastive Loss promotes consistency across sequence augmentations, and Cross-pseudo Supervision ensures uniform recommendations for users, whether their behaviour is labelled or not. Sequence-to-sequence loss enhances longer-term dependency modelling in user behaviour, leading to more accurate recommendations over extended sequences. The Diffusion Process captures dynamic user interests and latent item aspects. The total loss is a linear weighted sum with λ as the trade-off weight.

SASRec and Learning to Rank introduces an extra loss function in addition to the one utilized by SASRec [17]. Let $P^u = \{p_1, p_2, ..., p_{|\mathcal{V}|-1}, p_{|\mathcal{V}|}\}$ be a list of predicted relevance scores for each user u then:

$$\text{PairwiseLoss}(\mathcal{P}^u) = \sum_{i \subset |\mathcal{V}|} \sum_{j \in |\mathcal{V}|} \text{ReLU}(1 - (p_i - p_j)) \tag{4}$$

The ReLU function ensures that the loss is only calculated for cases where the difference between relevance scores is positive when one item is predicted to be more relevant. Let $\mathcal{O} = \{o_1, o_2, ..., o_{|\mathcal{V}|-1}, o_{|\mathcal{V}|}\}$ be a list of binary values indicating whether the user has observed the item then we get the overall ranking loss:

$$\mathcal{L}_{\text{RankingLoss}}(\mathcal{P}^u, \mathcal{O}^u) = \text{PairwiseLoss}(\mathcal{P}^u)\frac{1}{|\mathcal{O}^u|} \sum_{i \in |\mathcal{V}|} \sum_{j \in |\mathcal{V}|} (1 - (o_i + o_j)) \tag{5}$$

That is, we compute an average PairwiseLoss only for pairs of items where at least one of them has been observed by the user, either o_i or o_j. The total loss is a linear weighted sum:

$$\mathcal{L}_{\text{total}} = \mathcal{L}_{\text{SASRec}} + \lambda \mathcal{L}_{\text{RankingLoss}} \tag{6}$$

where λ represents the weight. The pairwise Learning-To-Rank method considers each user's preferences against a pair of items. Instead of predicting an isolated relevance score for each item, LTR ranks them relative to their neighbouring pairs [18].

SASRec and LambdaRank. The LambdaRank loss function is a pairwise classification loss. The gradient is defined directly by considering ranking loss. The gradient is scaled by the size of the change in NDCG if item i is swapped with j:

$$\mathcal{L}_{\text{LambdaRank}}(\mathcal{P}^u) = \sum_{i=1}^{|\mathcal{V}|} \sum_{j=i+1}^{|\mathcal{V}|} \text{PairwiseLoss}(\mathcal{P}^u) \Delta NDCG_{ij} \tag{7}$$

Similarly, the total loss is linearly weighted:

$$\mathcal{L}_{\text{total}} = \mathcal{L}_{\text{SASRec}} + \lambda \mathcal{L}_{\text{LambdaRank}} \tag{8}$$

LambdaRank is a listwise Learning-To-Rank method that models each user's preference to a list of items. LambdaRank directly optimizes the Normalized Discounted Cumulative Gain (NDCG) accuracy metric [18].

SASRec and Contrastive Loss. We implement a contrastive learning framework inspired by the work of Xie *et al.* [35]. Initially, we enhance the dataset through a set, $\mathcal{A}$, of data augmentations that include three methods tailored for a sequential recommendation: item cropping, item masking, and item reordering. We apply two randomly sampled augmentation methods $a_i \in \mathcal{A}$ *and* $a_j \in \mathcal{A}$ to each user's historical behaviour sequence, $\mathcal{S}_u$, and obtain two views of the sequence, denoted by $\mathcal{S}_u^{a_i}$ and $\mathcal{S}_u^{a_j}$, respectively. We treat $(\mathcal{S}_u^{a_i}, \mathcal{S}_u^{a_j})$ as a positive pair and treat the remaining augmented sequences within the same batch as negative samples, denoted by $\mathcal{S}^-$. The contrastive loss is then formulated as:

$$\mathcal{L}_{\text{contrastive}}^u(\mathbf{F}_u^{a_i}, \mathbf{F}_u^{a_j}) = -\log \frac{\exp(\text{sim}(\mathbf{F}_u^{a_i}, \mathbf{F}_u^{a_j}))}{\exp(\text{sim}(\mathbf{F}_u^{a_i}, \mathbf{F}_u^{a_j})) + \sum_{m \in \mathcal{S}^-} \exp(\text{sim}(\mathbf{F}_u^{a_i}, \mathbf{F}_m^-))} \tag{9}$$

where $\mathbf{F}_u^{a_i}$ and $\mathbf{F}_u^{a_j}$ were computed as:

$$\mathbf{F}_u^{a_i} = f_{\text{SASRec}}(\mathcal{S}_u^{a_i}, \mathcal{K}), \quad \mathbf{F}_u^{a_j} = f_{\text{SASRec}}(\mathcal{S}_u^{a_j}, \mathcal{K}) \tag{10}$$

and, $\mathbf{F_m}^-$ denotes the representations of the negative samples from $\mathcal{S}^-$. The objective to be optimized then becomes:

$$\mathcal{L}_{\text{total}} = \mathcal{L}_{\text{SASRec}} + \lambda \mathcal{L}_{\text{contrastive}} \tag{11}$$

A contrastive loss function encourages the model to learn similar representations for different views of the same sequence while maximizing the distance between representations from the augmented sequences of different users. Contrastive learning is a self-supervised learning paradigm that helps reduce the data sparsity problem by enabling training on massive unlabeled data [35]. The main objective of contrastive learning is to improve the representations of the users' behaviour sequences.

Cross-Pseudo Supervision. Cross-pseudo Supervision is a semi-supervised learning paradigm that uses a consistency regularization approach over two networks [7]. The main objective of Cross-pseudo Supervision is to ensure consistency of predictions from both labelled and unlabeled user sequences. We use the dual parallel cross-pseudo supervision network structure approach introduced by Chen *et al.* [7]:

$$(\mathcal{S}_u, \mathcal{K}) \rightarrow f_{\text{SASRec}}((\mathcal{S}_u, \mathcal{K}) : \theta_1) \rightarrow \mathcal{P}_1^u \rightarrow \mathcal{O}_1^u \tag{12}$$

$$\searrow f_{\text{SASRec}}((\mathcal{S}_u, \mathcal{K}) : \theta_2) \rightarrow \mathcal{P}_2^u \rightarrow \mathcal{O}_2^u \tag{13}$$

where two f_{SASRec} networks maintain the same structure, but their weights, i.e., θ_1 and θ_2, are initialized differently. The supervision loss:

$$\mathcal{L}_{\text{supervision}}^u = \frac{1}{|\mathcal{V}|}(\ell_{ce}(\mathcal{P}_1^u, \mathcal{O}_1^{u^*}) + \ell_{ce}(\mathcal{P}_2^u, \mathcal{O}_2^{u^*})) \tag{14}$$

is defined through the conventional cross-entropy loss function, $\ell_{ce}(\cdot)$, applied to the actual binary values $\mathcal{O}^{u^*}$, denoting whether the user has interacted with the item. The cross-pseudo supervision loss is bidirectional:

$$\mathcal{L}_{\text{crosspseudo}}^u = \frac{1}{|\mathcal{V}|}(\ell_{ce}(\mathcal{P}_1^u, \mathcal{O}_2^u) + \ell_{ce}(\mathcal{P}_2^u, \mathcal{O}_1^u)) \tag{15}$$

where the output, $\mathcal{O}_1^u$, from one network $f_{\text{SASRec}}(\theta_1)$ to supervise relevance scores $\mathcal{P}_2^u$ of the other network $f_{\text{SASRec}}(\theta_2)$, and the other one is from $f_{\text{SASRec}}(\theta_2)$ to $f_{\text{SASRec}}(\theta_1)$. This calculation takes place across the two parallel segmentation networks. The total loss is then:

$$\mathcal{L}_{\text{total}} = \mathcal{L}_{\text{supervision}} + \lambda \mathcal{L}_{\text{crosspseudo}} \tag{16}$$

Sequence to Sequence training strategy asks the model to predict the representation of the future sub-sequence given the earlier sequence's representations [22]. Given the earlier sequence $\mathcal{S}_u^{1:\frac{n}{2}} = \{v_1^u, v_2^u, \ldots, v_{\frac{n}{2}}^u\}$ and the future sequence $\mathcal{S}_u^{\frac{n}{2}+1:n} = \{v_{\frac{n}{2}+1}^u, v_{\frac{n}{2}+2}^u, \ldots, v_n^u\}$, we divide the sequence equally to ensure enough training signals in the future sequence, and obtain the model outputs for each sequence:

$$\mathbf{F}_u^{1:\frac{n}{2}} = f_{\text{SASRec}}(\mathcal{S}_u^{1:\frac{n}{2}}, \mathcal{K}), \quad \mathbf{F}_u^{\frac{n}{2}+1:n} = f_{\text{SASRec}}(\mathcal{S}_u^{\frac{n}{2}+1:n}, \mathcal{K}) \tag{17}$$

The sequence-to-sequence loss is defined as:

$$\mathcal{L}_{\text{seq2seq}}^u = -\ln \frac{\exp(\frac{1}{\sqrt{d}} \mathbf{F}_u^{1:\frac{n}{2}} \cdot \mathbf{F}_u^{\frac{n}{2}+1:n})}{\sum_{m \in \mathcal{B}} \exp(\frac{1}{\sqrt{d}} \mathbf{F}_m^{1:\frac{n}{2}} \cdot \mathbf{F}_m^{\frac{n}{2}+1:n})} \tag{18}$$

where $\mathcal{B}$ be a mini-batch of users from $\mathcal{U}$. The combined loss function is then:

$$\mathcal{L}_{\text{total}} = \mathcal{L}_{\text{SASRec}} + \lambda \mathcal{L}_{\text{seq2seq}} \tag{19}$$

Sequence-to-item training can be seen as myopic and can lead to non-diverse recommendations [22]. By looking at the longer-term future given the earlier sequence, we can avoid cases where the next immediate behaviour in the training data is irrelevant to the earlier sequence of behaviours.

Diffusion process explores modelling users' interests and preferences as a distribution rather than static vectors. We adopt the same strategy as Li *et al.* [21]. Let x_s be the noised target item representation that has undergone s diffusion steps. Then we use x_s to adjust the representations of each item in the users' sequence $\mathcal{S}_u$ to obtain the representation $\mathbf{Z}_{x_s}$ from exploiting the auxiliary semantic signals in x_s. Next we obtain $\hat{x}_0 = f_{\text{SASRec}}(\mathbf{Z}_{x_s})$ by enforcing that $\hat{x}_0$ is close to its target item embedding. A fixed vector may struggle to capture the inconsistent and dynamic interest of a user as well as the multiple latent aspects of an item [21]. The Diffusion process can model the multiple interests of users and multiple latent aspects of items as distributions under a unified framework.

4   Experimental Setup

4.1   Datasets

The study considers four benchmark datasets from different application domains. The Amazon Fashion dataset comprises more than 1.5 million customer reviews and ratings for over 100,000 fashion products on Amazon. The dataset also offers product details, including their titles, prices, and descriptions [24]. MovieLens is a large-scale benchmark dataset with over 20 million user ratings on 26,000 movies and TV shows [11]. Steam dataset contains over 600,000 user reviews on 30,000 games from the Steam website. The dataset also includes information about the games, such as the game title, price, and release date. The LastFm dataset contains over two million user profiles, and over 500 million song listens. The dataset includes information about the songs, such as the song title, artist, and release date.

4.2   Evaluation

We assess our models using two accuracy-based metrics. The Hit Rate at N (HR@N) evaluates the proportion of users for whom a query of size N includes the correct item [19]. The Normalized Discounted Cumulative Gain at N (NDCG@N) assigns higher importance to more relevant items at the top of the query [15]. Given the persistence of popularity bias within sequential recommendation models, evaluating the entire dataset across all users might not provide comprehensive insights. As a solution, we categorize users into two groups based on their preference for popular or unpopular items. We analyse the last item in the user's sequence to determine a user's preference for popular or unpopular items. If this item falls within the top $\alpha \in \{0, 0.1, 0.2, 0.3, 0.4, 0.5, 0.6, 0.7, 0.8, 0.9, 1.0\}$ fraction of items in the power law

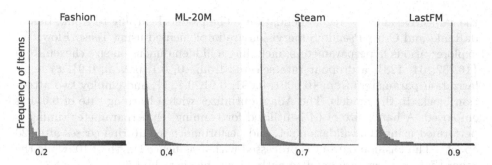

Fig. 1. Power law distributions for each dataset with an exemplified threshold value. The green area signifies popular items, while the red represents unpopular items. (Color figure online)

distribution, it signifies a preference for popular items. For instance, if $\alpha = 0.2$, then we take the top 20% of the items ranked according to their popularity and classify them as popular items. The remaining 80% of items would belong to the unpopular set. Figure 1 illustrates the power law distributions of each dataset, accompanied by an exemplified threshold value. The green area would be the popular items, and the red area corresponds to the unpopular items. We calculate HR@100 and NDCG@100 scores separately for users with popular and unpopular preferences. It's important to highlight that these metrics vary with changing α threshold values. A practical recommendation model should cater to both preferences, ensuring high performance for both categories [3,6]. That is, we expect to see a high performance for both popular and unpopular preference groups despite the change in threshold. If a recommendation model predominantly excels at lower threshold values, it indicates an inherent bias towards popular items. We ranked a relevant item against the entire item collection for each user and computed the aggregate score attained for each metric [20]. We conducted ten independent runs of the experiments and conducted paired t-tests to determine the significance of the SASRec variants over the epsilon-greedy algorithm.

4.3 Baselines and Implementation Details

The study uses three recommendation baselines: traditional, probabilistic and sequential recommendation models. Traditional: **Epsilon-greedy algorithm** is a simple exploration strategy widely used in reinforcement learning [31]. We set ϵ as the probability of selecting an item from the unpopular list. Probabilistic: **Bayesian Personalized Ranking (BPR)** optimizes matrix factorisation with implicit feedback using pairwise Bayesian Personalized Ranking [27]. Sequential: **Bert4Rec**, which enhances sequential recommendations using a bidirectional transformer architecture [30].

For BPR, BERT4Rec, SASRec, Contrastive loss, Sequence to Sequence, and Diffusion, we adopt hyperparameters and initialization strategies in line with the respective authors' recommendations. The models Learning to Rank, Lamb-daRank, and Cross-Pseudo Supervision are implemented using TensorFlow. We explore various hyperparameters, including a hidden dimension size chosen from {16, 32, 64, 128}, a dropout rate selected from {0, 0.1, 0.2, ..., 0.9}, ℓ_2 regularization parameters from {0.0001, 0.001, 0.01, 0.1, 1}, and employ two attention blocks in the models. The Adam optimizer with a learning rate of 0.001 is employed. A batch size of 64 is utilized for training. Hyperparameter tuning is performed using the validation set, and performance monitoring ceases after 20 epochs. The Fashion dataset is processed with a sequence length of 10, while the other three datasets are handled with a sequence length of 50.

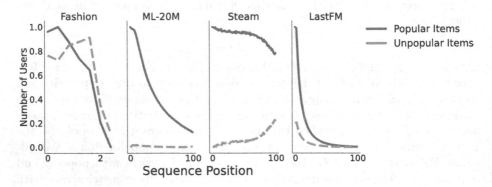

Fig. 2. Illustrates the count of users for each sequence position against two subsets of users: those who prefer popular or unpopular items. The earliest positions from left to right depict a declining number of users who prefer popular items while increasing number of users prefer unpopular items.

5   Results

5.1   Baselines and Transformer Varients

Table 1 provides a comprehensive overview of the outcomes attained across ten models, evaluated through two distinct metrics. In particular, we juxtapose the performance of our proposed models against that of established baselines such as the ϵ-greedy algorithm, BPR, Bert4Rec and SASRec. The ϵ-greedy algorithm notably outperforms most models, particularly for users inclined towards popular items. We designate an ϵ value of 0.2, representing the likelihood of selecting an item from the unpopular set. This choice results in the algorithm recommending popular items, almost resembling the naive PopRec baseline. As a result, SASRec and its variants perform worse than the ϵ-greedy algorithm but significantly better for users who prefer unpopular items. Considering the ML-20M dataset, SASRec achieved an HR@100 of 0.114 overall while reaching 0.128 for users who prefer popular items and performing less effectively for those inclined

Table 1. Accuracy metrics of four baseline models and six other architectural enhancements made on the unidirectional transformer model (SASRec). The bold values represent the best-performing model, while the grey values indicate the top two best-performing models.

Dataset Metrics	0.2-greedy	BPR	BERT4Rec	SASRec	LTR	LambdaRank	Contrastive	Seq2Seq	Cross-Pseudo	Diffusion
Fashion										
HR@100	**.1078**	.0411	.0414	.0367	.0117	.0392	.0389	.0159	.0104	.0417
NDCG@100	**.0373**	.0129	.0132	.0088	.0028	.0127	.0112	.0036	.0022	.0126
Popular										
HR@100	**.2606**	.0604	.0817	.0716	.0119	.0784	.0762	.0166	.0084	.0791
NDCG@100	**.0910**	.0182	.0279	.0179	.0030	.0273	.0234	.0034	.0017	.0262
unpopular										
HR@100	.0038	**.0373**	.0139	.0027	.0116	.0126	.0135	.0154	.0117	.0162
NDCG@100	.0007	.0123	.0032	**.0128**	.0027	.0028	.0029	.0036	.0025	.0034
ML-20M										
HR@100	.2429	.3086	.0949	.1137	.0157	.0042	.0091	.0010	**.3300**	.0101
NDCG@100	.0592	.0763	.0246	.0301	.0031	.0007	.0018	.0002	**.0907**	.0018
Popular										
HR@100	.2756	.3129	.1061	.1275	.0146	.0044	.0119	.0013	**.3752**	.0131
NDCG@100	.0672	.0774	.0277	.0338	.0028	.0007	.0023	.0002	**.1031**	.0021
unpopular										
HR@100	.0041	**.2774**	.0131	.0126	.0237	.0037	.0021	.0005	.0000	.0026
NDCG@100	.0009	**.0685**	.0024	.0024	.0051	.0006	.0005	.0001	.0000	.0010
Steam										
HR@100	.0010	.0097	.0095	.0037	.0019	**.0112**	.0091	.0010	.0003	.0101
NDCG@100	.0002	.0023	.0016	.0006	.0004	**.0019**	.0018	.0002	.0000	.0018
Popular										
HR@100	.0015	.0100	.0125	.0044	.0021	**.0148**	.0119	.0013	.0004	.0131
NDCG@100	.0003	.0022	.0020	.0007	.0004	**.0024**	.0023	.0002	.0001	.0021
unpopular										
HR@100	.0000	**.0089**	.0021	.0021	.0016	.0021	.0021	.0005	.000	.0026
NDCG@100	.0000	**.0023**	.0006	.0005	.0003	.0004	.0005	.0001	.000	.0010
LastFM										
HR@100	.1453	**.1822**	.0061	.0092	.0041	.0184	.0082	.0041	.0205	.0082
NDCG@100	.0339	**.0460**	.0013	.0019	.0007	.0039	.0033	.0007	.0118	.0016
Popular										
HR@100	.2141	**.2691**	.0061	.0107	.0031	.0229	.0092	.0031	.0306	.0107
NDCG@100	.0502	**.0678**	.0014	.0023	.0005	.0049	.0042	.0005	.0176	.0021
unpopular										
HR@100	.0062	.0062	.0062	.0062	.0062	**.0093**	.0062	.0062	.0000	.0031
NDCG@100	.0011	.0018	.0010	.0010	.0010	**.0019**	.0015	.0010	.0000	.0006

towards unpopular items with a score of 0.013. A similar trend holds across all the different SASRec variations. This observation implies that the prominent contribution to the overall HR@N and NDCG@N scores comes from recommendations of popular items. Consequently, these models optimize themselves over one group of users and perpetuate the popularity bias. This behaviour corresponds to the pattern observed in the 0.2-greedy Algorithm, where the optimal performance is concentrated among users who prefer popular items. Hence, this contributes to the algorithm's elevated overall HR@N and NDCG@N scores.

While it might be tempting to assume that popular items are inherently good recommendations, we show in Fig. 2 that this may not be true. Figure 2 illustrates the number of users who prefer popular items (indicated by the blue line) versus those inclined towards unpopular items (depicted by the orange line) relative to the position of the item in the sequence. The pattern reveals a decline in the count of popular items as we approach the most recently consumed items (from left to right). In contrast, the count of users who prefer unpopular items converges with those favouring popular items. This observation implies that while users exhibit a higher preference for popular items earlier in their consumption history, this inclination diminishes rapidly as their preference shifts towards unpopular items. Such a dynamic further explains the bias in sequential recommendation models. These models tend to be trained predominantly on popular items initially and consequently struggle to recommend unpopular items due to their relatively limited exposure to unpopular items.

5.2 ROC Curves

Given that our choice of the threshold value can be perceived as arbitrary, potentially sparking debates about whether popular items constitute 10% or even 50% of the items at the short head of the power law distribution, we introduce a novel evaluation scheme. We systematically alter the threshold value within the range of 0.0 to 1.0 in increments of 0.1. Subsequently, we construct ROC curves for users who prefer popular or unpopular items. This method provides a more comprehensive perspective on the model's behaviour across user preferences.

Figure 3 depicts the average ROC curve over all thresholds with the associated confidence interval. The top row plots in green represent ROC curves for users favouring popular items, and the bottom row plots in red represent users

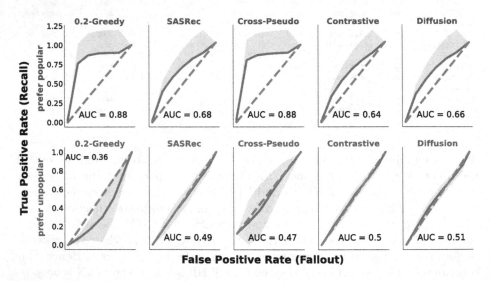

Fig. 3. Depicts the average ROC curve over all threshold values for the ML-20M dataset, with a blue shadow representing the confidence interval. The top plots in green represent the performance of each model on users who prefer popular items, while the bottom plots in red are for users who prefer unpopular items. Each model performs substantially better on popular items, achieving an AUC greater than 0.5. (Color figure online)

who prefer unpopular items. For users who prefer popular items, the curves consistently remain above the random classifier line for all models. Hence, all models effectively cater to the preferences of these users, with the ϵ-Greedy algorithm performing the best with an average AUC of 0.88. However, for users who prefer unpopular items (red plots), most models closely align with the random classifier line except for the ϵ-greedy algorithm with $\epsilon = 0.2$, which falls far below the random classifier line for most thresholds. This reaffirms that the ϵ-Greedy model struggles to accommodate the preference for unpopular items.

A practical recommendation model is expected to represent the differing preferences of users [3,6]. To understand which model best represents the two preferences considered in this paper, we calculate the average Recall and Fallout values for users with preference for popular or unpopular items and plot a unified ROC curve in Fig. 4. The ϵ-greedy algorithm with $\epsilon = 0.2$ consistently maintains superior performance across all datasets, while each model shows a slight upward trend above the random classifier line.

Given that this study exclusively employs the $\epsilon = 0.2$ for the greedy algorithm, other epsilon values might yield significantly better results. Consequently, traditional recommendation models could still maintain their position as the best-performing option. The inherent nature of the ϵ-greedy algorithm with $\epsilon = 0.2$ prevents it from effectively catering to users who prefer unpopular items. Our observation indicates that high AUC values are primarily attainable for low epsilon values. Consequently, the performance of a greedy algorithm is inherently constrained by the number of popular items. Even in scenarios where there is an extensive array of popular items, these items invariably constitute only a small fraction of the recommendation dataset. Hence the greedy algorithm is limited in its performance to a small subset of popular items. To clarify, while the algorithm may accurately rank 100% of the users' preferences for popular items, its performance remains suboptimal for users with a preference for unpopular items.

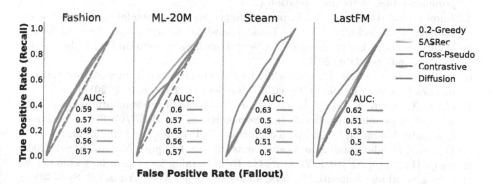

Fig. 4. Average Recall and Fallout values for users with preferences for popular and unpopular items across varying threshold values. A simple Epsilon = 0.2 greedy algorithm consistently outperforms other models across all datasets. While SASRec and its enhanced versions (Learning To Rank, Contrastive Learning, and Diffusion) closely approach the random classifier line.

Consequently, achieving a Hit Rate or NDCG score of close to one becomes an unattainable goal. As a result, a performance ceiling is established for this model. This limitation applies to SASRec and its variants, as their behaviour closely mirrors and performs below the greedy algorithms. As a result, the performance of SASRec and its modified versions remains restricted due to their inclination towards users who favour popular items.

6 Conclusion

This study demonstrates that a prevalent popularity bias endures regardless of architectural improvements when assessing sequential recommendation models across the entire item collection. Sequential recommendation models closely resemble the behaviour of the ϵ-greedy algorithm. As a result, a performance ceiling exists. The findings of this study indicate that the scarcity of interactions across all items hampers the sequential recommendation models' ability to effectively represent the distances and similarities among items. This scenario could be akin to a class imbalance issue and warrants further investigation in this avenue.

References

1. Abdollahpouri, H., Burke, R., Mobasher, B.: Controlling popularity bias in learning-to-rank recommendation. In: Proceedings of the Eleventh ACM Conference on Recommender Systems, RecSys 2017, pp. 42–46. Association for Computing Machinery, New York (2017)
2. Abdollahpouri, H., Mansoury, M.: Multi-sided exposure bias in recommendation. CoRR abs/2006.15772 (2020)
3. Abdollahpouri, H., Mansoury, M., Burke, R., Mobasher, B.: The unfairness of popularity bias in recommendation (2019)
4. Adomavicius, G., Kwon, Y.: Improving aggregate recommendation diversity using ranking-based techniques. IEEE Trans. Knowl. Data Eng. **24**(5), 896–911 (2012)
5. Bahdanau, D., Cho, K., Bengio, Y.: Neural machine translation by jointly learning to align and translate (2016)
6. Bellogín, A., Castells, P., Cantador, I.: Statistical biases in information retrieval metrics for recommender systems. Inf. Retrieval J. **20**, 606–634 (2017)
7. Chen, X., Yuan, Y., Zeng, G., Wang, J.: Semi-supervised semantic segmentation with cross pseudo supervision. In: Proceedings of the IEEE/CVF Conference on Computer Vision and Pattern Recognition (CVPR), pp. 2613–2622 (2021)
8. Dey, A.K.: Understanding and using context. Pers. Ubiquit. Comput. **5**, 4–7 (2001)
9. Fang, H., Zhang, D., Shu, Y., Guo, G.: Deep learning for sequential recommendation: algorithms, influential factors, and evaluations. ACM Trans. Inf. Syst. **39**(1), 1–42 (2020)
10. Fischer, E., Zoller, D., Dallmann, A., Hotho, A.: Integrating keywords into BERT4Rec for sequential recommendation. In: Schmid, U., Klügl, F., Wolter, D. (eds.) KI 2020. LNCS (LNAI), vol. 12325, pp. 275–282. Springer, Cham (2020). https://doi.org/10.1007/978-3-030-58285-2_23

11. Harper, F.M., Konstan, J.A.: The movielens datasets: History and context. ACM Trans. Interact. Intell. Syst. (TIIS) **5**(4), 1–19 (2015)
12. Hidasi, B., Karatzoglou, A.: Recurrent neural networks with top-k gains for session-based recommendations. In: Proceedings of the 27th ACM International Conference on Information and Knowledge Management, CIKM 2018, pp. 843–852. Association for Computing Machinery, New York (2018)
13. Jannach, D., Lerche, L., Kamehkhosh, I., Jugovac, M.: What recommenders recommend: an analysis of recommendation biases and possible countermeasures. User Model. User-Adap. Inter. **25**, 427–491 (2015)
14. Jannach, D., Ludewig, M.: When recurrent neural networks meet the neighborhood for session-based recommendation. In: Proceedings of the Eleventh ACM Conference on Recommender Systems, RecSys 2017, pp. 306–310. Association for Computing Machinery, New York (2017)
15. Järvelin, K., Kekäläinen, J.: Cumulated gain-based evaluation of IR techniques. ACM Trans. Inf. Syst. (TOIS) **20**(4), 422–446 (2002)
16. Breese, J.S., Heckerman, D., Kadie, C.M.: Empirical analysis of predictive algorithms for collaborative filtering. CoRR abs/1301.7363, pp. 43–52 (2013)
17. Kang, W.C., McAuley, J.: Self-attentive sequential recommendation. In: 2018 IEEE International Conference on Data Mining (ICDM), pp. 197–206 (2018)
18. Karatzoglou, A., Baltrunas, L., Shi, Y.: Learning to rank for recommender systems. In: Proceedings of the 7th ACM Conference on Recommender Systems, RecSys 2013, pp. 493–494. Association for Computing Machinery, New York (2013)
19. Karypis, G.: Evaluation of item-based top-n recommendation algorithms. In: Proceedings of the Tenth International Conference on Information and Knowledge Management, CIKM 2001, pp. 247–254. Association for Computing Machinery, New York (2001)
20. Krichene, W., Rendle, S.: On sampled metrics for item recommendation. In: Proceedings of the 26th ACM SIGKDD International Conference on Knowledge Discovery & Data Mining, KDD 2020, pp. 1748–1757. Association for Computing Machinery, New York (2020)
21. Li, Z., Sun, A., Li, C.: DiffuRec: a diffusion model for sequential recommendation (2023)
22. Ma, J., Zhou, C., Yang, H., Cui, P., Wang, X., Zhu, W.: Disentangled self-supervision in sequential recommenders. In: Proceedings of the 26th ACM SIGKDD International Conference on Knowledge Discovery & Data Mining, KDD 2020, pp. 483–491. Association for Computing Machinery, New York (2020)
23. Muthivhi, M., van Zyl, T., Wang, H.: Multi-modal recommendation system with auxiliary information. In: Pillay, A., Jembere, E., Gerber, A. (eds.) SACAIR 2022. CCIS, vol. 1734, pp. 108–122. Springer, Cham (2022). https://doi.org/10.1007/978-3-031-22321-1_8
24. Ni, J., Li, J., McAuley, J.: Justifying recommendations using distantly-labeled reviews and fine-grained aspects. In: Proceedings of the 2019 Conference on Empirical Methods in Natural Language Processing and the 9th International Joint Conference on Natural Language Processing (EMNLP-IJCNLP), pp. 188–197. Association for Computational Linguistics, Hong Kong (2019)
25. Park, Y.J., Tuzhilin, A.: The long tail of recommender systems and how to leverage it. In: Proceedings of the 2008 ACM Conference on Recommender Systems, RecSys 2008, pp. 11–18. Association for Computing Machinery, New York (2008)
26. Pei, C., Zhang, Y., Zhang, Y., et al.: Personalized re-ranking for recommendation. In: Proceedings of the 13th ACM Conference on Recommender Systems, RecSys 2019, pp. 3–11. Association for Computing Machinery, New York (2019)

27. Rendle, S., Freudenthaler, C., Gantner, Z., Schmidt-Thieme, L.: BPR: Bayesian personalized ranking from implicit feedback. arXiv preprint arXiv:1205.2618 (2012)
28. Rendle, S., Freudenthaler, C., Schmidt-Thieme, L.: Factorizing personalized Markov chains for next-basket recommendation. In: Proceedings of the 19th International Conference on World Wide Web, WWW 2010, pp. 811–820. Association for Computing Machinery, New York (2010)
29. Singer, U., Roitman, H., Eshel, Y., et al.: Sequential modeling with multiple attributes for watchlist recommendation in e-commerce. In: Proceedings of the Fifteenth ACM International Conference on Web Search and Data Mining, WSDM 2022, pp. 937–946. Association for Computing Machinery, New York (2022)
30. Sun, F., Liu, J., Wu, J., et al.: Bert4rec: sequential recommendation with bidirectional encoder representations from transformer. In: Proceedings of the 28th ACM International Conference on Information and Knowledge Management, CIKM 2019, pp. 1441–1450. Association for Computing Machinery, New York (2019)
31. Sutton, R.S., Barto, A.G.: Reinforcement Learning: An Introduction. MIT Press, Cambridge (2018)
32. Tenney, I., Das, D., Pavlick, E.: BERT rediscovers the classical NLP pipeline (2019)
33. Vaswani, A., et al.: Attention is all you need. In: Guyon, I., et al. (eds.) Advances in Neural Information Processing Systems, vol. 30. Curran Associates, Inc. (2017)
34. Wolf, T., Sanh, V., Chaumond, J., et al.: Hugging face transformers (2022)
35. Xie, X., Sun, F., Liu, Z., et al.: Contrastive learning for sequential recommendation. In: 2022 IEEE 38th International Conference on Data Engineering (ICDE), pp. 1259–1273 (2022)
36. Yang, Y., Huang, C., Xia, L., Huang, C., Luo, D., Lin, K.: Debiased contrastive learning for sequential recommendation. In: Proceedings of the ACM Web Conference 2023, WWW 2023, pp. 1063–1073. Association for Computing Machinery, New York (2023)
37. Yu, F., Liu, Q., Wu, S., Wang, L., Tan, T.: A dynamic recurrent model for next basket recommendation. In: Proceedings of the 39th International ACM SIGIR Conference on Research and Development in Information Retrieval, SIGIR 2016, pp. 729–732. Association for Computing Machinery, New York (2016)
38. Zhang, T., Zhao, P., Liu, Y., et al.: Feature-level deeper self-attention network for sequential recommendation. In: IJCAI, pp. 4320–4326 (2019)

A Comparative Study of Over-Sampling Techniques as Applied to Seismic Events

Mpho Mokoatle[1,2]([⊠]) [iD], Toshka Coleman[1], and Paul Mokilane[1] [iD]

[1] Council for Scientific and Industrial Research Cluster: Next Generation Enterprises and Institutions, Data Science, Pretoria, South Africa
Mmokoatle@csir.co.za
[2] University of Pretoria, Pretoria, South Africa
https://www.csir.co.za/

Abstract. The likelihood that an earthquake will occur in a specific location, within a specific time frame, and with ground motion intensity greater than a specific threshold is known as a **seismic hazard**. Predicting these types of hazards is crucial since doing so can enable early warnings, which can lessen the negative effects. Research is currently being executed in the field of machine learning to predict seismic events based on previously recorded incidents. However, because these events happen so infrequently, this presents a class imbalance problem to the machine learning or deep learning learners. As a result, this study provided a comparison of the performance of popular over-sampling techniques that seek to even out class imbalance in seismic events data. Specifically, this work applied SMOTE, SMOTENC, SMOTEN, BorderlineSMOTE, SVMSMOTE, and ADASYN to an open source Seismic Bumps dataset then trained several machine learning classifiers with stratified K-fold cross-validation for seismic hazard detection. The SVMSMOTE algorithm was the best over-sampling method as it produced classifiers with the highest overall accuracy, F1 score, recall, and precision of 100%, respectively, whereas the ADASYN over-sampling methodology showed the lowest performance in all the reported metrics of all the models. To our understanding, no research has been done comparing the effectiveness of the aforementioned over-sampling techniques for tasks involving seismic events.

1 Introduction

Seismic hazards, such as earthquakes, pose significant risks to human life, infrastructure, and the environment. Accurate prediction of these events is crucial for implementing early warning systems and taking preventive measures to minimize their impact. Additionally, accurate prediction of these events informs crucial decisions on infrastructure development, land-use planning, and disaster preparedness, guiding the implementation of building codes and engineering practices to withstand earthquakes [3, 21].

Over the years, researchers have turned to machine learning techniques to forecast seismic events based on historical data. However, due to the rare occurrence of such events, imbalanced data presents a challenge to these machine-learning models. One specific challenge is the class imbalance problem, where the positive instances (seismic events) are significantly outnumbered by the negative instances (non-seismic events) [9]. This class imbalance can lead to biased models and limited predictive performance due to the scarcity of positive instances available for learning and generalization [11].

Several studies [4,19,31] have explored the application of machine learning techniques for seismic hazard prediction. Some of these studies have employed the publicly available Seismic Bumps dataset, which records energy readings and bump counts in a coal mine in Poland. Researchers have utilized various classifiers, including Naïve Bayes, Support Vector Machine (SVM), and neural networks, to tackle this prediction problem. While existing research has shown promising results in utilizing machine learning for seismic hazard prediction, one critical gap remains unaddressed. Little attention has been given to the comparison and evaluation of different over-sampling techniques as a potential solution to the class imbalance problem in this domain. Over-sampling methods create synthetic instances of the minority class or modify existing instances to balance the class distribution [6]. The comparative study proposed in this research aims to fill this gap by assessing the performance of popular over-sampling techniques applied to seismic events data. Specifically, this study applies SMOTE, SMO-TENC, SMOTEN, BorderlineSMOTE, SVMSMOTE, and ADASYN (which are described in Sect. 2.2) to the Seismic Bumps dataset. To evaluate the efficacy of these over-sampling approaches, various machine learning classifiers were trained using K-fold cross-validation. By analyzing the impact of these over-sampling methods, the study seeks to identify the most effective technique for handling the class imbalance and improving the performance of seismic event prediction models, helping researchers and practitioners in the field to make informed decisions when building predictive models for rare events like seismic hazard.

2   Related Work

2.1   Review of Existing Research Studies Focused on Seismic Hazard Prediction Using Machine Learning

Numerous research studies have delved into the application of machine learning techniques for detecting seismic hazards, with several of them utilizing the Seismic Bumps dataset employed in our own study. A noteworthy paper [19] uses the dataset and proposed a new approach that incorporates negation handling in the Naïve Bayes classifier to improve accuracy. Experimental results show that the proposed approach achieves a higher accuracy of 76.98% compared to the traditional Naïve Bayes classifier without negation handling (64.5%) and the native MATLAB Naïve Bayes classifier without negation handling (65.09%). Another research study [31] using the dataset developed a prediction model for detecting periods of increased seismic activity that pose a threat to miners in coal mines.

Various classification models, including Random Forest, Naïve Bayes Classifier, Logistic Regression, and Support Vector Machine (SVM), are applied to evaluate their performance. The experimental results demonstrate that the Random Forest and SVM models achieve the highest accuracy, with 92.2%, respectively, for the Seismic Bumps dataset.

Additionally, this dataset was used in another study [4] where the authors proposed two deep temporal convolution neural network (CNN) models: dilated causal temporal convolution network (DCTCNN) and CNN long short-term memory hybrid model (CNN-LSTM). DCTCNN utilizes dilated CNN kernels, a causal strategy, and residual connections, while CNN-LSTM combines the advantages of CNN and LSTM. Both models are designed to extract long-term historical features from monitoring seismic data. The proposed models were evaluated using two real-life coal mine seismic datasets and compared with a traditional time series prediction method, two classic machine learning algorithms, and two standard deep learning networks. The results demonstrate that DCTCNN and CNN-LSTM outperform the other algorithms, indicating their effectiveness in completing the seismic prediction task. However, the issue of class imbalance is overlooked in these studies [4,19,31] and not addressed through any over-sampling, undersampling or resampling methods, resulting in models that may not adequately capture the characteristics of seismic events. Furthermore, different methods for over-sampling to address this have not yet been compared in tasks predicting seismic hazards. The study by [15] employed resampling for the Seismic Bumps dataset, but chose a single resampling method without comparing it against other possible methods for addressing its class imbalance. Previous studies have demonstrated the shortcomings of traditional machine learning algorithms when faced with imbalanced datasets, including decreased accuracy, sensitivity, and precision in detecting seismic events.

2.2 Overview of Over-Sampling Techniques for Class Imbalance

Over-sampling techniques have gained attention as a potential solution to address the class imbalance problem [6,16]. These techniques aim to re-balance the class distribution by generating synthetic instances of the minority class (seismic events) or modifying existing instances to amplify their representation. By artificially increasing the number of positive instances, over-sampling methods aim to provide a more balanced training dataset, enabling machine learning models to better capture the characteristics of seismic events [6]. We provide descriptions below of the over-sampling techniques used in this study.

SMOTE: Synthetic Minority Over-Sampling Technique. SMOTE is a popular over-sampling approach introduced in 2002. It works by over-sampling the minority class through creating synthetic instances rather than over-sampling with replacement. The SMOTE algorithm works by first identifying a minority class sample and choosing five nearest neighbors of the minority class sample. Then, for example, if the amount of over-sampling that is required is 200%, then two nearest neighbors will be selected from the initial five nearest neighbors. Finally, the synthetic instance is generated by drawing a line between the two nearest neighbors and the initial minority class sample [2, 14].

BorderlineSMOTE. The BorderlineSMOTE algorithm is a variant of the SMOTE algorithm and seeks to address the shortcomings of SMOTE. For instance, if there are minority class samples that are outliers and appears in the majority class, a synthetic instance will be created using the samples of the majority class. BorderlineSMOTE alleviates this problem by classifying any minority sample as noise if all its neighbors are the majority class samples. The noise minority samples will be ignored when creating new synthetic instances. Additionally, the BorderlineSMOTE algorithm classifiers a few samples as border points that majority and minority samples as neighbors and creates synthetic instances completely from these border points [6, 25, 28].

Adaptive Synthetic (ADASYN). ADASYN is a universal over-sampling method. For each of the minority samples it first determines the impurity of the neighbourhood by taking the ratio of the majority samples in the neighbourhood and k. The higher the ratio, the more synthetic examples will be created for that instance [7].

Synthetic Minority Over-Sampling TEchnique-Nominal Continuous (SMOTENC). Since the dataset used in this study is a combination of continuous and categorical features, a SMOTE variant named SMOTENC was applied as it is known to generalize well across mixed-data. SMOTENC's algorithm involves the computation of the median and the nearest neighbors and also populates the synthetic instance using the same approach as in SMOTE [2, 8].

SVMSMOTE. The primary distinction between SVMSMOTE and other SMOTE is that the technique would use the Support Vector Machine (SVM) algorithm to determine the mis-classification in the Borderline-SMOTE rather than K-nearest neighbours [20].

2.3 Previous Applications of Over-Sampling Techniques for Class Imbalance

Several studies have investigated the effectiveness of over-sampling techniques. According to some studies, the inclusion of over-sampling techniques has shown to achieve better True Positive (TP) rates and improve model performance [10]. One study introduced two new minority over-sampling methods, Borderline-SMOTE1 and Borderline-SMOTE2, which focused on over-sampling only the minority examples near the borderline. The experiments showed that these approaches achieved better True Positive (TP) rate and F-value compared to

SMOTE and random over-sampling methods for the minority class [6]. Furthermore, in the prediction of environmental complaints related to construction projects, an over-sampling-based method was developed using imbalanced empirical data. The method involved over-sampling techniques combined with machine learning algorithms to predict complaints due to environmental pollutants. The study reported performance improvements ranging from 8% to 23% using the over-sampling-based method compared to non-over-sampling approaches [32]. In the domain of Medical Artificial Intelligence, SMOTE was also used in a study [30] to address class imbalance in detecting COVID-19 cases. The authors used SMOTE to generate synthetic samples of the minority class (COVID-19 cases) and improve classifier performance. This study demonstrated the effectiveness of SMOTE in the specific domain of COVID-19 detection. Another research study [18] focused on determining the effective rate of minority class over-sampling using five over-sampling methods, including SMOTE, SVMSMOTE, and BorderlineSMOTE. The study aimed to maximize the performance of machine learning models and found that different datasets required different effective over-sampling rates. Additionally, a comparative analysis examined the performance of SMOTE and ADASYN, concluding that both techniques were effective in handling class imbalance. However, the performance varied depending on the classifier and the minority class [24]. Drawing from these two studies, we note that the effectiveness of these over-sampling techniques can vary depending on dataset characteristics, imbalance severity, and the choice of machine learning algorithms. Therefore, conducting experiments and evaluating these techniques within the context of specific applications is crucial to determine the most suitable approach.

In the context of seismic hazard detection, the application of over-sampling techniques to address class imbalance is relatively limited. Limited studies have explored the use of these techniques specifically for seismic event prediction. However, insights gained from their application in other domains highlight the potential benefits of employing over-sampling techniques to improve the performance of seismic hazard detection models. Thus, the objective of this study is to provide a comparative assessment of over-sampling methods. Specifically, this study applies SMOTE, SMOTENC, SMOTEN, BorderlineSMOTE, SVMSMOTE, and ADASYN to the Seismic Bumps dataset.

3   Materials and Methods

The dataset, the pre-processing procedures, the over-sampling methods, as well as an overview of the performance indicators used to assess the efficacy of the various over-sampling methods are all described in this section (Fig. 1).

3.1   Data Description and Pre-processing

In this study, the Seismic Bumps dataset from Kaggle, which is available to the public, was used. This dataset, made up of 2584 observations and 19 columns,

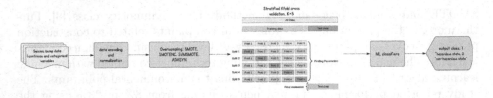

Fig. 1. This figure illustrates the strategy taken in this study to evaluate the effectiveness of various over-sampling strategies.

contains energy readings and bump counts that were captured during the preceding shifts in a coal mine in Poland. All input features were used in this work, with the exception of "nbumps6," "nbumps7," and "nbumps86," which exclusively contained zero values. The omission of these features was crucial to prevent their conversion to null during normalization. The target class is a binary variable where "1" indicates a high energy seismic bump occurred in the following shift (a "hazardous state"), and "0" indicates no high energy seismic bumps in the following shift (a "non-hazardous state"). This target variable had 2414 *non-hazardous state* and only 170 *hazardous state*, which posed a significant class-imbalance problem.

For data processing, the scikit-learn LabelEncoder was used to encode the non-numeric variables then, a data normalization procedure (min-max scaling) was performed to transform the numeric columns to a standard scale to ensure that the features with high values do not dominate the learning process.

3.2   Machine Learning Models

To compare the effectiveness of the different over-sampling methods, seven different machine learning models from the scikit-learn library [22] were used: KNeighborsClassifier, DecisionTreeClassifier, RandomForestClassifier, AdaBoostClassifier, MLPClassifier, GaussianNB, and LogisticRegression. Table 1 provides a description of all the models used.

3.3   Stratified *K*-Fold Cross Validation

This study evaluated the classification performance of the models with the over-sampling techniques after performing 5-fold stratified cross-validation. Stratified *K*-Fold cross validation is a variation of the standard *K*-Fold cross-validation. As opposed to splitting the data at random, stratified *k*-fold cross-validation ensures that the ratio of the target classes is the same in each fold as it is across the entire dataset. This method is particularly useful in small datasets with class imbalance problems [23,33].

3.4   Performance Measures

Accuracy, precision, recall, and F1-score measurements were used to assess how well the over-sampling techniques worked in conjunction with the machine learn-

Table 1. This table gives a description of the machine learning classifiers

ML model	Description
KNeighbors	This algorithm makes predictions on new data by assuming that data with similar characteristics cluster together [5]. In this work, the default metric *minkowski* [29] was used to compute the distance
DecisionTree	This algorithm is applicable to both classification and regression. Its structure resembles a tree and is composed of leaf nodes, internal nodes, branches, and root nodes [12]. In this work, the split's quality was assessed using the *gini* default criterion
RandomForest	A type of bagging estimator that fits several decision trees on various sub-samples of the data then uses averaging to increase the predictive performance [26]. As with the decision tree, in this work, the *gini* criterion was used to measure the quality of a split
AdaBoost	AdaBoost is a boosting technique that trains classifiers sequentially so as to minimise the errors produced by earlier learners [27].
MLP	Multi-layer perceptrons (MLPs), are a sort of feed-forward networks in which data is only sent in one direction. This model was trained with 100 neurons, ReLU activation function, and Adam optimizer [13] .
GaussianNB	Gaussian Naive Bayes (NB) is a classification algorithm based on the probabilistic method and Gaussian distribution. GaussianNB assumes that each independent variable has an independent capacity of predicting the target class [17]
LogisticRegression	An algorithm that uses a number of independent variables to predict a binary target class [1]

ing classifiers. These metrics are all defined below in terms of true positives (TP), true negatives (TN), false negatives (FN), and false positives (FP). The mean and standard deviation are reported for each metric.

$$accuracy = \frac{TP + TN}{TP + TN + FP + FN} \tag{1}$$

$$precision = \frac{TP}{TP + FP} \tag{2}$$

$$recall = \frac{TP}{TP + FN} \tag{3}$$

$$F1score = \frac{2TP}{2TP + FP + FN} \tag{4}$$

4 Experimental Results

The results of all the models are shown (Table 2, 3, 4 and 5) in terms of their average accuracy, F1-score, recall and precision over 5 k-folds.

When no over-sampling techniques were used on the data, the machine learning models' accuracy was quite good. The confusion matrices, however, showed a considerable bias towards the minority class and some models only predicted one class label (Table 2 and Fig. 2, 3, 4).

The ADASYN over-sampling approach produced the least accurate classifiers (49–65%), while the SVMSMOTE over-sampling method produced the best classifiers with an average accuracy of 98–100% and F1 score of 98–100% (Table 3, Fig. 5).

Table 2. Mean accuracy

	KNeighbors	DecisionTree	RandomForest	AdaBoost	MLP	GaussianNB	LogisticRegression
No over-sampling	0.93 ± 0.003	0.81 ± 0.07	0.92 ± 0.01	0.83 ± 0.11	0.93 ± 0.003	0.10 ± 0.02	0.93 ± 0.003
ADASYN	0.52 ± 0.03	0.59 ± 0.04	0.65 ± 0.04	0.62 ± 0.05	0.49 ± 0.11	0.49 ± 0.03	0.52 ± 0.12
BorderlineSMOTE	0.83 ± 0.05	0.86 ± 0.06	0.89 ± 0.05	0.82 ± 0.06	0.84 ± 0.08	0.84 ± 0.07	0.83 ± 0.07
SMOTE	0.75 ± 0.05	0.75 ± 0.06	0.82 ± 0.07	0.78 ± 0.06	0.77 ± 0.06	0.78 ± 0.05	0.75 ± 0.06
SVMSMOTE	0.98 ± 0.01	1.0 ± 0.00	1.0 ± 0.00	1.0 ± 0.00	0.99 ± 0.01	0.99 ± 0.00	0.99 ± 0.01
SMOTENC	0.76 ± 0.02	0.81 ± 0.04	0.89 ± 0.03	0.81 ± 0.04	0.80 ± 0.02	0.56 ± 0.09	0.71 ± 0.02

Table 3. Mean F1 score

	KNeighbors	DecisionTree	RandomForest	AdaBoost	MLP	GaussianNB	LogisticRegression
No over-sampling	0.0 ± 0.0	0.10 ± 0.08	0.0 ± 0.0	0.0 ± 0.0	0.0 ± 0.0	0.12 ± 0.01	0.0 ± 0.0
ADASYN	0.55 ± 0.07	0.60 ± 0.04	0.64 ± 0.03	0.63 ± 0.03	0.45 ± 0.07	0.64 ± 0.03	0.54 ± 0.12
BorderlineSMOTE	0.83 ± 0.04	0.86 ± 0.05	0.88 ± 0.04	0.83 ± 0.05	0.84 ± 0.06	0.84 ± 0.06	0.83 ± 0.06
SMOTE	0.76 ± 0.03	0.76 ± 0.04	0.84 ± 0.04	0.80 ± 0.04	0.77 ± 0.04	0.77 ± 0.03	0.76 ± 0.03
SVMSMOTE	0.98 ± 0.01	1.0 ± 0.0	1.0 ± 0.0	1.0 ± 0.0	0.99 ± 0.01	0.99 ± 0.01	0.99 ± 0.01
SMOTENC	0.76 ± 0.02	0.83 ± 0.03	0.89 ± 0.04	0.82 ± 0.03	0.79 ± 0.03	0.69 ± 0.04	0.71 ± 0.02

Since the F1 scores were inadequate when no over-sampling technique was used, as was expected, the average recall scores and precision were similarly quite poor as well when no over-sampling technique was applied to the data. Additionally, across all over-sampling techniques, SMVSMOTE still maintained

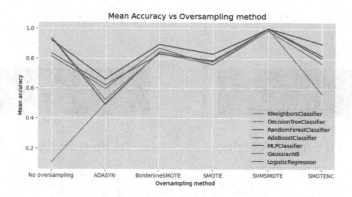

Fig. 2. Mean accuracy vs over-sampling method

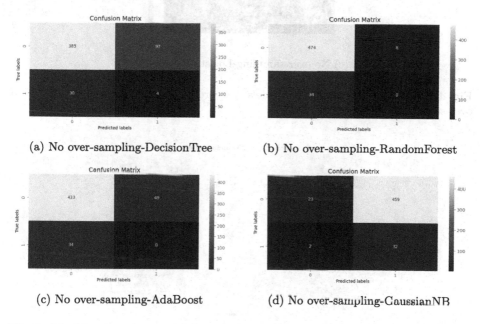

Fig. 3. Machine learning models' confusion matrices without the use of an over-sampling technique

Table 4. Mean recall

	KNeighbors	DecisionTree	RandomForest	AdaBoost	MLP	GaussianNB	LogisticRegression
No over-sampling	0.0 ± 0.0	0.11 ± 0.06	0.0 ± 0.0	0.0 ± 0.0	0.0 ± 0.0	0.93 ± 0.07	0.0 ± 0.0
ADASYN	0.62 ± 0.16	0.62 ± 0.09	0.66 ± 0.09	0.66 ± 0.06	0.43 ± 0.07	0.93 ± 0.08	0.60 ± 0.19
BorderlineSMOTE	0.84 ± 0.04	0.91 ± 0.02	0.91 ± 0.03	0.86 ± 0.02	0.83 ± 0.02	0.81 ± 0.04	0.83 ± 0.04
SMOTE	0.78 ± 0.03	0.86 ± 0.04	0.87 ± 0.02	0.83 ± 0.03	0.75 ± 0.06	0.72 ± 0.06	0.75 ± 0.05
SVMSMOTE	1.0 ± 0.0	1.0 ± 0.0	1.0 ± 0.0	1.0 ± 0.0	1.0 ± 0.0	1.0 ± 0.0	1.0 ± 0.0
SMOTENC	0.74 ± 0.05	0.86 ± 0.03	0.91 ± 0.02	0.84 ± 0.03	0.77 ± 0.06	0.99 ± 0.01	0.69 ± 0.05

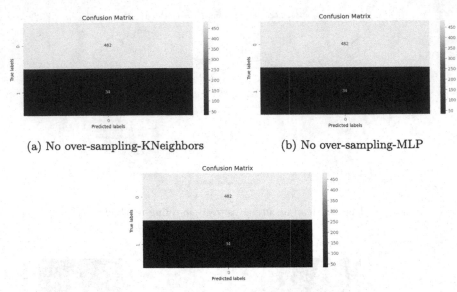

(a) No over-sampling-KNeighbors   (b) No over-sampling-MLP

(c) No over-sampling-LogisticRegression

Fig. 4. Confusion matrices of the models that predicted a single class label when no over-sampling technique was used

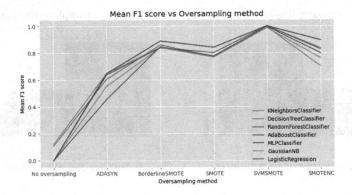

Fig. 5. Mean F1 vs over-sampling method

Table 5. Mean precision

	KNeighbors	DecisionTree	RandomForest	AdaBoost	MLP	GaussianNB	LogisticRegression
No over-sampling	0.0 ± 0.0	0.04 ± 0.04	0.0 ± 0.0	0.0 ± 0.0	0.0 ± 0.0	0.06 ± 0.01	0.0 ± 0.0
ADASYN	0.51 ± 0.03	0.59 ± 0.04	0.68 ± 0.06	0.61 ± 0.07	0.51 ± 0.14	0.49 ± 0.02	0.52 ± 0.09
BorderlineSMOTE	0.84 ± 0.09	0.82 ± 0.08	0.87 ± 0.07	0.81 ± 0.10	0.88 ± 0.11	0.87 ± 0.10	0.84 ± 0.11
SMOTE	0.76 ± 0.08	0.72 ± 0.06	0.81 ± 0.10	0.78 ± 0.11	0.80 ± 0.12	0.84 ± 0.11	0.78 ± 0.10
SVMSMOTE	**0.97 ± 0.01**	**1.0 ± 0.0**	**1.0 ± 0.0**	**1. 0 ± 0.0**	**0.99 ± 0.01**	**0.98 ± 0.0**	**0.98 ± 0.02**
SMOTENC	0.78 ± 0.05	0.79 ± 0.07	0.87 ± 0.06	0.81 ± 0.07	0.84 ± 0.07	0.54 ± 0.06	0.73 ± 0.05

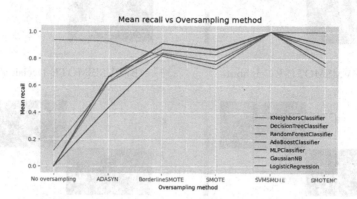

Fig. 6. Mean recall vs over-sampling method

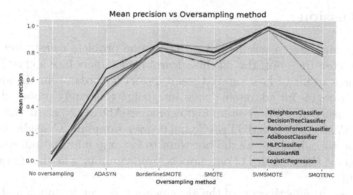

Fig. 7. Mean precision vs over-sampling method

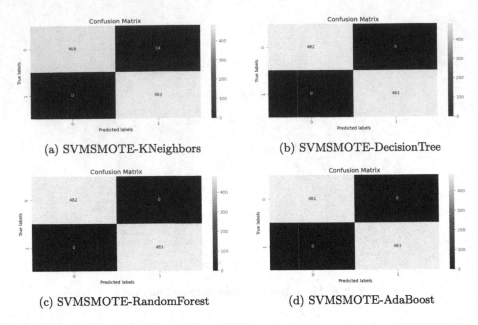

Fig. 8. Confusion matrices of some the best over-sampling technique: SVMSMOTE

the highest average recall and precision scores of 100% respectively, whereas ADASYN had the lowest recall and precision (Table 4, 5, and Fig. 6, 7).

The confusion matrices of the best performing over-sampling method is shown (Fig. 8).

5   Discussion

This study presented a seismic hazard prediction problem using machine learning. More often than not, the machine learning classifiers face a class imbalance problem as a result of how rarely these types of hazards occur. In an effort to examine strategies for addressing class inequality, this study proposed to evaluate the influence of five over-sampling techniques-ADASYN, BorderlineSMOTE, SMOTE, SVMSMOTE, SMOTENC, and no over-sampling. Considering that the dataset was too small and we did not want to lose any information, which would have impacted the accuracy of our models, over-sampling was chosen rather than under-sampling.

To assess the effectiveness of the aforementioned over-sampling methods, this work trained several machine learning techniques, including KNeighbors, DecisionTree, RandomForest, AdaBoost, MLP, GaussianNB, and LogisticRegression. Then, the average accuracy, F1 score, recall, and precision for each classifier and over-sampling technique were reported.

In the first run, all seven models were tested using a dataset that had not been over-sampled and several observations were made: all the learning algorithms-

aside from the GaussianNB model-returned very high accuracies. The F1 scores, recall, and precision scores, however, were incredibly poor. This was an apparent indication that the class imbalance problem posed a significant issue.

In the second run of the experiments, the dataset was over-sampled with the above over-sampling techniques. The SVMSMOTE algorithm was the best over-sampling method as it provided the highest overall accuracy, F1 score, recall and precision whereas the ADASYN over-sampling methodology showed the lowest performance in all the reported metrics of all the models.

Although using different methodologies, the dataset used in this work has previously been applied in prior studies [4,15,31] and some of the observations made in this work are in agreement with those in the literature.

For example, prior to over-sampling, the accuracy in [15] was consistently good, but the F1 scores, precision, and recall were subpar. Similar to our work, the performance metrics only increased after accounting for class imbalance. The sole significant distinction between this work [15] and ours is that only one sampling strategy was suggested and put to the test while our study examined several other strategies.

A study [31] also used the dataset that was used in this work and a key difference between our work and theirs is that the models were biased towards a single class as the authors did not account for class imbalance.

In another paper, the authors [4] employed the same dataset as in our study, but used a different response variable to describe the seismic hazard problem. For instance, the authors of this work employed the *energy* variable as the response variable, which signifies the total energy of seismic bumps reported within a previous shift, whereas our study used the categorical variable *hazardous state/ or non hazardous* state as the response variable. As a result, this problem was automatically transformed into a regression problem rather than being modelled as a classification problem.

6   Conclusion

Predicting seismic hazards is essential for reducing the negative effects of earthquakes, such as casualties and property loss. Early warning systems can be created, enabling prompt evacuation and disaster response preparation, by precisely predicting seismic hazards. However, machine learning or deep learning models face a class imbalance issue because seismic events occur infrequently. This study therefore offered a comparative evaluation of widely used over-sampling techniques for the problem of seismic events. This study discovered that classifiers for seismic hazard can perform noticeably better when machine learning techniques incorporating SVMSMOTE are used.

References

1. Bisong, E., Bisong, E.: Logistic regression. In: Building Machine Learning and Deep Learning Models on Google Cloud Platform: A Comprehensive Guide for Beginners, pp. 243–250 (2019)

2. Chawla, N.V., Bowyer, K.W., Hall, L.O., Kegelmeyer, W.P.: SMOTE: synthetic minority over-sampling technique. J. Artif. Intell. Res. **16**, 321–357 (2002)
3. Cutter, S.L.: Vulnerability to environmental hazards. Prog. Hum. Geogr. **20**(4), 529–539 (1996)
4. Geng, Y., Su, L., Jia, Y., Han, C.: Seismic events prediction using deep temporal convolution networks. J. Electr. Comput. Eng. **2019** (2019)
5. Guo, G., Wang, H., Bell, D., Bi, Y., Greer, K.: KNN model-based approach in classification. In: Meersman, R., Tari, Z., Schmidt, D.C. (eds.) OTM 2003. LNCS, vol. 2888, pp. 986–996. Springer, Heidelberg (2003). https://doi.org/10.1007/978-3-540-39964-3_62
6. Han, H., Wang, W.-Y., Mao, B.-H.: Borderline-SMOTE: a new over-sampling method in imbalanced data sets learning. In: Huang, D.-S., Zhang, X.-P., Huang, G.-B. (eds.) ICIC 2005. LNCS, vol. 3644, pp. 878–887. Springer, Heidelberg (2005). https://doi.org/10.1007/11538059_91
7. He, H., Bai, Y., Garcia, E.A., Li, S.: ADASYN: adaptive synthetic sampling approach for imbalanced learning. In: 2008 IEEE International Joint Conference on Neural Networks (IEEE World Congress on Computational Intelligence), pp. 1322–1328. IEEE (2008)
8. Islahulhaq, W.W., Ratih, I.D.: Classification of non-performing financing using logistic regression and synthetic minority over-sampling technique-nominal continuous (SMOTE-NC). Int. J. Adv. Soft Comput. Appl. **13**, 115–128 (2021)
9. Japkowicz, N., Stephen, S.: The class imbalance problem: a systematic study. Intell. Data Anal. **6**(5), 429–449 (2002)
10. Kalaycioglu, O., Akhanli, S.E., Mentese, E.Y., Kalaycioglu, M., Kalaycioglu, S.: Using machine learning algorithms to identify predictors of social vulnerability in the event of an earthquake: Istanbul case study. Nat. Hazards Earth Syst. Sci. Discuss. **2022**, 1–32 (2022)
11. Kiani, J., Camp, C., Pezeshk, S.: On the application of machine learning techniques to derive seismic fragility curves. Comput. Struct. **218**, 108–122 (2019). https://doi.org/10.1016/j.compstruc.2019.03.004. https://www.sciencedirect.com/science/article/pii/S0045794918318650
12. Kotsiantis, S.B.: Decision trees: a recent overview. Artif. Intell. Rev. **39**, 261–283 (2013)
13. Kruse, R., Mostaghim, S., Borgelt, C., Braune, C., Steinbrecher, M.: Multi-layer perceptrons. In: Computational Intelligence. TCS, pp. 53–124. Springer, Cham (2022). https://doi.org/10.1007/978-3-030-42227-1_5
14. Maldonado, S., López, J., Vairetti, C.: An alternative smote oversampling strategy for high-dimensional datasets. Appl. Soft Comput. **76**, 380–389 (2019)
15. Menon, A.P., Varghese, A., Joseph, J.P., Sajan, J., Francis, N.: Performance analysis of different classifiers for earthquake prediction: Pace. IJIRT **2**, 142–146 (2020)
16. Mohammed, R., Rawashdeh, J., Abdullah, M.: Machine learning with oversampling and undersampling techniques: overview study and experimental results. In: 2020 11th International Conference on Information and Communication Systems (ICICS), pp. 243–248. IEEE (2020)
17. Mugdha, S.B.S., et al.: A Gaussian Naive Bayesian classifier for fake news detection in Bengali. In: Hassanien, A.E., Bhattacharyya, S., Chakrabati, S., Bhattacharya, A., Dutta, S. (eds.) Emerging Technologies in Data Mining and Information Security. AISC, vol. 1300, pp. 283–291. Springer, Singapore (2021). https://doi.org/10.1007/978-981-33-4367-2_28

18. Naim, F.A., Hannan, U.H., Humayun Kabir, M.: Effective rate of minority class over-sampling for maximizing the imbalanced dataset model performance. In: Gupta, D., Polkowski, Z., Khanna, A., Bhattacharyya, S., Castillo, O. (eds.) Proceedings of Data Analytics and Management. LNDECT, vol. 91, pp. 9–20. Springer, Singapore (2022). https://doi.org/10.1007/978-981-16-6285-0_2

19. Netti, K., Radhika, Y.: An efficient Naïve Bayes classifier with negation handling for seismic hazard prediction. In: 2016 10th International Conference on Intelligent Systems and Control (ISCO), pp. 1–4. IEEE (2016)

20. Nguyen, H.M., Cooper, E.W., Kamei, K.: Borderline over-sampling for imbalanced data classification. Int. J. Knowl. Eng. Soft Data Paradigms 3(1), 4–21 (2011)

21. Nicolis, O., Plaza, F., Salas, R.: Prediction of intensity and location of seismic events using deep learning. Spat. Stat. 42, 100442 (2021)

22. Pedregosa, F., et al.: Scikit-learn: machine learning in python. J. Mach. Learn. Res. 12, 2825–2830 (2011)

23. Prusty, S., Patnaik, S., Dash, S.K.: SKCV: stratified K-fold cross-validation on ML classifiers for predicting cervical cancer. Front. Nanotechnol. 4, 972421 (2022)

24. Rachburee, N., Punlumjeak, W.: Oversampling technique in student performance classification from engineering course. Int. J. Electr. Comput. Eng. 11(4), 3567 (2021)

25. Revathi, M., Ramyachitra, D.: A modified borderline smote with noise reduction in imbalanced datasets. Wirel. Pers. Commun. 121, 1659–1680 (2021)

26. Rigatti, S.J.: Random forest. J. Insur. Med. 47(1), 31–39 (2017)

27. Schapire, R.E.: Explaining AdaBoost. In: Schölkopf, B., Luo, Z., Vovk, V. (eds.) Empirical Inference, pp. 37–52. Springer, Heidelberg (2013). https://doi.org/10.1007/978-3-642-41136-6_5

28. Shen, W., Fan, W., Chen, C.: An electric vehicle charging pile fault diagnosis system using Borderline-SMOTE and LightGBM. In: Tenth International Symposium on Precision Mechanical Measurements, vol. 12059, pp. 615–622. SPIE (2021)

29. Singh, A., Yadav, A., Rana, A.: K-means with three different distance metrics. Int. J. Comput. Appl. 67(10) (2013)

30. Turlapati, V.P.K., Prusty, M.R.: Outlier-SMOTE: a refined oversampling technique for improved detection of COVID-19. Intell.-Based Med. 3, 100023 (2020)

31. Verma, L.K., Kishore, N., Jharia, D.: Predicting dangerous seismic events in active coal mines through data mining. Int. J. Appl. Eng. Res. 12(5), 567–571 (2017)

32. Wang, D., Liang, Y., Yang, X., Dong, H., Tan, C.: A safe zone smote oversampling algorithm used in earthquake prediction based on extreme imbalanced precursor data. Int. J. Pattern Recogn. Artif. Intell. 35(13), 2155013 (2021)

33. Widodo, S., Brawijaya, H., Samudi, S.: Stratified K-fold cross validation optimization on machine learning for prediction. Sinkron: jurnal dan penelitian teknik informatika 7(4), 2407–2414 (2022)

Author Index

Printed in the United States
by Baker & Taylor Publisher Services